Systems & Control: Foundations & Applications

Bert van Keulen

$\mathcal{H}_\infty$-Control for Distributed Parameter Systems: A State-Space Approach

Birkhäuser
Boston • Basel • Berlin

Bert van Keulen
Department of Mathematics
University of Groningen
The Netherlands

Library of Congress Cataloging In-Publication Data
Keulen, Bert van.
H [infinity]-control for distributed parameter systems / a state space approach / Bert van Keulen.
p. cm. -- (Systems & control)
On t.p. "[infinity]" appears as the infinity symbol.
Includes bibliographical references.
ISBN 0-8176-3709-5 (acid-free).
1. H [infinity symbol] control. I. Distributed paramter systems.
3. State-space methods. I. Title. II. Series.
QA402.3.K42 1993 93-24351
003'.78--dc20 CIP

Printed on acid-free paper

Birkhäuser

ISBN 0-8176-3709-5
ISBN 3-7643-3709-5
Typeset by the Author in LATEX.
Printed and bound by Quinn-Woodbine, Woodbine, NJ.
Printed in the U.S.A.

9 8 7 6 5 4 3 2 1

Contents

Preface

Control of distributed parameter systems is a fascinating and challenging topic, from both a mathematical and an applications point of view. The same can be said about $\mathcal{H}_\infty$-control theory, which has become very popular lately. I am therefore pleased to present in this book a complete treatment of the state-space solution to the $\mathcal{H}_\infty$-control problem for a large class of distributed parameter systems.

The class of distributed parameter systems considered in this book allows for a certain degree of unboundedness of the input and output operators and is usually referred to as the Pritchard-Salamon class. In addition to deriving the state-space solution to the $\mathcal{H}_\infty$-control problem for this class of systems, it is my aim to show the very nice system theoretic properties of the class.

Some basic knowledge about infinite-dimensional systems theory will be assumed: the book is intended for graduate students and other researchers interested in infinite-dimensional systems theory and $\mathcal{H}_\infty$-control.

Most of the research that lead to this book was performed when I was a PhD student in the Systems & Control group of the Department of Mathematics of the University of Groningen in The Netherlands. My supervisor was Ruth Curtain and, since this book would not have appeared without her advice and support, I owe her many thanks.

It is my pleasure to acknowledge here the help of all the others who in some way or another contributed to this book. In particular, I would like to thank the members of the Systems & Control group in Groningen for the pleasant (working) atmosphere. I thank Professors Bensoussan, Kaashoek and Pritchard for their comments on the thesis that forms the core of this book, the Systems and Control Theory Network in The Netherlands for the high level courses and the extra funding, and the staff of Birkhauser Boston for excellent cooperation. Finally, and most importantly of all, I thank my family and my friends for their interest and their support over the last few years.

Bert van Keulen
June 1993

$\mathcal{H}_\infty$-Control for Distributed Parameter Systems: A State-Space Approach

Chapter 1

Introduction

Due to the fact that there have been many recent papers, books and PhD-theses on the $\mathcal{H}_\infty$-control problem, there are few members of the systems & control community who have not at least heard of this problem. The reason for all the research in this field is that a lot of interesting control problems can be formulated in a general $\mathcal{H}_\infty$-design framework (see e.g. Francis [34] and references therein); in particular, we mention several robustness optimization problems and the so-called mixed-sensitivity problem. In order to explain what $\mathcal{H}_\infty$-control is about, we shall introduce the problem in Section 1.1 below, without mentioning all the technical details, and we shall give some examples. Furthermore, we shall discuss the relevant literature on the state-space approach to this problem, both for finite-dimensional systems and distributed parameter systems (we shall often also use the term infinite-dimensional systems). In Section 1.2 we shall introduce our class of distributed parameter systems, again without going into too much detail, and some examples will be given in order to motivate the choice of this class. In Section 1.3 an outline of the rest of the book is given.

1.1 $\mathcal{H}_\infty$-control

The standard problem

First of all, we sketch the so-called *standard $\mathcal{H}_\infty$-control problem*, following the book of Francis [34]. Suppose that we have some (infinite-dimensional) plant Σ_G with two inputs and two outputs as in Figure 1.1. Denoting the transfer function of Σ_G by G, we can also describe the system by

$$\begin{pmatrix} z \\ y \end{pmatrix} = G \begin{pmatrix} w \\ u \end{pmatrix} = \begin{pmatrix} G_{11} & G_{12} \\ G_{21} & G_{22} \end{pmatrix} \begin{pmatrix} w \\ u \end{pmatrix}.$$

Sometimes Σ_G (or G) is called the *generalized plant.* As usual, we call u the *control input*, w the *disturbance input*, y is the *measured output* and z is the

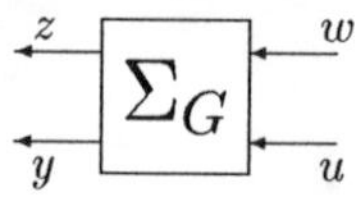

Figure 1.1: The generalized plant Σ_G.

to-be-controlled output. Furthermore, suppose that we have a controller Σ_K (with transfer function K) that takes as an input y and produces the output u, as in Figure 1.2. The problem is to find a controller Σ_K that stabilizes Σ_G

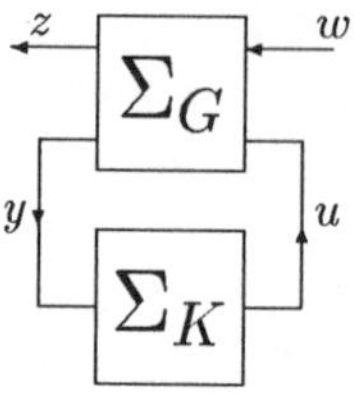

Figure 1.2: The closed-loop system $\Sigma_{G_{zw}}$.

such that the influence of the disturbance w on the to-be-controlled output z is minimized in some sense. Clearly we need to clarify 'stability' and 'influence of w on z' (details about the exact formulation can be found in the next chapters). In our approach to the problem we shall always assume that we have a state-space representation of the plant, denoted here by

$$\Sigma_G \left\{ \begin{array}{rcllll} \dot{x} &=& Ax &+& B_1 w &+& B_2 u \\ z &=& C_1 x &+& D_{11} w &+& D_{12} u \\ y &=& C_2 x &+& D_{21} w &+& D_{22} u, \end{array} \right. \tag{1.1}$$

and a state-space representation of the controller, denoted by

$$\Sigma_K \left\{ \begin{array}{rclll} \dot{p} &=& Mp &+& Ny \\ u &=& Lp &+& Ry, \end{array} \right. \tag{1.2}$$

where x and p are the states of the plant and the controller and A, B_i, C_i, D_{ij}, M, N, L and R are certain linear operators. By stability of the closed-loop system $\Sigma_{G_{zw}}$ in Figure 1.2, we shall mean the exponential stability of $\Sigma_{G_{zw}}$ on the extended state-space. Now if the closed-loop system is stable, it follows that the closed-loop transfer function from w to z denoted by G_{zw} is an element of the Hardy space $\mathcal{H}_\infty(\mathcal{L}(W, Z))$ (the space of functions with values in $\mathcal{L}(W, Z)$ and analytic and bounded on $\mathbb{C}^+$). The influence of w on z can be measured in terms of the $\mathcal{H}_\infty$-norm of the transfer function, which equals the supremum over all frequencies of $\|G_{zw}(i\omega)\|_{\mathcal{L}(W,Z)}$. A time-domain characterization of this norm follows from the fact that for zero initial conditions the closed-loop system defines a bounded linear map from $w(\cdot) \in L_2(0,\infty;W)$ to $z(\cdot) \in L_2(0,\infty;Z)$ and the norm of this map precisely equals the $\mathcal{H}_\infty$-norm of the transfer function. Consequently, the standard $\mathcal{H}_\infty$-control problem may be considered as a general disturbance attenuation problem.

In the case that the state of the system is available for controlling the system (i.e. $y = x$ in (1.1)), we shall use the term *$\mathcal{H}_\infty$-control with state-feedback*. In the general case the term *$\mathcal{H}_\infty$-control with measurement-feedback* or *output-feedback* will be used. Choosing a particular state-space representation of the plant is one of the elements of the so-called *state-space approach* to the $\mathcal{H}_\infty$-control problem. Another element is that one usually considers the *sub-optimal $\mathcal{H}_\infty$-control problem*, which is the problem of finding a controller that stabilizes the plant and is such that the norm of the closed-loop map is smaller than some pre-specified bound γ. The solution to the sub-optimal $\mathcal{H}_\infty$-control problem constitutes a first step toward solving the *optimal $\mathcal{H}_\infty$-control problem*. The optimal solution can be obtained (approximately) by a procedure that is usually referred to as γ-iteration.

Some examples of $\mathcal{H}_\infty$-control problems

Next, we show how several $\mathcal{H}_\infty$-control problems can be formulated as standard $\mathcal{H}_\infty$-control problems, by choosing a suitable generalized plant Σ_G (see also Francis [34] and Glover and McFarlane [63]). We shall discuss two robust control problems and the so-called mixed-sensitivity problem.

Generally speaking, a robust control problem is the problem of finding a controller that not only stabilizes a certain *nominal* plant P, but also a set of perturbations P_Δ of this nominal plant. Depending on the class of perturbations, one can try to find 'optimally robust' controllers that allow for as many perturbations within this class as possible. To illustrate this point, we first consider the case of robustness against *additive* perturbations. Let P (or $P(\cdot)$) be the transfer function of a nominal plant and suppose that the perturbation P_Δ is of the form $P_\Delta = P + \Delta$, where Δ is a transfer function in L_∞ such that P_Δ has the same number of unstable poles as P (in this case we call Δ an *admissible perturbation*). Actually, for infinite-dimensional systems the formulation should be more precise, as in Curtain et al. [11, 12], but those details will be omitted here. Consider the control configuration of Figure 1.3. Under certain extra conditions, it can be shown that the following are equivalent:

- K stabilizes $P_\Delta = P + \Delta$ for all admissible perturbations $\Delta \in L_\infty$ satisfying $\|\Delta\|_\infty < \epsilon$
- K stabilizes P and $\|K(I - PK)^{-1}\|_\infty \leq 1/\epsilon$.

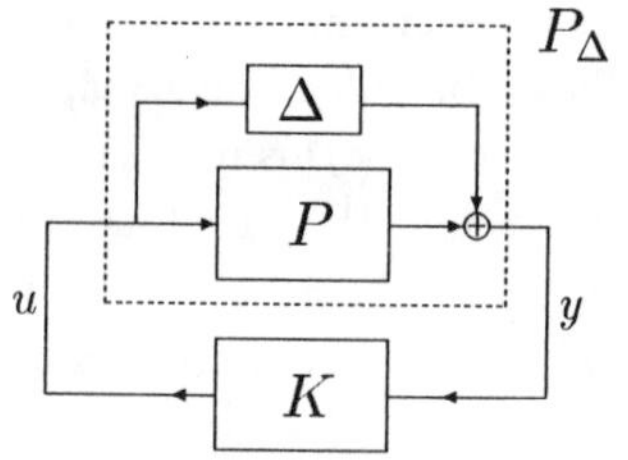

Figure 1.3: Additive perturbations.

This defines, in a natural way, the robustness optimization problem as the problem of finding a stabilizing controller K that allows for as large as possible ϵ. Then the corresponding optimal controller stabilizes P_Δ for as many admissible perturbations as possible. The point is that for the choice of the generalized plant

$$G = \begin{pmatrix} G_{11} & G_{12} \\ G_{21} & G_{22} \end{pmatrix} = \begin{pmatrix} 0 & I \\ I & P \end{pmatrix},$$

the closed-loop transfer function is given by $G_{zw} = K(I - PK)^{-1}$, so that the robustness optimization problem corresponds to the standard $\mathcal{H}_\infty$-control problem of finding a stabilizing controller K such that $\|G_{zw}\|_\infty$ is minimal.

It should be noted that in practice it is often better to shape the admissible perturbation by some weighting function, but this does not cause extra difficulties in the above procedure: in this case the expression of the generalized plant should be adapted (see e.g. Francis [34]).

Usually, one calls the type of perturbation above *unstructured*. Other examples of unstructured perturbations are *multiplicative* perturbations and *coprime factor* perturbations (see the work of Glover and McFarlane [63], Curtain [11, 12], Özbay [65] and Curtain and Zwart [17] for both finite- and infinite-dimensional formulations). Those problems can be cast in the standard $\mathcal{H}_\infty$-control framework as well.

In many practical situations one has explicit a priori knowledge of the structure of the perturbation of a system (one could think of physical parameters of a system varying in a pre-specified range) and it may be useful to exploit that knowledge in the perturbation model. Keeping this in mind, Hinrichsen and Pritchard [39] and Doyle [24] have presented a theory of *structured perturbations* of the state-space realization of a finite-dimensional linear system. This theory has been extended by Pritchard and Townley [74, 91] to allow for infinite-dimensional systems in the Pritchard-Salamon class. Using their results, we shall describe the corresponding robust optimization problem and show how it can be formulated as a standard $\mathcal{H}_\infty$-control problem. The exposition shall be on a somewhat formal level: the technical details will be omitted. Suppose that we have a nominal plant given by a state-space description of the form

$$P \quad \dot{x}(t) = Ax(t) + Bu(t) \; ; \quad t > 0,$$

where A is the generator of a C_0-semigroup on a complex Hilbert space $\mathcal{H}$ and B is a possibly unbounded input operator on the input space U such that (A, B) is stabilizable. Furthermore, suppose that the perturbed plant is of the form

$$P_\Delta \quad \dot{x}(t) = (A + D\Delta E)x(t) + Bu(t) \; ; \quad t > 0,$$

where D is a possibly unbounded input operator defined on a Hilbert space U_0, E is a possibly unbounded output operator mapping into U_0 and the perturbation $\Delta \in \mathcal{L}(U_0)$ represents an unknown disturbance. As explained in Townley [91], this type of structured perturbation may represent perturbations of boundary data in boundary control systems or, for instance, perturbations due to neglected delay terms in delay equations. Now if F is a possibly unbounded ouput operator such that $A + BF$ is exponentially stable we can define the (complex) *stability radius* $r_\mathbb{C}(F)$ by

$$r_\mathbb{C}(F) := \sup_{r \geq 0}\{r \mid \|\Delta\| \leq r \;\text{ implies stability of } A + BF + D\Delta E\}.$$

In other words, if the state-feedback $u = Fx$ is stabilizing for the nominal model, then $r_\mathbb{C}(F)$ is the bound on the size of the perturbation Δ such that the perturbed plant is still stable. This defines a robustness optimization problem in a natural way: we can try to find a stabilizing F such that $r_\mathbb{C}(F)$ is maximal, so that F is stabilizing for perturbations $D\Delta E$ with $\|\Delta\|$ as large as possible. This problem is usually referred to as the stability radius optimization problem. The point is that there exists an interesting relationship between $r_\mathbb{C}(F)$ and the transfer function

$$G_F(s) = E(sI - A - BF)^{-1}D \in \mathcal{H}_\infty(\mathcal{L}(U_0)),$$

namely

$$r_\mathbb{C}(F) = \left(\sup_{\omega \in \mathbb{R}} \|G_F(i\omega)\|_{\mathcal{L}(U_0)}\right)^{-1}.$$

Therefore, the problem can be considered as a standard $\mathcal{H}_\infty$-control problem with state-feedback, using the following state-space description of the generalized plant

$$\Sigma_G \begin{cases} \dot{x}(t) &= Ax(t) + Dw(t) + Bu(t) \\ z(t) &= Ex(t) \\ y(t) &= x(t). \end{cases}$$

It follows that the stability radius optimization problem is equivalent to finding a stabilizing state-feedback $u(\cdot) = Fx(\cdot)$ such that the corresponding closed-loop norm $\|E(\cdot I - A - BF)^{-1}D\|_\infty$ is as small as possible (we note that in general the minimum need not exist). One of the results in this book is that there is no advantage in allowing for dynamic state-feedbacks. Pritchard and Townley [74] related this problem to the solvability of $\mathcal{H}_\infty$-type Riccati equations. The state-feedback solution of Chapter 4 represents an improvement of this result.

The last example of an $\mathcal{H}_\infty$-control problem that we discuss here is usually considered as the most important one: the *mixed-sensitivity minimization problem* (see e.g. the work of Verma and Jonckheere [93] and Kwakernaak [56] for the finite-dimensional case and Özbay [65] and references therein for some infinite-dimensional examples). It is an extension of the weighted-sensitivity minimization problem that was introduced by Zames [103]. The mixed-sensitivity minimization problem has received a lot of attention in the $\mathcal{H}_\infty$-control literature because it is of important practical value for controller design. The reason for this is that the problem can be considered as the combination of a robust control problem and a performance optimization problem (thus the term *robust performance problem*, used by Doyle et al. [25]). The problem is to find a stabilizing controller with transfer function K which minimizes the $\mathcal{H}_\infty$-norm

$$\left\| \begin{pmatrix} W_1(I-PK)^{-1} \\ W_2K(I-PK)^{-1} \end{pmatrix} \right\|_\infty,$$

where P is the transfer function of a given plant and W_1 and W_2 are certain (usually stable) weighting transfer functions. The first component of the criterion, $W_1(I-PK)^{-1}$, is called the *weighted sensitivity function* and the second component, $W_2K(I-PK)^{-1}$, is usually called the *weighted control sensitivity function*. As mentioned above, the mixed sensitivity problem can be motivated in terms of performance optimization and robustness optimization. Indeed, minimizing the $\mathcal{H}_\infty$-norm of the weighted sensitivity function corresponds to the optimization of the tracking properties of the controller.

Furthermore, minimizing the $\mathcal{H}_\infty$-norm of the control sensitivity function corresponds to the optimization of the (additive) robustness of the controller (see also above). We note that the choice of the weights W_1, W_2 is an important issue and that this choice typically depends on the kind of application (for details about choosing the weights we refer to Kwakernaak [56] and Doyle et al [25]). We also note that in some papers the mixed sensitivity problem is defined differently: the second component $W_2K(I - PK)^{-1}$ may be replaced by $W_2PK(I - PK)^{-1}$. This last choice corresponds to robustness optimization with respect to *multiplicative* perturbations. Defining the transfer function G of the generalized plant by

$$G := \left(\begin{array}{c|c} W_1 & W_1P \\ 0 & W_2 \\ \hline I & P \end{array}\right),$$

it follows that the closed-loop transfer function G_{zw} is given by

$$G_{zw} = \begin{pmatrix} W_1(I - PK)^{-1} \\ W_2K(I - PK)^{-1} \end{pmatrix}.$$

Hence, the mixed-sensitivity minimization problem corresponds to the standard $\mathcal{H}_\infty$-control problem of finding a stabilizing controller K such that the $\mathcal{H}_\infty$-norm of G_{zw} is minimal. In Chapter 6 we shall consider this problem for a class of delay systems, using the theory that will be covered in the next chapters.

Summary of results on $\mathcal{H}_\infty$-theory relevant for the book

We conclude this section with a summary of known results on the state-space approach to the standard $\mathcal{H}_\infty$-control problem, both for finite and infinite-dimensional systems.

The standard $\mathcal{H}_\infty$-control problem was first formulated for finite-dimensional linear time-invariant systems and several approaches to solve this problem were adopted (frequency-domain, state-space and combinations of these). One of the main contributions to solving the problem is the paper by Doyle, Glover, Khargonekar and Francis [26] (see also Glover and Doyle [37], where some of the a priori assumptions in [26] are weakened). The authors used the state-space approach to the problem and their solution to the (sub-optimal, regular) $\mathcal{H}_\infty$-control problem can be seen as comprising three main parts:

- The first part consists of the *state-feedback* result: the existence of a controller that solves the problem implies the existence of a solution to a certain Riccati equation and this solution can be used to construct a state-feedback controller that also solves the problem.

- The second part is the *measurement-feedback* result: if there exists a controller that solves the problem, a solution to a second 'dual' Riccati equation (with a certain coupling condition) can be obtained from the state-feedback result, using some duality arguments and an important auxiliary result, which is referred to as *Redheffer's Lemma* in this book. The solution of this Riccati equation, together with the solution to the state-feedback one, can be used to construct a measurement-feedback controller that solves the $\mathcal{H}_\infty$-control problem.

- The third part is the *controller parametrization.* Using the results of the first two parts, a parametrization of all controllers that solve the (sub-optimal) $\mathcal{H}_\infty$-control problem can be given.

In [87], Tadmor obtained a different proof of the result of Doyle et al. in [26]. One difference between his approach and the one in [26] is that he solved the state-feedback part by considering a certain *sup-inf*-problem (this idea was also introduced by Weiland [94] and Başar and Bernhard [6]). Stoorvogel [86] (who solved the *singular* $\mathcal{H}_\infty$-control problem for finite-dimensional systems) gave an elegant proof of the measurement-feedback result, based on the approaches of Tadmor [87] and Glover et al. [26].

State-space results for the $\mathcal{H}_\infty$-control control problem for *infinite-dimensional systems* were obtained by Pritchard and Townley [74], Tadmor [88], van Keulen et al. [51], van Keulen [47, 48, 50], Ichikawa [41, 40], Bensoussan and Bernhard [7], Barbu [5] and McMillan and Triggiani [64]. For a frequency domain approach to the $\mathcal{H}_\infty$-control problem for infinite-dimensional systems we refer to Tannenbaum [90], Özbay [65] and references therein.

Pritchard and Townley [74] were the first to consider the infinite-dimensional $\mathcal{H}_\infty$-control problem with state-feedback, in connection with the stability radius optimization problem. The class of systems in their paper is usually referred to as the Pritchard-Salamon class (see Section 1.2 and Chapter 2). The papers of van Keulen et al. [51], Bensoussan and Bernhard [7], Barbu [5] and McMillan and Triggiani [64] are also devoted to the state-feedback problem only; [51, 7] treat the bounded input/output case and in [5, 64] boundary control systems are considered. The measurement-feedback problem is solved by Tadmor [88], van Keulen [47, 48, 50] and Ichikawa [40] for several classes of infinite-dimensional systems. Van Keulen [47, 48] and Ichikawa solved the problem for semigroup control systems with bounded input and output operators, whereas Tadmor considered time-varying systems (also with bounded inputs and outputs). Finally, van Keulen [50] solved the problem for the Pritchard-Salamon class, using the procedures that were developed in [51, 47, 48, 49]. This last result, which forms the core of the book, is a complete generalization of the solution to the finite-dimensional $\mathcal{H}_\infty$-control problem described above to the Pritchard-Salamon class. In particular, all the weak a priori assumptions of Glover and Doyle [37] have been generalized to their natural infinite-dimensional counterparts.

1.2 Unbounded inputs and outputs

As mentioned above, we shall extend the finite-dimensional results for the standard $\mathcal{H}_\infty$-control problem of Doyle et al. [26] and Glover and Doyle [37] to the Pritchard-Salamon class. This is a class of *infinite-dimensional systems* that allows for *unbounded input and output operators* (the precise definition is postponed until the next chapter). In this section we shall give some motivation for considering this particular class of systems. First of all, we shall give two examples of infinite-dimensional systems that are modelled using either *bounded* or *unbounded* input and output operators, depending on the kind of control and observation. The first example is a parabolic partial differential equation (PDE) with either distributed or boundary control and the second example is a delay differential equation, with or without delays in the observation (both examples are borrowed from Pritchard and Salamon [72]). These systems can *formally* be described by equations of the form

$$\Sigma_G \begin{cases} \dot{x}(t) = Ax(t) + Bu(t), \quad x(0) = x_0 \\ y(t) = Cx(t) + Du(t). \end{cases} \tag{1.3}$$

Here, the state $x(\cdot)$, the control $u(\cdot)$ and the observation $y(\cdot)$ are all functions of time with values in certain Hilbert spaces. Furthermore, A is the infinitesimal generator of a C_0-semigroup, B is the (possibly unbounded) input operator, C is the (possibly unbounded) output operator and D is the feedthrough operator. One usually represents the system Σ_G by the so-called *mild solution* to the differential equation (1.3):

$$\Sigma_G \begin{cases} x(t) = S(t)x_0 + \int_0^t S(t-s)Bu(s)ds \\ y(t) = Cx(t) + Du(t), \end{cases} \tag{1.4}$$

where $S(\cdot)$ is the C_0-semigroup generated by A (we refer to [16, 17, 69] for a discussion about the concept of mild solutions). This mild solution is always well-defined for bounded input and output operators, but in the unbounded case one must be more careful. For both the parabolic PDE and the delay differential equation we consider the bounded case first.

Example 1.1 (*parabolic PDE: bounded case*)
The temperature distribution in a simple model of a heated rod may be described by

$$\begin{cases} \frac{\partial z}{\partial t}(\xi,t) = \frac{\partial^2 z}{\partial \xi^2}(\xi,t) + b(\xi)u(t), & t>0\,; \quad 0<\xi<1, \\ \frac{\partial z}{\partial \xi}(0,t)=0, \quad \frac{\partial z}{\partial \xi}(1,t)=0 & t>0, \\ y(t) = \int_0^t c(\xi)z(\xi,t)d\xi & t>0, \\ z(\xi,0)=z_0(\xi) & 0<\xi<1 \quad \text{(the intitial condition)}, \end{cases}$$

where both $b(\cdot)$ and $c(\cdot)$ are elements of $L_2(0,1)$. Here $z(t,\xi)$ represents the temperature profile at time t, $b(\xi)u(t)$ represents the addition of heat along the rod (according to the shape of $b(\cdot)$) and $y(t)$ is some averaged measurement of the temperature. Choosing the state-space $\mathcal{H} = L_2(0,1)$ with the state function

$$x(t) = z(t,\xi) \qquad t>0\,; \quad 0<\xi<1,$$

the PDE can be modelled as a system of the form (1.4) with $D=0$, i.e.

$$\begin{cases} x(t) &=& S(t)x_0 + \int_0^t S(t-s)Bu(s)ds \\ y(t) &=& Cx(t). \end{cases}$$

Here $S(\cdot)$ is a C_0-semigroup on $\mathcal{H}$ and its infinitesimal generator A, the input operator B and the output operator C are defined by

$$\begin{aligned} D(A) &= \{x \in \mathcal{H} \mid x, \tfrac{dx}{d\xi} \text{ are absolutely continuous}, \\ &\quad\; \tfrac{d^2x}{d\xi^2} \in \mathcal{H} \text{ and } \tfrac{dx}{d\xi}(0) = \tfrac{dx}{d\xi}(1) = 0\}, \\ Ax &= \frac{d^2x}{d\xi^2}, \\ Bu &= b(\cdot)u, \\ Cx &= \langle c(\cdot), x(\cdot) \rangle_{\mathcal{H}}\,. \end{aligned}$$

Furthermore, the initial condition $x_0 \in \mathcal{H}$ is given by $x_0(\xi) = z(\xi,0)$. It is easy to see that both the input and the output operator are linear and bounded: $B \in \mathcal{L}(\mathbb{R},\mathcal{H})$ and $C \in \mathcal{L}(\mathcal{H},\mathbb{R})$. For smooth input functions $u(\cdot)$ and for initial conditions $x_0 \in D(A)$, the state function $x(\cdot)$ is differentiable (with respect of the topology of $\mathcal{H}$) so that (1.4) can be expressed as (1.3) with $D=0$ (for technical details we refer to the following chapter). □

Example 1.2 (*delay differential equation: bounded case*)
Consider the delay differential equation

$$\begin{cases} \dot{z}(t) &= A_{f1}z(t) + A_{f2}z(t-2) + B_f u(t), \\ \\ y(t) &= C_f z(t), \end{cases} \tag{1.5}$$

where $z(t) \in \mathbb{R}^n, u(t) \in \mathbb{R}^m, y(t) \in \mathbb{R}^p, A_{f1}, A_{f2} \in \mathbb{R}^{n \times n}, B_f \in \mathbb{R}^{n \times m}$ and $C_f \in \mathbb{R}^{p \times n}$ (the index f stands for 'finite-dimensional'). This delay differential equation is a very simple example of a retarded functional differential equation and we shall show how it can be modelled as a system of the form (1.4). Define the product Hilbert space $M_2 = \mathbb{R}^n \times L_2(-2, 0\,; \mathbb{R}^n)$ (with the obvious inner product). For every initial condition $x_0 = (\eta_0, \phi_0) \in M_2$ there exists a unique solution of (1.5) satisfying

$$\lim_{t \downarrow 0} z(t) = \eta_0, \qquad z(\tau) = \phi_0(\tau); \quad -2 < \tau < 0.$$

In order to give the state of the system we define the solution segment

$$z_t(\tau) = z(t + \tau)\,; \qquad -2 \leq \tau \leq 0\,;$$

(the past of $z(\cdot)$ up to 2 time units). Choosing the state-space $\mathcal{H} = M_2$ with state function

$$x(t) = (z(t), z_t) \in M_2,$$

it follows that the system (1.5) can be reformulated as a system of the form (1.4), where $S(\cdot)$ is a C_0-semigroup on the Hilbert space $\mathcal{H}$, the input and output operators are given by

$$\begin{aligned} Bu &= (B_f u, 0) \ \text{ for } \ u \in \mathbb{R}^m \\ Cx &= C_f \eta \ \text{ for } \ x = (\eta, \phi) \in \mathcal{H} \end{aligned}$$

and the feedthrough operator is given by $D = 0$. Hence, $B \in \mathcal{L}(\mathbb{R}^m, \mathcal{H})$ and $C \in \mathcal{L}(\mathcal{H}, \mathbb{R}^p)$: the input and output operators are bounded. □

We have seen that the systems of Examples 1.1,1.2 can be modelled as semigroup control systems of the form (1.4), with bounded input and output operators. This class of systems has nice system theoretic properties and allows for extensions of important finite-dimensional ideas (see e.g. the work of Curtain and Pritchard [16] and Curtain and Zwart [17] and references therein). Furthermore, many partial differential equations (such as in Example 1.1) can be modelled in this framework, as well as many delay equations (such as in Example 1.2). However, some infinite-dimensional systems are most naturally modelled using *unbounded* input and output operators. In PDE-models this unboundedness can be a consequence of point or boundary control and observation. Delay equations with delays in the inputs or outputs lead to unbounded control and observation operators as well. To give an idea of both situations we shall modify Examples 1.1 and 1.2: we shall consider the parabolic PDE of example 1.1 with boundary control, rather than distributed control and we show what happens to the delay differential equation of Example 1.1 if the output $y(\cdot)$ depends on the past $z_t(\cdot)$, rather than on $z(t)$.

Example 1.3 (*parabolic PDE: unbounded case*)
The temperature distribution of a heated rod with *boundary* heat control may be described by

$$\begin{cases} \dfrac{\partial z}{\partial t}(\xi,t) = \dfrac{\partial^2 z}{\partial \xi^2}(\xi,t), & t>0\,; \quad 0<\xi<1, \\ \dfrac{\partial z}{\partial \xi}(0,t) = u(t), \quad \dfrac{\partial z}{\partial \xi}(1,t)=0 & t>0, \\ y(t) = \int_0^t c(\xi)z(\xi,t)d\xi & t>0, \\ z(\xi,0) = z_0(\xi) & 0<\xi<1 \ \text{(intitial condition)}, \end{cases}$$

where $c(\cdot) \in L_2(0,1)$ (as before, the output $y(t)$ is some averaged measurement of the temperature). Formally, the PDE may be expressed as

$$\frac{\partial z}{\partial t}(\xi,t) = \frac{\partial^2 z}{\partial \xi^2}(\xi,t) - \delta_0 u(t),$$

where δ_0 denotes the Dirac delta impulse at $\xi = 0$ (actually, this can be done in a rigorous way, using distribution theory, see Salamon [81]). In fact, choosing the state-space $\mathcal{H} = L_2(0,1)$ with state function

$$x(t) = z(t,\xi) \qquad t>0\,; \quad 0<\xi<1$$

as in Example 1.1, this equation can formally be modelled as a system of the form (1.4) with the same C_0-semigroup $S(\cdot)$, the same (bounded) output operator C and with B given by $Bu = -\delta_0 u$. The difference with Example 1.1 is that now $b(\cdot)$ is not an L_2-function, but a distribution. Hence $B \notin \mathcal{L}(\mathbb{R},\mathcal{H})$: we say that the input operator is unbounded. In the next chapter we shall see how one can interpret the system equations (1.4) in a rigorous manner. The idea is to introduce another Hilbert space $\mathcal{V}$ that contains $\mathcal{H}$ and is such that $B \in \mathcal{L}(U,\mathcal{V})$. Furthermore, B and C satisfy certain 'admissibility' properties which should be used in the interpretation of (1.4). □

Example 1.4 (*delay differential equation: unbounded case*)
Consider the delay differential equation

$$\begin{cases} \dot{z}(t) &= A_{f1}z(t) + A_{f2}z(t-2) + B_f u(t), \\ \\ y(t) &= C_f z(t-1), \end{cases} \tag{1.6}$$

where $z(t) \in \mathbb{R}^n, u(t) \in \mathbb{R}^m, y(t) \in \mathbb{R}^p$ and the matrices A_{f1}, A_{f2}, B_f and C_f are as in Example 1.2. The only difference with Example 1.2 is that now there is a delay in the output equation for $y(t)$. Hence the state equation can still be given by

$$x(t) = S(t)x_0 + \int_0^t S(t-s)Bu(s)ds,$$

where $S(\cdot)$ is the same C_0-semigroup on the same state-space $\mathcal{H} = M_2 = \mathbb{R}^n \times L_2(-2,0\,;\mathbb{R}^n)$, $B \in \mathcal{L}(\mathbb{R}^m, \mathcal{H})$ is the same (bounded) input operator and x_0 is the same initial condition as in Example 1.2. Defining $< \delta_x, \phi >= \phi(x)$ for $\phi \in C(-2,0\,;\mathbb{R}^n)$, we can express the delayed output equation for y as in formula

$$y(t) = Cx(t),$$

where C is given by

$$Cx = C_f < \delta_{-1}, \phi >= C_f\phi(-1) \quad \text{for} \quad x = (\eta, \phi).$$

However, we see that $C \notin \mathcal{L}(\mathcal{H}, \mathbb{R}^p)$, because an element of $L_2(-2,0\,;\mathbb{R}^n)$ does not have any continuity properties in general: C is unbounded with respect to $\mathcal{H}$. Due to an 'admissibility' property of B, however, the second component of $x(t)$ is an element of $C(-2,0\,;\mathbb{R}^n)$ for all $t \geq 0$ if the initial condition is smooth. Furthermore, C satisfies another 'admissibility' property so that this can somehow be extended to initial conditions in $\mathcal{H} = M_2$ and formula (1.4) makes sense. The notions of admissibility for input and output operators will be further explained in the next chapter.

Finally, we note that one of the simplest delay systems one could think of, namely the system represented by the transfer function

$$\exp(-s)C_f(sI - A_f)^{-1}B_f,$$

where A_f, B_f and C_f are matrices of appropriate dimensions, can be described in the above formulation (take $A_{f1} = A_f$ and $A_{f2} = 0$ in formula (1.6)). It does not, however, admit a bounded formulation as in Example 1.2. We shall come back to this particular example in Chapter 6. □

As mentioned above, the class of infinite-dimensional systems with *bounded* inputs and outputs has nice system theoretic properties and several interesting results have been obtained for this class. The attractive aspect of having such a class of systems is of course that one can obtain, at once, nice results for each element of the class. However, as we have seen in Examples 1.3,1.4, when modelling infinite-dimensional systems it is sometimes necessary to introduce *unbounded* input and output operators. Thus it would be convenient to have a *class* of systems that allows for such unboundedness.

In [72, 73] Pritchard and Salamon introduced an interesting class of systems in order to solve the linear quadratic control problem for several infinite-dimensional systems with unbounded inputs and outputs. This class of systems is now commonly referred to as the *Pritchard-Salamon class.* The reason for its popularity is that it is an extension of the 'bounded class' including many 'unbounded' examples, such that most of the nice system theoretic properties of the bounded class are retained. In particular, its structure is rich enough to solve the linear quadratic control problem (see Pritchard and Salamon [72, 73] and van Keulen [49]) and the $\mathcal{H}_\infty$-control problem, as shown in this book. As far as the unbounded examples are concerned, several PDE-models with point or boundary control, c.q. observation (including Example 1.3) fall within the Pritchard-Salamon framework (some examples can be found in Pritchard and Salamon [72]; we refer to Bontsema [9] and Curtain [11] for an interesting beam example). Other unbounded examples in the Pritchard-Salamon class are neutral systems with output delays (including Example 1.4) and retarded systems with delays in inputs and outputs (see Pritchard and Salamon [72, 73] and Salamon [80]).

Since the introduction of the Pritchard-Salamon class in [72, 73], several researchers have explored this class and succeeded in revealing many of its important mathematical and system theoretic properties. A rather recent reference related to this topic is [15], where Curtain et al. have given a nice overview containing many results about perturbation theory, transfer functions and stability theory for Pritchard-Salamon systems. One of the aims of the book is to demonstrate the nice properties of the Pritchard-Salamon class, while adding several new results in this respect. In particular, we shall see that the Pritchard-Salamon class is closed under (dynamic output) feedback perturbations and that it has very nice duality properties (these are needed in the derivation of the second 'dual' Riccati equation in the solution to the $\mathcal{H}_\infty$-control problem). Consequently, the contribution of this book is two-fold: we extend the finite-dimensional results for the $\mathcal{H}_\infty$-control problem to the Pritchard-Salamon class, but while doing so, we also show that this class has just the right system theoretic properties that allow for such an extension. At this point we should mention that our class of systems by no means contains all the infinite-dimensional systems for which Riccati equations are well-posed (see e.g. Lasiecka and Triggiani [57] and Bensoussan et al. [8] and references therein). However, these other classes do not share the nice perturbation and duality properties of the Pritchard-Salamon class. In the standard $\mathcal{H}_\infty$-control problem these properties are needed in order to allow for both input operators to be unbounded (for the control and the disturbance) and output operators to be unbounded (for the observation and the to-be-controlled output). Therefore, our class seems to be the largest class of infinite-dimensional state-space systems for which the $\mathcal{H}_\infty$-control problem with both unbounded input operators and unbounded output operators will have a solution in terms of two coupled Riccati equations.

1.3 Organization of the book

The rest of the book is organized as follows.

In Chapter 2 we introduce the Pritchard-Salamon class and we shall present many interesting preliminary results for this class, including frequency domain results, perturbation results, duality theory, stability theory, some theory about dynamic output feedback stabilization and a note on Riccati equations.

In Chapter 3 we consider the indefinite linear quadratic control problem (LQ-problem) for Pritchard-Salamon systems. The main result relates the solvability of the LQ-problem to the solvability of a Riccati equation and a frequency domain inequality, in the spirit of the well-known Kalman-Yakubovich-Popov Lemma (see Popov [71] and Willems [100]). This result, which is interesting in its own right, is a perfect generalization of the 'bounded' result of Louis and Wexler [62] and it is needed in the solution to the $\mathcal{H}_\infty$-control problem.

Chapter 4 is devoted to the $\mathcal{H}_\infty$-control problem with state-feedback. As explained above, this is the first step in the solution to the measurement-feedback problem. The solution is given using the natural analogues of the weak a priori assumptions for the finite-dimensional case in Glover and Doyle [37].

The main result of the book is given in Chapter 5. There we show that the (sub-optimal) $\mathcal{H}_\infty$-control problem for Pritchard-Salamon systems is solvable if and only if two coupled Riccati equations have stabilizing solutions. Furthermore, in the case that the problem is solvable we give a parametrization of all sub-optimal controllers. As for the state-feedback case, all the weak finite-dimensional assumptions can be replaced by their natural infinite-dimensional counterparts.

In Chapter 6 we shall clarify some of the developed theory by use of several examples. In particular, we shall examine some system theoretic properties of delay systems of the form $\exp(-s)C_f(sI - A_f)^{-1}B_f$ (see also Example 1.4) and we shall consider a mixed-sensitivity problem for these systems. In the last section of Chapter 6 we summarize the contributions of the book and we point out some directions for future research.

Some of the technical details in the proofs of several results have been deferred to an appendix. For the reader's convenience, the appendix also contains a number of known results about semigroup control systems.

Chapter 2

Pritchard-Salamon systems

In this chapter we introduce the same class of systems already considered in [72, 73, 74, 80, 15] (the Pritchard-Salamon class) and state a number of important results for this class. In particular, we shall consider frequency domain results, perturbation theory, duality theory, stabilizability and detectability concepts, the notion of dynamic output-feedback and Riccati equations for Pritchard-Salamon systems.

2.1 Notation

First of all, we introduce some notation that will be used throughout the book:

- For any normed space X, the norm is denoted by $\|\cdot\|_X$. If a sequence $x_n \in X$ converges to $x \in X$ as $n \to \infty$, we shall denote this by $x_n \stackrel{X}{\to} x$.
- For any Hilbert space X the inner product is denoted by $<\cdot,\cdot>_X$.
- For two Hilbert spaces X and Y,

  $$X \hookrightarrow Y$$

  means that $X \subset Y$, X is dense in Y and the canonical injection $X \to Y$ denoted by $x \mapsto x$ is continuous. In particular, there exists some constant c such that for all $x \in X$ there holds $\|x\|_Y \leq c\|x\|_X$.
- For Banach spaces X, Y, $\mathcal{L}(X,Y)$ denotes the Banach space of bounded linear operators from X to Y and $\mathcal{L}(X,X)$ is denoted by $\mathcal{L}(X)$.
- If a linear operator T has a unique extension to a bounded linear operator, this extension will be denoted by $\overline{T}$.

- If $T \in \mathcal{L}(X,Y)$ for some Hilbert spaces X, Y, then $T^* \in \mathcal{L}(Y,X)$ denotes the Hilbert adjoint (another kind of adjoint of a linear operator shall be introduced in Section 2.5, as well as the notion of dual spaces and some less standard notation).

- If $T \in \mathcal{L}(X,Y)$ for some Hilbert spaces X, Y, then the induced operator norm of T is denoted by $\|T\|$. If $T \in \mathcal{L}(X)$ for some Hilbert space X, then $r_\sigma^X(T)$ denotes its spectral radius. The resolvent set of a linear operator $A : D(A) \subset X \to X$ shall be denoted by $\rho^X(A)$ or $\rho(A)$.

- For $T \in \mathcal{L}(X)$, where X is some Hilbert space, we call T *coercive* if $T = T^*$ and there exists an $\epsilon > 0$ such that for all $x \in X$, $< Tx, x >_X \geq \epsilon < x, x >_X$ (this implies that T has a bounded inverse).

- If $S(\cdot)$ is a C_0-semigroup on two Hilbert spaces X and Y, then its infinitesimal generator shall be denoted by using the corresponding space as a superscript, e.g. A^X and A^Y.

- The growth bound of a C_0-semigroup $S(\cdot)$ on a Hilbert space X shall be denoted by $\omega_{S(\cdot)}^X$. If $\omega_{S(\cdot)}^X < 0$, $S(\cdot)$ is exponentially stable on X.

- If $S(\cdot)$ is a C_0-semigroup on a Hilbert space X with infinitesimal generator A^X, then $D(A^X)$ denotes the Hilbert space with the (graph) inner product defined by

$$< x, y >_{D(A^X)} = < x, y >_X + < A^X x, A^X y >_X .$$

We note that for $\sigma > \omega_{S(\cdot)}^X$ there holds $(\sigma I - A^X)^{-1} X = D(A^X)$ and the norm on $D(A^X)$ given by $\|(\sigma I - A^X) \cdot \|_X$ is equivalent to the (graph) norm of $D(A^X)$.

- For a separable Hilbert space X we denote for $-\infty \leq a \leq b \leq \infty$

$$L_p(a, b\,; X) = \{f : (a, b) \to X| < f(\cdot), x >_X$$

is Lebesgue measurable for all $x \in X$ and

$$\|f(\cdot)\|_{L_p(a,b\,;X)} = \left(\int_a^b \|f(t)\|_X^p \, dt \right)^{1/p} < \infty\}; 1 \leq p < \infty$$

By $L_p^{loc}(0, \infty; X)$ we shall denote the set of functions $f(\cdot)$ such that $f(\cdot) \in L_p(0, T; X)$ for all $0 < T < \infty$ (for Lebesgue (or Bochner) integration theory on Hilbert/Banach spaces we refer to [38]; the appendix of [17] is also a good reference).

- For $-\infty \leq a \leq b \leq \infty$ we define

$$\begin{array}{lcl} C(a,b\,;X) & = & \{f:(a,b)\to X \mid f(\cdot) \text{ is continuous}\} \\ CD(a,b\,;X) & = & \{f:(a,b)\to X \mid f(\cdot) \text{ is continuously differentiable}\} \\ CCD(a,b\,;X) & = & \{f:(a,b)\to X \mid f \in CD(a,b\,;X) \text{ with compact} \\ & & \qquad\qquad\qquad\qquad\qquad\qquad\qquad \text{support}\}. \end{array}$$

- $\mathbb{C}_\xi^+ = \{s \in \mathbb{C} \,|\mathrm{Re}(s) > \xi\}$ and $\mathbb{C}^+ = \mathbb{C}_0^+$.
- $\mathbb{C}_\xi^- = \{s \in \mathbb{C} \,|\mathrm{Re}(s) < \xi\}$ and $\mathbb{C}^- = \mathbb{C}_0^-$.
- For a complex Banach space Z, we consider the Hardy spaces

$$\mathcal{H}_\infty(\mathbb{C}_\xi^+, Z) = \{F : \mathbb{C}_\xi \to Z \mid F \text{ is holomorphic on } \mathbb{C}_\xi^+ \text{ and}$$

$$\|F(\cdot)\|_{\mathcal{H}_\infty(\mathbb{C}_\xi^+,Z)} = \sup_{s\in\mathbb{C}_\xi^+} \|F(s)\|_Z < \infty\},$$

$$\mathcal{H}_\infty(Z) = \mathcal{H}_\infty(\mathbb{C}_0^+, Z)$$

and

$$\mathcal{H}_2(Z) = \{F : \mathbb{C} \to Z \mid F \text{ is holomorphic on } \mathbb{C}^+ \text{ and}$$

$$\|F(\cdot)\|_{\mathcal{H}_2(Z)} = \frac{1}{\sqrt{2\pi}} \sup_{x>0} \left(\int_{-\infty}^{\infty} \|F(x+iy)\|_Z^2 dy \right)^{1/2} < \infty\}$$

(see e.g. [78, chapter 4]).

2.2 Definitions

In this section we shall give the definition of the Pritchard-Salamon class. As explained in the introduction, this class of systems extends the class of semigroup control systems with bounded input and output operators, by allowing for certain unboundedness of the input and output operators. In [72], the authors introduce a semigroup control system in the following manner.
Let $\mathcal{W}, \mathcal{H}$ and $\mathcal{V}$ be real separable Hilbert spaces, satisfying

$$\mathcal{W} \hookrightarrow \mathcal{H} \hookrightarrow \mathcal{V} \tag{2.1}$$

and suppose that $S(\cdot)$ is a C_0-semigroup on $\mathcal{V}$ which restricts to C_0-semigroups on $\mathcal{H}$ and $\mathcal{W}$. The infinitesimal generators of $S(\cdot)$ on $\mathcal{W}, \mathcal{H}$ and $\mathcal{W}$ are denoted by $A^{\mathcal{V}}, A^{\mathcal{H}}$ and $A^{\mathcal{W}}$, respectively. Furthermore, let the input space U and the output space Y also be real separable Hilbert spaces and suppose that the input operator is given by $B \in \mathcal{L}(U, \mathcal{V})$, the output operator is given by $C \in \mathcal{L}(\mathcal{W}, Y)$ and the feedthrough operator is given by $D \in \mathcal{L}(U, Y)$. We would like to give meaning to a system of the form

$$\begin{cases} x(t) &= S(t)x_0 + \int_0^t S(t-s)Bu(s)ds \\ y(t) &= Cx(t) + Du(t), \end{cases} \tag{2.2}$$

where $u(\cdot) \in L_2^{loc}(0, \infty; U)$ and $x_0 \in \mathcal{V}$ (see also the 'strong' formulation in (1.3)). We note that for the *bounded* case in Section 1.2 this formulation corresponds to $\mathcal{W} = \mathcal{H} = \mathcal{V}$. Often $\mathcal{H}$ can be considered as the 'natural state-space' of the system (see [72] and the continuation of Examples 1.3,1.4 below). Identifying $\mathcal{H}$ with its dual, it can also be used as a pivot space in order to define the duals of $\mathcal{W}$ and $\mathcal{V}$ (see Section 2.5). Since in general $B \notin \mathcal{L}(U, \mathcal{H})$ and $C \notin \mathcal{L}(\mathcal{H}, Y)$, we say that both B and C are unbounded with respect to this state-space. Due to the unboundedness of B, the integral term of the state $x(\cdot)$ given by

$$\int_0^t S(t-s)Bu(s)ds$$

is in principle a function with values only in the 'larger space' $\mathcal{V}$ (the integral is taken with respect to the topology of $\mathcal{V}$). Furthermore, in general $S(t)x_0$ also has values in $\mathcal{V}$, so that $y(\cdot) = Cx(\cdot) + Du(\cdot)$ need not be well-defined. In order to deal with this problem we first give a definition.

Definition 2.1
Let $\mathcal{W}, \mathcal{V}, U, Y$ be real separable Hilbert spaces as above and suppose that

$$\mathcal{W} \hookrightarrow \mathcal{V} \tag{2.3}$$

and that $S(\cdot)$ is a C_0-semigroup on $\mathcal{V}$ which restricts to a C_0-semigroup on $\mathcal{W}$.

(*i*) An operator $B \in \mathcal{L}(U, \mathcal{V})$ is called an *admissible input operator (with respect to* $(\mathcal{W}, \mathcal{V})$ *for* $S(\cdot)$*)*, if there exist $t_1 > 0$ and $\beta > 0$ such that

$$\int_0^{t_1} S(t_1 - s)Bu(s)ds \in \mathcal{W} \text{ and}$$

$$\left\| \int_0^{t_1} S(t_1 - s)Bu(s)ds \right\|_{\mathcal{W}} \le \beta \|u(\cdot)\|_{L_2(0,t_1;U)} \tag{2.4}$$

for all $u(\cdot) \in L_2(0, t_1; U)$.

(*ii*) An operator $C \in \mathcal{L}(\mathcal{W}, Y)$ is called an *admissible output operator* (*with respect to* $(\mathcal{W}, \mathcal{V})$ *for* $S(\cdot)$), if there exist $t_1 > 0$ and $\gamma > 0$ such that

$$\|CS(\cdot)x\|_{L_2(0,t_1;Y)} \leq \gamma \|x\|_{\mathcal{V}} \text{ for all } x \in \mathcal{W}. \tag{2.5}$$

Remark 2.2

If it is clear which semigroup and what spaces are involved, we shall omit the additional statements 'for $S(\cdot)$' and 'with respect to $(\mathcal{W}, \mathcal{V})$' .

If the properties (2.4) (or (2.5), respectively) hold for some t_1 and β (or t_1 and γ, respectively) then they hold for any $t_1 > 0$ and some $\beta > 0$ (or some $\gamma > 0$, respectively). Furthermore, for $u(\cdot) \in L_2^{loc}(0, \infty; U)$, $\bar{x}(t)$ defined by

$$\bar{x}(t) = \int_0^t S(t-s)Bu(s)ds,$$

is continuous with respect to the topology on $\mathcal{W}$ (see e.g. [15]).

Property (2.5) implies that the linear map from $\mathcal{W}$ to $L_2(0, T; Y)$, $x \mapsto CS(\cdot)x$ has a unique bounded extension from $\mathcal{V}$ to $L_2(0, T; Y)$. We shall denote this *observability map* by $x \mapsto \overline{CS}(\cdot)x$ for $x \in \mathcal{V}$. It follows from [72, Remark 3.5 (i)] that if $S(\cdot)$ is exponentially stable on $\mathcal{V}$, then for all $x \in \mathcal{V}$

$$\overline{CS}(\cdot)x \in L_2(0, \infty; Y) \text{ and } \|\overline{CS}(\cdot)x\|_{L_2(0,\infty;Y)} \leq \bar{\gamma}\|x\|_{\mathcal{V}} \tag{2.6}$$

(a similar result is proved in Appendix C.1, Lemma C.1). Finally, we note that the concepts of admissible input operator and admissible output operator are dual to each other, in the sense that B is an admissible input operator with respect to $(\mathcal{W}, \mathcal{V})$ for $S(\cdot)$ if and only if B' is an admissible output operator with respect to $(\mathcal{V}', \mathcal{W}')$ for $S'(\cdot)$ (for the precise formulation of this result we refer to Section 2.5 below).

Using Definition 2.1 and Remark 2.2, we can give meaning to the system in (2.2) in the case that B and C are admissible input and output operators, respectively. The output equation should be interpreted as the function

$$y(\cdot) = \overline{CS}(\cdot)x_0 + C\int_0^{\cdot} S(\cdot - s)Bu(s)ds + Du(\cdot)$$

(strictly speaking, there is no pointwise expression for $y(\cdot)$, due to the terms $\overline{CS}(\cdot)x_0$ and $Du(\cdot)$). We note that if B is an admissible input operator with respect to $(\mathcal{W}, \mathcal{V})$ as in Definition 2.1, then it follows from the fact that $\mathcal{W} \hookrightarrow \mathcal{H}$ that it is also admissible with respect to $(\mathcal{H}, \mathcal{V})$ (in Definition 2.1 $\mathcal{W}$ may be replaced by $\mathcal{H}$). Similarly, if C is an admissible output operator with respect to $(\mathcal{W}, \mathcal{V})$, it follows from the fact that $\mathcal{H} \hookrightarrow \mathcal{V}$ that it is also admissible with respect to $(\mathcal{W}, \mathcal{H})$. In some applications B (or C) may be such that the admissibility condition with respect to $(\mathcal{W},\mathcal{V})$ is not satisfied, whereas the admissibility condition with respect to $(\mathcal{H},\mathcal{V})$ (or $(\mathcal{W},\mathcal{H})$, respectively) is. In this case it may still be possible to give meaning to (2.2), but the theory becomes more complicated. A thorough analysis of that theory is beyond the scope of the book, but we shall mention some aspects. In a series of papers [81, 82, 95, 96, 97], Salamon and Weiss introduced the notions of *abstract linear systems* and *well-posedness* for linear systems. The concept of abstract linear systems captures the intrinsic characteristics of a linear system which are relevant for state-space analysis and synthesis of input-output systems. A linear system of the form (2.2) may be called well-posed if it fits in the framework of an abstract linear system. It can be shown that if B and C are admissible with respect to $(\mathcal{W},\mathcal{V})$ as in Definition 2.1, then the system (2.2) is well-posed in the above sense. A Pritchard-Salamon system has by definition such B and C:

Definition 2.3
Let $\mathcal{W},\mathcal{V},U,Y$ be real separable Hilbert spaces as above and suppose that $\mathcal{W}\hookrightarrow\mathcal{V}$ and that $S(\cdot)$ is a C_0-semigroup on $\mathcal{V}$ which restricts to a C_0-semigroup on $\mathcal{W}$. Let $B\in\mathcal{L}(U,\mathcal{V})$ and $C\in\mathcal{L}(\mathcal{W},Y)$ be admissible input and output operators with respect to $(\mathcal{W},\mathcal{V})$ for $S(\cdot)$, respectively, and suppose that $D\in\mathcal{L}(U,Y)$. The system given by

$$\left\{\begin{array}{lcl} x(t) & = & S(t)x_0+\int_0^t S(t-s)Bu(s)ds \\ \\ y(t) & = & Cx(t)+Du(t), \end{array}\right. \tag{2.7}$$

where $x_0\in\mathcal{V}, t\geq 0$ and $u(\cdot)\in L_2^{loc}(0,\infty;U)$ is called a *Pritchard-Salamon system (with respect to* $(\mathcal{W},\mathcal{V})$ *for* $S(\cdot)$) and is denoted by $\Sigma(S(\cdot),B,C,D)$. If, in addition, we have

$$D(A^{\mathcal{V}})\hookrightarrow\mathcal{W}, \tag{2.8}$$

we call the system a *smooth* Pritchard-Salamon system.

As before, if it is clear what spaces are involved, we shall omit the additional statement 'with respect to $(\mathcal{W},\mathcal{V})$ for $S(\cdot)$'.

Remark 2.4
In [15] it is shown that the smoothness condition $D(A^{\mathcal{V}})\hookrightarrow\mathcal{W}$ is implied by the assumption that $D(A^{\mathcal{V}})\subset\mathcal{W}$. Furthermore, if this assumption is satisfied it follows that the resolvent sets of $A^{\mathcal{V}}$ and $A^{\mathcal{W}}$ are the same. However, the growth bounds $\omega_{S^{\mathcal{W}}(\cdot)}$ and $\omega_{S^{\mathcal{V}}(\cdot)}$ need not be the same (a counterexample is given in [15]).

Remark 2.5
It is clear from Remark 2.2 that for a Pritchard-Salamon system $\Sigma(S(\cdot), B, C, D)$ of the form (2.7) the map G defined by

$$(Gu)(\cdot) := C\int_0^{\cdot} S(\cdot - s)Bu(s)ds + Du(\cdot).$$

is a linear time-invariant map from $L_2^{loc}(0, \infty; U)$ to $L_2^{loc}(0, \infty; Y)$. We call $\Sigma(S(\cdot), B, C, D)$ a *realization* for G, although it should of course be noted that not every linear time-invariant map from $L_2^{loc}(0, \infty; U)$ to $L_2^{loc}(0, \infty; Y)$ can be represented by a Pritchard-Salamon system. In the rest of the book, we shall sometimes use the notation Σ_G instead of $\Sigma(S(\cdot), B, C, D)$ if we want to express that the linear map corresponding to the system with zero initial conditions is given by G.

Remark 2.6
The class of systems introduced by Pritchard and Salamon in [72] satisfies all the assumptions in Definition 2.3. We have not incorporated their intermediate space $\mathcal{H}$ in the definition, because it is not relevant for the results in the next sections. The definition of a Pritchard-Salamon system in this form was given in [15]. Assumption (2.8) is used in [72] in the solution to the standard linear quadratic control problem for this class of systems. It is satisfied by all known examples that satisfy (2.4) and (2.5) as long as the spaces $\mathcal{W}$ and $\mathcal{V}$ are chosen appropriately (see [72]) and so it is not very restrictive.

The reason for considering the Pritchard-Salamon class, rather than systems with more unbounded input and output operators, is that it has many of the nice structural properties in common with finite-dimensional systems, such as closedness under (dynamic) output-feedback and several 'duality' properties. This structure is explored in the next sections and is exploited in the derivation of the solutions to the (nonstandard) linear quadratic control problem in Chapter 3 and the $\mathcal{H}_\infty$-control problem in Chapters 4 and 5.

Next, we show how the systems of Examples 1.3 and 1.4 can both be formulated as Pritchard-Salamon systems. The problem is always to find the right choices for $\mathcal{W}$ and $\mathcal{V}$, so that the input and output operators satisfy the admissibility conditions. The reformulation of the examples as Pritchard-Salamon systems should give an impression of this. As before, the results are borrowed from [72].

Example 2.7 (*continuation of Example* 1.3)
In order to formulate the parabolic PDE with boundary control of Example

1.3 as a Pritchard-Salamon system, we first consider a more general type of parabolic system. Let A be a self-adjoint operator on a separable Hilbert space $\mathcal{H}$ and suppose that it has compact resolvent and that its spectrum consists of strictly decreasing real eigenvalues $\lambda_n\ ; n \in \mathbb{N}$ with eigenvectors $\phi_n \in \mathcal{H}$, $\|\phi_n\|_{\mathcal{H}} = 1$. In this case $\{\phi_n\ ; n \in \mathbb{N}\}$ forms an orthonormal basis of $\mathcal{H}$ so that for all $x \in \mathcal{H}$,

$$\sum_{n=0}^{\infty} < x, \phi_n >_{\mathcal{H}}^2 < \infty \quad \text{and} \quad x = \sum_{n=0}^{\infty} < x, \phi_n >_{\mathcal{H}} \phi_n.$$

A can be represented as

$$\begin{aligned} D(A) &= \{x \in \mathcal{H} \mid \textstyle\sum_{n=0}^{\infty} \lambda_n^2 < x, \phi_n >_{\mathcal{H}}^2 < \infty\} \\ Ax &= \textstyle\sum_{n=0}^{\infty} \lambda_n < x, \phi_n >_{\mathcal{H}} \phi_n \end{aligned} \tag{2.9}$$

and the C_0-semigroup $S(\cdot)$ generated by A is given by

$$S(t)x = \sum_{n=0}^{\infty} \exp(\lambda_n t) < x, \phi_n >_{\mathcal{H}} \phi_n. \tag{2.10}$$

Now let β_n and γ_n be positive sequences satisfying $0 < \beta_n \leq 1 \leq \gamma_n < \infty$ and suppose that $\mathcal{W}$ and $\mathcal{V}$ are determined by

$$\mathcal{W} = \{x \in \mathcal{H} \mid \sum_{n=0}^{\infty} \gamma_n < x, \phi_n >_{\mathcal{H}}^2 < \infty\}$$

$$\mathcal{V}' = \{x \in \mathcal{H} \mid \sum_{n=0}^{\infty} \beta_n^{-1} < x, \phi_n >_{\mathcal{H}}^2 < \infty\},$$

with the obvious inner products. Here we assume that $\mathcal{H}$ is identified with its dual, so that $\mathcal{V}' \subset \mathcal{H} = \mathcal{H}' \subset \mathcal{V}$ (see also Section 2.5 about duality). This means that $\mathcal{V}$ can be represented as a space of sequences

$$\mathcal{V} = \{x \in \mathbb{R}^{\mathbb{N}} \mid \sum_{n=0}^{\infty} \beta_n x_n^2 < \infty\}$$

and the injection $\mathcal{H} \subset \mathcal{V}$ is given by identifying $x \in \mathcal{H}$ with the sequence $\{< x, \phi_n >_{\mathcal{H}}\ ; n \in \mathbb{N}\}$. Finally, let $B \in \mathcal{L}(\mathbb{R}, \mathcal{V})$ and $C \in \mathcal{L}(\mathcal{W}, \mathbb{R})$ be given by

$$Bu = \{b_n u\ ; n \in \mathbb{N}\} \quad \text{and} \quad Cx = \sum_{n=0}^{\infty} c_n < x, \phi_n >_{\mathcal{H}},$$

where the sequences $\{b_n\ ; n \in \mathbb{N}\}$ and $\{c_n\ ; n \in \mathbb{N}\}$ are such that

$$\sum_{n=0}^{\infty} \beta_n b_n^2 < \infty \quad \text{and} \quad \sum_{n=0}^{\infty} \gamma_n^{-1} c_n^2 < \infty. \tag{2.11}$$

It is not difficult to show that B and C are admissible with respect to $(\mathcal{W}, \mathcal{V})$ in the sense of Definition 2.1 if

$$\sum_{n=n_0}^{\infty} \frac{\gamma_n b_n^2}{|\lambda_n|} < \infty \quad \text{and} \quad \sum_{n=n_0}^{\infty} \frac{c_n^2}{\beta_n |\lambda_n|} < \infty, \tag{2.12}$$

where $n_0 = 1+\max\{n \in \mathbb{N} \mid \lambda_n \geq 0\}$. Furthermore, given sequences $b_n, c_n, \lambda_n \in \mathbb{R}$ such that λ_n is strictly decreasing and $\lambda_n \downarrow -\infty$, there exist sequences β_n, γ_n such that the inequalities (2.11)-(2.12) are satisfied if and only if

$$\sum_{n=n_0}^{\infty} \frac{b_n c_n}{|\lambda_n|^{1/2}} < \infty. \tag{2.13}$$

(see [72, Lemma 4.4]). This last result particularly shows that B and C cannot be 'too unbounded' with respect to $\mathcal{H}$. Furthermore, it shows that B can be 'more unbounded' as long as C is 'less unbounded' and vice versa. We mention that in addition to (2.4)-(2.5), the condition $D(A^{\mathcal{V}}) \hookrightarrow \mathcal{W}$ is also satisfied, provided that γ_n, β_n are chosen appropriately.

Finally, we can formulate the parabolic PDE with boundary control of Example 1.3 as a Pritchard-Salamon system. Choosing $\mathcal{H} = L_2(0,1)$ it follows that A is indeed self-adjoint, that it has compact resolvent and that it is of the form (2.9), where $\lambda_0 = 0$; $\phi_0 = 1$ and $\lambda_n = -n^2\pi^2$; $\phi_n(\xi) = \sqrt{2}\cos(n\pi\xi)$ for $n \geq 1$. Recall that the input and output operators were given by

$$Cx = \int_0^1 c(\xi)x(\xi)d\xi, \quad c(\cdot) \in \mathcal{H}.$$

and

$$Bu = -\delta u.$$

Hence we get $c_n =< c(\cdot), \phi_n >_{\mathcal{H}}$; $n \in \mathbb{N}$ and $b_0 = -1$; $b_n = -\sqrt{2}$ for $n \geq 1$ and condition (2.13) is satisfied if and only if

$$\sum_{n=1}^{\infty} \frac{|c_n|}{n} < \infty. \tag{2.14}$$

Since $c(\cdot) \in \mathcal{H}$, we have $\sum_{n=1}^{\infty} c_n^2 < \infty$ and so condition (2.14) is satisfied. In fact, we can choose the sequences γ_n and β_n by $\gamma_n = 1$ and $\beta_n = n^{-2}$, $n \geq 1$. This corresponds to the choice $\mathcal{W} = \mathcal{H}$ and $\mathcal{V}' = W^{1,2}(0,1)$ (the Sobolev space of absolute continuous functions in $L_2(0,1)$ whose derivative is in $L_2(0,1)$). Hence, we have modelled Example 1.3 as a Pritchard-Salamon system.

Finally, we note that we could allow for more unboundedness of the operator C, because c_n is only required to satisfy (2.14). However, C cannot be taken to represent a point observation, because this would correspond to $c_n = \sqrt{2}$, $n \geq 1$ and then condition (2.14) would be violated. $\square$

Example 2.8 (*continuation of Example* 1.4)
In order to model the delay differential equation of Example 1.4 as a Pritchard-Salamon system, we first introduce a much more general functional differential equation (cf. [72, 80]), so that the system of Example 1.4 will be merely a special case (in fact, here we consider a system of the neutral type, rather than of the retarded type). Consider the system given by

$$\begin{cases} \frac{d}{dt}(z(t) - Mz_t) & = & Lz_t + B_0u(t), \\ \\ y(t) & = & C_0z_t, \end{cases} \tag{2.15}$$

where $z(t) \in \mathbb{R}^n, u(t) \in \mathbb{R}^m, y(t) \in \mathbb{R}^p$ and z_t is the solution segment defined by

$$z_t(\tau) = z(t + \tau)\ ; \quad -h \le \tau \le 0\ ; \quad h > 0.$$

We have $B_0 \in \mathbb{R}^{n \times m}$ and L, M, C_0 are bounded linear functionals from $C(-h, 0\ ; \mathbb{R}^n)$ into $\mathbb{R}^n$ and $\mathbb{R}^p$ respectively. It is easy to see that the delay differential equation of Example 1.4 can be reformulated as in (2.15), using

$$\begin{aligned} h &= 2 \\ M &= 0 \\ L &= A_{f1} < \delta_0, \cdot > + A_{f2} < \delta_{-2}, \cdot > \\ B_0 &= B_f \\ C_0 &= C_f < \delta_{-1}, \cdot >, \end{aligned}$$

where $< \delta_x, \phi >= \phi(x)$ for $\phi \in C(-1, 0\ ; \mathbb{R}^n)$.

Under some conditions (see [72, Example 4.1]), the system given by (2.15) has a unique solution $z(t)$; $t \ge -h$, for every input $u(\cdot) \in L_2^{loc}(0, \infty; \mathbb{R}^m)$ and every initial condition satisfying

$$\lim_{t \downarrow 0}(z(t) - Mz_t) = \eta_0, \quad z(\tau) = \phi_0(\tau); \quad -h < \tau < 0,$$

where $x_0 = (\eta_0, \phi_0) \in M_2 = \mathbb{R}^n \times L_2(-h, 0\ ; \mathbb{R}^n)$. Moreover, the evolution of the state

$$x(t) = (z(t) - Mz_t, z_t) \in M_2$$

of the system can be described by

$$x(t) = S(t)x_0 + \int_0^t S(t - s)Bu(s)ds,$$

where $B \in \mathcal{L}(\mathbb{R}^m, M_2)$ maps $u \in \mathbb{R}^m$ into the pair $Bu = (B_0 u, 0)$ and $S(t) \in \mathcal{L}(M_2)$ is the C_0-semigroup generated by the operator A given by

$$\begin{aligned} D(A) &= \{x = (\eta, \phi) \in M_2 \mid \phi \in W^{1,2}, \eta = \phi(0) - M\phi\}, \\ Ax &= (L\phi, \dot{\phi}). \end{aligned}$$

Here $W^{1,2}$ denotes the Sobolev space $W^{1,2}(-h, 0\,; \mathbb{R}^n)$ of absolute continuous functions in $L_2(-h, 0\,; \mathbb{R}^n)$ whose derivative is in $L_2(-h, 0\,; \mathbb{R}^n)$.
Now $D(A)$ can be considered as a Hilbert space by choosing the inner product

$$< (\eta, \phi), (\bar{\eta}, \bar{\phi}) >_{D(A)} = < \phi, \bar{\phi} >_{W^{1,2}}$$

and it follows that $S(\cdot)$ restricts to a C_0-semigroup on $D(A)$.
The output $y(t) = C_0 z_t$ of the system may formally be described by

$$y(t) = Cx(t) = C\left(z(t) - Mz_t, z_t\right),$$

where the output operator C is given by

$$C : D(A) \to \mathbb{R}^p, \quad Cx = C(\eta, \phi) = C_0 \phi.$$

We recall that by assumption C_0 is a bounded linear map from $C(-h, 0\,; \mathbb{R}^n)$ to $\mathbb{R}^p$ so that $C \in \mathcal{L}(D(A), \mathbb{R}^p)$ but in general $C \notin \mathcal{L}(M_2, \mathbb{R}^p)$. This last situation of course occurs in Example 1.4, because there $C_0 = C_f < \delta_{-1}, \cdot >$, which is not bounded on $L_2(-2, 0\,; \mathbb{R}^n)$.

Using the fact that $S(\cdot)$ restricts to a C_0-semigroup on $D(A)$, a natural choice for $\mathcal{W}$ and $\mathcal{V}$ is $\mathcal{W} = D(A)$ and $\mathcal{V} = M_2$, because then $C \in \mathcal{L}(\mathcal{W}, \mathbb{R}^p)$ and $B \in \mathcal{L}(\mathbb{R}^m, \mathcal{V})$ and we can choose $U = \mathbb{R}^m$ and $Y = \mathbb{R}^p$. In [72] it is explained that now B and C are both admissible in the sense of Definition 2.1, so that the neutral functional differential equation of Example 1.4 can indeed be modelled as a Pritchard-Salamon system. Finally, we note that condition (2.8) is trivially satisfied because of $\mathcal{W} = D(A)$, so that in fact we have a smooth Pritchard-Salamon system. □

Remark 2.9

We have seen two examples of infinite-dimensional systems with unbounded inputs and outputs that can be formulated as Pritchard-Salamon systems: a class of parabolic PDE's with Neumann boundary control in Example 2.7 and the class of neutral functional differential equations of Example 2.8. In [72] the authors show that also certain hyperbolic PDE's can be modelled as Pritchard-Salamon systems. However, these systems are in general only exponentially stabilizable if there is some internal damping (we shall need a stabilizability assumption for both the linear quadratic control problem in Chapter 3 and the $\mathcal{H}_\infty$-control problem in Chapters 4,5). Furthermore, it should be noted that Dirichlet boundary control usually leads to input operators that are 'too unbounded' for the Pritchard-Salamon framework. In [73] it is shown that also the class of retarded functional differential equations with delays in both the input and the output falls within the Pritchard-Salamon framework.

The section is concluded with some differentiability properties for Pritchard-Salamon systems in the case that the initial condition and the input are smooth. In some sense, we obtain a 'strong' formulation of a system of the form (2.7). The result is similar to the known results for the bounded case (see Appendix B.3). The difference is of course caused by the fact that the topologies of $\mathcal{W}$ and $\mathcal{V}$ are not the same. The proof of the result is similar to the proofs of [81, Lemma 2.3, Lemma 2.5] and is therefore deleted.

Lemma 2.10
Let $\Sigma(S(\cdot), B, C, D)$ be a Pritchard-Salamon system as in (2.7), suppose that $x_0 \in D(A^{\mathcal{V}})$ and let $u(\cdot) \in CD(0,T;U)$ for some $T > 0$. Then $x(\cdot)$ is continuously differentiable with respect to the topology of $\mathcal{V}$ and for all $t \in (0,T)$ we have $x(t) \in D(A^{\mathcal{V}})$ and

$$\dot{x}(t) = A^{\mathcal{V}}x(t) + Bu(t) = S(t)(A^{\mathcal{V}}x_0 + Bu(0)) + \int_0^t S(t-s)B\dot{u}(s)ds.$$

If, in addition, $x_0 \in \mathcal{W}$ and $A^{\mathcal{V}}x_0 + Bu(0) \in \mathcal{W}$, then $x(\cdot)$ is also differentiable with respect to the topology of $\mathcal{W}$ and the derivative is given by the same formula.

2.3 Frequency domain results

As a matter of course, we shall also be interested in frequency domain properties of Pritchard-Salamon systems. Below we shall give a definition of Fourier and Laplace transforms, but first we make a remark about the *complexification* of a real Hilbert space H. Formally, the elements of the complexification H^c of H are of the form $x + iy$, where $x, y \in H$ and the inner product on H is determined by

$$< x_1 + iy_1, x_2 + iy_2 >_{H^c} :=$$

$$< x_1, x_2 >_H + < y_1, y_2 >_H + i(< y_1, x_2 >_H - < x_1, y_2 >_H).$$

Any linear operator T with domain $D(T) \in H$ defines a linear operator T^c with domain $D(T^c) = D(T) + iD(T) \subset H^c$ and

$$T^c(x + iy) = Tx + iTy \text{ for all } x, y \in D(T).$$

Henceforth, we shall use complexifications where necessary, without using the superscript c.
Suppose that H is some Hilbert space and let $x(\cdot) \in L_2(0,\infty;H)$. The Fourier transform of $x(\cdot)$ denoted by $\hat{x}(\cdot)$ is defined by

$$\hat{x}(i\omega) = \underset{T\to\infty}{\text{l.i.m.}} \int_0^T \exp(-i\omega t)x(t)dt,$$

where l.i.m. stands for limit in (quadratic) mean. It is well known (Plancherel's Theorem, see e.g. [78, section 4.8]) that the function determined by $\omega \mapsto \hat{x}(i\omega)$ is an element of $L_2(\mathbb{R},H)$ and

$$\|x(\cdot)\|^2_{L_2(0,\infty;H)} = (1/2\pi)\|\hat{x}(\cdot)\|^2_{L_2(\mathbb{R};H)}. \tag{2.16}$$

Furthermore, we define the Laplace transform $\hat{x}(\cdot)$ of $x(\cdot)$, by

$$\hat{x}(s) = \int_0^\infty \exp(-st)x(t)dt, \ \mathrm{Re}(s) > 0.$$

It follows from Fatou's Theorem (see e.g. [78, section 4.6]) and the Paley-Wiener Theorem (see e.g. [78, section 4.8]) that $\hat{x}(\cdot) \in \mathcal{H}_2(X)$, $\hat{x}(s)$ converges to $\hat{x}(i\omega)$ almost everywhere on the imaginary axis (nontangential limit, see [78, section 4.6]) and

$$\|x(\cdot)\|^2_{L_2(0,\infty;H)} = \|\hat{x}(\cdot)\|^2_{\mathcal{H}_2(X)} \tag{2.17}$$

We shall introduce the concept of transfer function for Pritchard-Salamon systems as in [15, Definition 2.14]. Let $\Sigma_G = \Sigma(S(\cdot),B,C,D)$ be a Pritchard-Salamon system of the form (2.7) so that G denotes the linear map from $L_2^{loc}(0,\infty;U)$ to $L_2^{loc}(0,\infty;Y)$ given by

$$(Gu)(\cdot) = C\int_0^{\cdot} S(\cdot - s)Bu(s)ds + Du(\cdot). \tag{2.18}$$

Define Ω as the subset of functions in L_2 with at most exponential growth:

$$\Omega := \{u \in L_2^{loc}(0,\infty;U) \mid \exists\gamma \in \mathbb{R} \text{ s.t. } u(\cdot)e^{-\gamma\cdot} \in L_2(0,\infty;U)\}$$

and for $u(\cdot) \in \Omega$ the *energy bound*

$$\lambda(u) := \inf\{\lambda \in \mathbb{R} \mid u(\cdot)e^{-\lambda\cdot} \in L_2(0,\infty;U)\}.$$

Definition 2.11
A holomorphic function $\tilde{G} : \mathbb{C}^+_\xi \to \mathcal{L}(U,Y)$ is called a *transfer function* of $\Sigma(S(\cdot),B,C,D)$ if for all $u(\cdot) \in \Omega$ there holds

$$\widehat{Gu}(s) = \tilde{G}(s)\hat{u}(s) \text{ for all } s \in \mathbb{C}^+_{\max(\xi,\lambda(u))}.$$

It is clear that if $\tilde{G}_1 : \mathbb{C}^+_{\xi_1} \to \mathcal{L}(U,Y)$ and $\tilde{G}_2 : \mathbb{C}^+_{\xi_2} \to \mathcal{L}(U,Y)$ are two transfer functions of $\Sigma(S(\cdot), B, C, D)$, then $\tilde{G}_1(s) = \tilde{G}_2(s)$ for $s \in \mathbb{C}^+_{\max(\xi_1,\xi_2)}$. Hence the appropriate phrase '*the* transfer function of a Pritchard-Salamon system'. However, first we should show that a Pritchard-Salamon system indeed *has* a transfer function. Using the results in [15], we shall show that $\Sigma(S(\cdot), B, C, D)$ has a transfer function and we shall derive an expression for the transfer function in terms of $A^{\mathcal{V}}, A^{\mathcal{W}}, B, C$ and D. Let $\xi > \max(\omega_{S^{\mathcal{W}}(\cdot)}, \omega_{S^{\mathcal{V}}(\cdot)})$. It follows from [15, Lemma 2.12] that for all $s \in \mathbb{C}^+_\xi$ we have $(sI - A^{\mathcal{V}})^{-1}B \in \mathcal{L}(U, \mathcal{W})$ and that $C(sI - A^{\mathcal{W}})^{-1} \in \mathcal{L}(\mathcal{W}, Y)$ has a continuous extension to a linear map $\overline{C(sI - A^{\mathcal{W}})^{-1}} \in \mathcal{L}(\mathcal{V}, Y)$. In fact, for all $s \in \mathbb{C}^+_\xi$ we have

$$\|(sI - A^{\mathcal{V}})^{-1}B\|_{\mathcal{L}(U,\mathcal{W})} \leq \frac{c_1}{(\mathrm{Re}(s) - \xi)^{\frac{1}{2}}} \tag{2.19}$$

and

$$\|C(sI - A^{\mathcal{W}})^{-1}x\|_Y \leq \frac{c_2\|x\|_{\mathcal{V}}}{(\mathrm{Re}(s) - \xi)^{\frac{1}{2}}} \tag{2.20}$$

for all $x \in \mathcal{W}$. For $s \in \mathbb{C}^+_\xi$, we define

$$\tilde{G}(s) := C(sI - A^{\mathcal{V}})^{-1}B + D = \overline{C(sI - A^{\mathcal{W}})^{-1}}B + D \tag{2.21}$$

(the last equality follows from the proof of [15, Theorem 3.3]). Consider G given by (2.18) and let $\eta > \xi$. It follows from [15, Proposition 2.15] that if $u(\cdot) \in L_2^{loc}(0,\infty;U)$ is such that $\exp(-\xi\cdot)u(\cdot) \in L_2(0,\infty;U)$, then $\exp(-\eta\cdot)y(\cdot) = \exp(-\eta\cdot)(Gu)(\cdot) \in L_1(0,\infty;Y) \cap L_2(0,\infty;Y)$. Furthermore,

$$\hat{y}(s) = \tilde{G}(s)\hat{u}(s) \text{ for all } s \in \mathbb{C}^+_\xi \tag{2.22}$$

and

$$\tilde{G}(\cdot) \in \mathcal{H}_\infty(\mathbb{C}^+_\xi, \mathcal{L}(U,Y))$$

and so $\tilde{G}(\cdot)$ given by (2.21) is the transfer function of Σ_G.
It follows from (2.19) that for $\mathrm{Re}(s) \to \infty$

$$\|C(sI - A^{\mathcal{V}})^{-1}B\|_{\mathcal{L}(U,Y)} \to 0$$

so that

$$\lim_{\mathrm{Re}(s)\to\infty} \tilde{G}(s) = D. \tag{2.23}$$

We conclude this section with a result that will be very useful in the sequel. It is a direct consequence of the property $\mathcal{W} \hookrightarrow \mathcal{V}$.

Lemma 2.12
Let $\mathcal{W}$ and $\mathcal{V}$ be Hilbert spaces such that $\mathcal{W} \hookrightarrow \mathcal{V}$ and suppose that $S(\cdot)$ is a C_0-semigroup on both spaces. For any $\sigma > max(\omega_{S^{\mathcal{W}}(\cdot)}, \omega_{S^{\mathcal{V}}(\cdot)})$, there holds

$$(\sigma I - A^{\mathcal{V}})^{-1} \mid_{\mathcal{W}} = (\sigma I - A^{\mathcal{W}})^{-1} \tag{2.24}$$

and we have

$$D(A^{\mathcal{W}}) \hookrightarrow D(A^{\mathcal{V}}). \tag{2.25}$$

Proof
We know that $S(\cdot)$ is a C_0-semigroup on both $\mathcal{V}$ and $\mathcal{W}$ and that $\mathcal{W} \hookrightarrow \mathcal{V}$. Hence [69, Theorem 4.5.5] implies that $D(A^{\mathcal{W}})$ is the part of $D(A^{\mathcal{V}})$ in $\mathcal{W}$. Therefore,

$$(\sigma I - A^{\mathcal{V}})^{-1}(\sigma I - A^{\mathcal{W}})x = x$$

for all $x \in D(A^{\mathcal{W}})$. This implies that for all $y \in \mathcal{W}$

$$(\sigma I - A^{\mathcal{V}})^{-1}(\sigma I - A^{\mathcal{W}})(\sigma I - A^{\mathcal{W}})^{-1}y = (\sigma I - A^{\mathcal{W}})^{-1}y$$

and this shows (2.24).

Next we prove (2.25). Since $D(A^{\mathcal{W}})$ is the part of $D(A^{\mathcal{V}})$ in $\mathcal{W}$, there holds $D(A^{\mathcal{W}}) \subset D(A^{\mathcal{V}})$ and for all $w \in D(A^{\mathcal{W}})$ we have $\|w\|^2_{D(A^{\mathcal{V}})} = \|A^{\mathcal{V}}w\|^2_{\mathcal{V}} + \|w\|^2_{\mathcal{V}} \leq const(\|A^{\mathcal{W}}w\|^2_{\mathcal{W}} + \|w\|^2_{\mathcal{W}}) = const\|w\|^2_{D(A^{\mathcal{W}})}$, so that the injection from $D(A^{\mathcal{W}})$ to $D(A^{\mathcal{V}})$ is continuous. Finally, we prove that $D(A^{\mathcal{W}})$ is dense in $D(A^{\mathcal{V}})$: let x be an element of $D(A^{\mathcal{V}})$ and define $v := (\sigma I - A^{\mathcal{V}})x \in \mathcal{V}$ for $\sigma > \max(\omega_{S^{\mathcal{W}}(\cdot)}, \omega_{S^{\mathcal{V}}(\cdot)})$. Because of the dense inclusion $\mathcal{W} \hookrightarrow \mathcal{V}$ there exists a sequence $v_n \in \mathcal{W}$ such that $v_n \stackrel{\mathcal{V}}{\rightarrow} v$ as $n \to \infty$. Hence, $\|(\sigma I - A^{\mathcal{V}})^{-1}(v_n - v)\|_{\mathcal{V}} \to 0$ as $n \to \infty$. Since $(\sigma I - A^{\mathcal{V}})^{-1} \mid_{\mathcal{W}} = (\sigma I - A^{\mathcal{W}})^{-1}$, the sequence $w_n := (\sigma I - A^{\mathcal{V}})^{-1}v_n \in D(A^{\mathcal{W}})$ satisfies $\|w_n - x\|_{D(A^{\mathcal{V}})} \to 0$ as $n \to \infty$. ∎

2.4 Perturbation results

In this section we give a number of perturbation results for Pritchard-Salamon systems. Most of these results can be found in [15, section 4]; only the new results will be proved.

Lemma 2.13
Let $\Sigma(S(\cdot), B, C, D)$ be a Pritchard-Salamon system of the form (2.7) and let $F \in \mathcal{L}(\mathcal{W}, U)$ be an admissible output operator with respect to $(\mathcal{W}, \mathcal{V})$ and $H \in \mathcal{L}(Y, \mathcal{V})$ an admissible input operator with respect to $(\mathcal{W}, \mathcal{V})$ for this system.

(*i*) *There exists a unique* C_0*-semigroup* $S_{BF}(\cdot)$ *on* $\mathcal{W}$ *such that for all* $x \in \mathcal{W}$

$$S_{BF}(t)x = S(t)x + \int_0^t S(t-s)BFS_{BF}(s)xds. \tag{2.26}$$

Furthermore, $S_{BF}(\cdot)$ *extends to a* C_0*-semigroup on* $\mathcal{V}$*,* $\Sigma(S_{BF}(\cdot), B, C, D)$ *is a Pritchard-Salamon system and*

$$S_{BF}(t)x = S(t)x + \int_0^t S(t-s)B\overline{FS_{BF}}(s)xds =$$

$$S(t)x + \int_0^t S_{BF}(t-s)B\overline{FS}(s)xds \quad \textit{for all } x \in \mathcal{V}. \tag{2.27}$$

In addition, F *is an admissible output operator for* $S_{BF}(\cdot)$ *and* H *is an admissible input operator for* $S_{BF}(\cdot)$.

(*ii*) *The generators of* $S_{BF}(\cdot)$ *on* $\mathcal{W}$ *and* $\mathcal{V}$ *satisfy*

$$D(A^{\mathcal{V}}_{BF}) = D(A^{\mathcal{V}}) \quad \textit{with equivalent graph norms}$$

$$D(A^{\mathcal{W}}_{BF}) = \{x \in D(A^{\mathcal{V}}) \cap \mathcal{W}) \mid A^{\mathcal{V}}_{BF}x \in \mathcal{W}\},$$

$$A^{\mathcal{W}}_{BF}x = A^{\mathcal{V}}_{BF}x \textit{ for all } x \in D(A^{\mathcal{W}}_{BF}),$$

$$A^{\mathcal{V}}_{BF}x = A^{\mathcal{V}}x + B\overline{F(\sigma I - A^{\mathcal{W}})^{-1}}(\sigma I - A^{\mathcal{V}})x \textit{ for all } x \in D(A^{\mathcal{V}}),$$

where σ *is any number with real part larger than the growth bounds of* $S(\cdot)$ *and* $S_{BF}(\cdot)$ *on* $\mathcal{W}$ *and* $\mathcal{V}$*. If, in addition, assumption* (2.8) *is satisfied, there holds*

$$A^{\mathcal{V}}_{BF}x = (A^{\mathcal{V}} + BF)x \textit{ for all } x \in D(A^{\mathcal{V}}_{BF}) = D(A^{\mathcal{V}}).$$

(*iii*) *Let* $F_1, F_2 \in \mathcal{L}(\mathcal{W}, U)$ *be admissible output operators with respect to* $(\mathcal{W}, \mathcal{V})$ *for* $S(\cdot)$*. There holds*

$$S_{B(F_1+F_2)}(\cdot) = (S_{BF_1})_{BF_2}(\cdot). \tag{2.28}$$

Proof

Almost all of the results in this lemma can be found in [15]. Here we prove that the graph norms of $D(A^{\mathcal{V}})$ and $D(A^{\mathcal{V}}_{BF})$ are equivalent and we derive the expressions for $A^{\mathcal{V}}_{BF}$. As explained in [15], we can apply the Laplace transform to (2.27) to obtain for all $x \in \mathcal{V}$

$$(\sigma I - A^{\mathcal{V}}_{BF})^{-1}x = (\sigma I - A^{\mathcal{V}})^{-1}x + (\sigma I - A^{\mathcal{V}})^{-1}B\overline{F(\sigma I - A^{\mathcal{W}}_{BF})^{-1}}x$$

and

$$(\sigma I - A^{\mathcal{V}})^{-1}x = (\sigma I - A^{\mathcal{V}}_{BF})^{-1}x - (\sigma I - A^{\mathcal{V}}_{BF})^{-1}B\overline{F(\sigma I - A^{\mathcal{W}})^{-1}}x,$$

where σ is any number with real part larger than the growth bounds of $S(\cdot)$ and $S_{BF}(\cdot)$ on $\mathcal{W}$ and $\mathcal{V}$. Defining $T_1 := I + \overline{BF(\sigma I - A^{\mathcal{W}}_{BF})^{-1}} \in \mathcal{L}(\mathcal{V})$ and $T_2 := I - \overline{BF(\sigma I - A^{\mathcal{W}})^{-1}} \in \mathcal{L}(\mathcal{V})$, we can reformulate the above equations as

$$(\sigma I - A^{\mathcal{V}})x = T_1(\sigma I - A^{\mathcal{V}}_{BF})x \quad \text{and} \quad (\sigma I - A^{\mathcal{V}}_{BF})x = T_2(\sigma I - A^{\mathcal{V}})x,$$

for all $x \in D(A^{\mathcal{V}}) = D(A^{\mathcal{V}}_{BF})$. Since $T_1, T_2 \in \mathcal{L}(\mathcal{V})$ it follows that the graph norms of $D(A^{\mathcal{V}})$ and $D(A^{\mathcal{V}}_{BF})$ are equivalent. Manipulation of the last equation shows that $A^{\mathcal{V}}_{BF}x = A^{\mathcal{V}}x + \overline{BF(\sigma I - A^{\mathcal{W}})^{-1}}(\sigma I - A^{\mathcal{V}})x$ for all $x \in D(A^{\mathcal{V}})$. Finally, we prove that if (2.8) is satisfied, it follows that for all $x \in D(A^{\mathcal{V}})$

$$\overline{F(\sigma I - A^{\mathcal{W}})^{-1}}(\sigma I - A^{\mathcal{V}})x = Fx. \tag{2.29}$$

Since $\mathcal{W} \hookrightarrow \mathcal{V}$, we have $D(A^{\mathcal{W}}) \hookrightarrow D(A^{\mathcal{V}})$ (see Lemma 2.12) and because (2.29) is satisfied for all $x \in D(A^{\mathcal{W}})$, the result follows by taking a sequence $x_n \in D(A^{\mathcal{W}})$ converging to x in the topology of $D(A^{\mathcal{V}})$. ■

Using the perturbation results in Lemma 2.13, we shall investigate the notion of preliminary feedback '$u = Fx + v$' for Pritchard-Salamon systems. This notion will be used a lot in the sequel: in Section 2.7 to formulate the concept of dynamic feedback precisely, in Chapter 3 to 'remove a cross term' in the quadratic criterion of the linear quadratic control problem and in Chapter 4 to remove a similar cross term in the 'sup-inf criterion' related to the $\mathcal{H}_\infty$-control problem.

Lemma 2.14
Let $\Sigma(S(\cdot), B, C, D)$ be a Pritchard-Salamon system of the form (2.7) *and let $F \in \mathcal{L}(\mathcal{W}, U)$ be an admissible output operator for this system.*

(*i*) *Suppose that $v(\cdot) \in L_2^{loc}(0,\infty; U)$ and define*

$$x_F(t) := S_{BF}(t)x_0 + \int_0^t S_{BF}(t-s)Bv(s)ds \tag{2.30}$$

and

$$u(t) := Fx_F(t) + v(t). \tag{2.31}$$

Then $u(\cdot) \in L_2^{loc}(0,\infty; U)$ and $x(t)$ given by

$$x(t) = S(t)x_0 + \int_0^t S(t-s)Bu(s)ds \tag{2.32}$$

satisfies

$$\left.\begin{aligned} &x(t) = x_F(t) \text{ for all } t \geq 0 \\ &Cx(t) = Cx_F(t) \text{ for almost all } t \geq 0. \end{aligned}\right\} \tag{2.33}$$

(*ii*) *Suppose that* $u(\cdot) \in L_2^{loc}(0,\infty;U)$ *and define* $x(\cdot)$ *by* (2.32) *and* $v(\cdot)$ *by*

$$v(t) := -Fx(t) + u(t). \tag{2.34}$$

Then $v(\cdot) \in L_2^{loc}(0,\infty;U)$ *and* $x_F(t)$ *given by* (2.30) *satisfies*

$$\left.\begin{array}{l} x(t) = x_F(t) \text{ for all } t \geq 0 \\ \\ Cx(t) = Cx_F(t) \text{ for almost all } t \geq 0. \end{array}\right\} \tag{2.35}$$

(*iii*) *Let* $v(\cdot) \in L_2^{loc}(0,\infty;U)$ *and define* $x_F(\cdot)$ *as in* (2.30) *and* $u(\cdot)$ *as in* (2.31), *i.e.* $u(\cdot) = Fx_F(\cdot) + v(\cdot)$. *Then* $v(\cdot)$ *satisfies* (2.34), *i.e.* $v(\cdot) = -Fx(\cdot) + u(\cdot)$. *Conversely, let* $u(\cdot) \in L_2^{loc}(0,\infty;U)$ *and define* $x(\cdot)$ *as in* (2.32) *and* $v(\cdot)$ *as in* (2.34). *Then* $u(\cdot)$ *satisfies* (2.31), *i.e.* $u(\cdot) = Fx_F(\cdot) + v(\cdot)$.

Proof

Proof of (*i*):

The expression $Fx_F(\cdot)$ in $u(\cdot) := Fx_F(\cdot) + v(\cdot)$ should be interpreted as explained in and after Remark 2.2:

$$u(\cdot) = \overline{FS}_{BF}(\cdot)x_0 + F\int_0^{\cdot} S_{BF}(\cdot - s)Bv(s)ds + v(\cdot) \in L_2^{loc}(0,\infty;U),$$

where we have used the fact that F is an admissible output operator with respect to $(\mathcal{W},\mathcal{V})$ for $S_{BF}(\cdot)$ (see Lemma 2.13).

Substituting $u(\cdot)$ in (2.32) gives

$$x(t) = S(t)x_0 + \int_0^t S(t-s)B\overline{FS}_{BF}(s)x_0ds + \int_0^t S(t-s)Bv(s)ds +$$

$$\int_0^t S(t-s)BF\left[\int_0^s S_{BF}(s-\tau)Bv(\tau)d\tau\right]ds. \tag{2.36}$$

It follows from Lemma 2.13 that the two terms of $x_F(t)$ in (2.30) can be reformulated as

$$S_{BF}(t)x_0 = S(t)x_0 + \int_0^t S(t-s)B\overline{FS}_{BF}(s)x_0ds$$

and

$$\int_0^t S_{BF}(t-s)Bv(s)ds = \int_0^t S(t-s)Bv(s)ds +$$

$$\int_0^t \left[\int_0^{t-s} S(t-s-\sigma)B\overline{FS}_{BF}(\sigma)Bv(s)d\sigma\right]ds.$$

Comparing this with (2.36), we see that to prove that $x(t) = x_F(t)$, we only have to show that

$$\int_0^t S(t-s)BF\left[\int_0^s S_{BF}(s-\tau)Bv(\tau)d\tau\right]ds =$$

$$\int_0^t \left[\int_0^{t-s} S(t-s-\sigma)B\overline{FS}_{BF}(\sigma)Bv(s)d\sigma\right]ds. \tag{2.37}$$

This means that we have to somehow 'get F inside the integral' (the other operators cause no problems, since they are bounded with respect to $\mathcal{V}$). It follows from [15, Theorem 3.3] that if $v(\cdot)$ is a *step function*, we have

$$F\int_0^s S_{BF}(s-\tau)Bv(\tau)d\tau = \int_0^s \overline{FS}_{BF}(s-\tau)Bv(\tau)d\tau. \tag{2.38}$$

We note that in [72] a result similar to (2.38) was proved under the extra assumption that (2.8) is satisfied. Using (2.38) and Fubini's Theorem to interchange the order of integration, it is straightforward to show that (2.37) is satisfied if $v(\cdot)$ is a step function. Next, we shall prove the general case, i.e. we prove (2.37) for $v(\cdot) \in L_2(0,T;U)$, T arbitrary. We introduce a sequence of step functions $v_n(\cdot)$ that converge to $v(\cdot)$ in the $L_2(0,T;U)$-norm. Since B is an admissible input operator for $S_{BF}(\cdot)$ and $F \in \mathcal{L}(\mathcal{W},U)$ it follows that for $n \to \infty$ there holds

$$F\left[\int_0^s S_{BF}(s-\tau)Bv_n(\tau)d\tau\right] \xrightarrow{U} F\left[\int_0^s S_{BF}(s-\tau)Bv(\tau)d\tau\right],$$

so that for the left hand side of (2.37) we have

$$\int_0^t S(t-s)BF\left[\int_0^s S_{BF}(s-\tau)Bv_n(\tau)d\tau\right]ds \xrightarrow{\mathcal{V}}$$

$$\int_0^t S(t-s)BF\left[\int_0^s S_{BF}(s-\tau)Bv(\tau)d\tau\right]ds. \tag{2.39}$$

On the other hand, it is easy to see that

$$\int_0^t \left[\int_0^{t-s} S(t-s-\sigma)B\overline{FS}_{BF}(\sigma)Bv_n(s)d\sigma\right]ds \xrightarrow{\mathcal{V}}$$

$$\int_0^t \left[\int_0^{t-s} S(t-s-\sigma)B\overline{FS}_{BF}(\sigma)Bv(s)d\sigma\right]ds \tag{2.40}$$

and so the result follows from (2.39) and (2.40). We note that in [72, Lemma 2.5] a result similar to (2.37) is proved, using different arguments.

Finally, we prove the second part of (2.33). This seems to follow trivially from the first part of (2.33). However, $Cx(t) = Cx_F(t)$ should be interpreted in the sense of Remark 2.2, i.e. in the L_2-sense. It is easy to see that $Cx(t) = Cx_F(t)$ holds if $x_0 \in \mathcal{W}$. The general result can be obtained by introducing a sequence $x_{0n} \in \mathcal{W}$ such that $x_{0n} \xrightarrow{\mathcal{V}} x_0$ as $n \to \infty$ and using the admissibility of B and C: let $T > 0$ be arbitrary and define

$$x_{Fn}(t) := S_{BF}(t)x_{0n} + \int_0^t S_{BF}(t-s)Bv(s)ds,$$

$$u_n(\cdot) := Fx_{Fn}(\cdot) + v(\cdot)$$

and

$$x_n(\cdot) := S(\cdot)x_{0n} + \int_0^{\cdot} S(\cdot - s)Bu_n(s)ds.$$

Since F is an admissible output operator for $S_{BF}(\cdot)$ and

$$Fx_{Fn}(\cdot) - Fx_F(\cdot) = \overline{FS}_{BF}(\cdot)(x_{0n} - x_0),$$

we have

$$\|Fx_{Fn}(\cdot) - Fx_F(\cdot)\|_{L_2(0,T;U)} \to 0, \tag{2.41}$$

so that $\|u_n(\cdot) - u(\cdot)\|_{L_2(0,T;U)} \to 0$. Furthermore, using the fact that C and B are both admissible, it follows from (the proof of) Appendix B.6, formula (B.20) that

$$\|Cx_n(\cdot) - Cx(\cdot)\|_{L_2(0,T;Y)} \to 0 \text{ as } n \to \infty. \tag{2.42}$$

Furthermore, it is easy to see that just as in (2.41)

$$\|Cx_{Fn}(\cdot) - Cx_F(\cdot)\|_{L_2(0,T;Y)} \to 0, \tag{2.43}$$

because C is an admissible output operator for $S_{BF}(\cdot)$.
Finally, $x_{0n} \in \mathcal{W}$ implies that $Cx_n(\cdot) = Cx_{Fn}(\cdot)$ and so the result follows from (2.42) and (2.43).

Proof of (ii):
Again, using the interpretation of Remark 2.2, we have

$$v(t) = -\overline{FS}(t)x_0 - F\int_0^t S(t-s)Bu(s)ds + u(t).$$

The rest of the proof of this part is similar to the proof of (i) and therefore deleted.

Proof of (iii):
This follows immediately from (i) and (ii). ∎

2.5 Duality theory

An important tool that will be used in the rest of this book is the notion of 'duality'. Some of the results in this sections can be found in [80, 72, 17]. For a short introduction about representations of dual spaces, duality pairings and pivot spaces we refer to [2]. Here we shall use some of the (rather standard) terminology of [2].

In principle, the dual space of a real Hilbert space X is just the Hilbert space of bounded linear functionals, i.e. $\mathcal{L}(X,\mathbb{R})$. We shall denote $\mathcal{L}(X,\mathbb{R})$ by X^d, the canonical isomorphism from X^d to X is given by i_X and the inner product on X^d is given by

$$< f,g >_{X^d} = < i_X f, i_X g >_X \quad \text{for all } f,g \in X^d. \tag{2.44}$$

The duality pairing $< \cdot,\cdot >_{<X^d,X>}$ is given by

$$< f,x >_{<X^d,X>} = f(x) \quad \text{for all } f \in X^d \text{ and } x \in X$$

and so the following holds

$$< f,x >_{<X^d,X>} = f(x) = < i_X f, x >_X \quad \text{for all } f \in X^d \text{ and } x \in X. \tag{2.45}$$

Now suppose that X' is a real Hilbert space and that there is an isometry j from X' onto X^d. This means that

$$X' \xrightarrow{j} X^d \xrightarrow{i_X} X \tag{2.46}$$

and $i_X j$ is an isometry from X' onto X. The pair $\{X', j\}$ is called a *representation* of X^d and the corresponding duality pairing $< \cdot,\cdot >_{<X',X>}$ is defined by the bilinear form

$$< x',x >_{<X',X>} := < jx',x >_{<X^d,X>} = (jx')(x) = < i_X jx', x >_X \tag{2.47}$$

for all $x' \in X'$ and $x \in X$.

In this case, we may also call X' the dual space of X (this is standard abuse of terminology).

We say that X is *identified with its dual* if the representation for X^d is chosen to be $\{X, (i_X)^{-1}\}$ (so that $X' = X$). It follows from (2.47) that in this case the pairing corresponds to the inner product on X. If X is identified with its dual, X is sometimes called a *pivot space* (cf. [2, section 3.5]).

Usually, it is assumed that whatever choice is made for the representation $\{X', j\}$ of $\mathcal{L}(X,\mathbb{R})$, the representation of the dual of this dual space (i.e.

$(X')^d = \mathcal{L}(X', \mathbb{R})$) is given by $\{X, c_X\}$, where c_X is the isometry from X to $(X')^d$ given by

$$c_X = (i_{X'})^{-1} j^{-1} (i_X)^{-1} \tag{2.48}$$

(we could also say that $X'' = X$). This implies that

$$< x'', x' >_{<X'',X'>} = < x'', x' >_{<X,X'>} = < c_X x'', x' >_{<(X')^d,X'>} = c_X x''(x')$$

$$= (i_{X'})^{-1} j^{-1} (i_X)^{-1} x''(x') = < j^{-1}(i_X)^{-1} x'', x' >_{X'} = < (i_X)^{-1} x'', j x' >_{X^d}$$

$$= < x'', i_X j x' >_X = j x'(x'') = < x', x'' >_{<X',X>} . \tag{2.49}$$

We identify the bidual of X with X itself and the pairing $< \cdot, \cdot >_{<X,X'>}$ is given by $< x'', x' >_{<X,X'>} = < x', x'' >_{<X',X>}$.

Now we explain the terminology of pivot spaces. We first note that if we have two real Hilbert spaces X and Y with $X \hookrightarrow Y$, then $Y^d \hookrightarrow X^d$. Here, we identify an element $g \in Y^d = \mathcal{L}(Y, \mathbb{R})$ with $g \mid_X \in X^d = \mathcal{L}(X, \mathbb{R})$. Similarly, if $f \in X^d$ satisfies $|f(x)| \leq const\|x\|_Y$ for all $x \in X$, then the unique continuous extension of f to Y is also denoted by f. Furthermore, for all $g \in Y^d$ and $x \in X$ we have

$$< g, x >_{<Y^d,Y>} = g(x) = < g, x >_{<X^d,X>} . \tag{2.50}$$

Now if Y is identified with its dual, we can define a space X' which is a representation of the dual of X such that $X \hookrightarrow Y = Y' \hookrightarrow X'$, as follows. Recall that i_Y denotes the canonical isomorphism from Y^d to Y. Let X' be the completion of $Y' = Y$ with respect to the inner product

$$< x_1', x_2' >_{X'} = < (i_Y)^{-1} x_1', (i_Y)^{-1} x_2' >_{X^d} \quad \text{for } x_1', x_2' \in Y = Y'. \tag{2.51}$$

We have

$$\begin{array}{ccc} Y^d & \hookrightarrow & X^d \\ \downarrow i_Y & & \\ Y' = Y & \hookrightarrow & X' \end{array} \tag{2.52}$$

According to (2.51), for all $g \in Y^d$

$$\|i_Y(g)\|_{X'} = \|(i_Y)^{-1} i_Y(g)\|_{X^d} = \|g\|_{X^d}. \tag{2.53}$$

Thus, it follows that i_Y has a unique extension $\overline{i_Y}$ to X^d and $\overline{i_Y}$ is an isometry from X^d to X'. Furthermore, we can choose the pair $\{X', (\overline{i_Y})^{-1}\}$ as a representation of X^d so that the the duality pairing $< \cdot, \cdot >_{<X',X>}$ is given by (compare with (2.47))

$$< x', x >_{<X',X>} = (\overline{i_Y})^{-1} x'(x)$$

and for all $(x', x) \in Y \times X$ we have

$$< x', x >_{<X',X>} = (\overline{i_Y})^{-1} x'(x) = (i_Y)^{-1} x'(x) = < x', x >_Y \tag{2.54}$$

(in the last step, apply (2.45) with X replaced by Y). Finally, $X \hookrightarrow Y = Y' \hookrightarrow X'$, showing that Y really is a pivot space.

If, in addition, we have a third real Hilbert space Z with $Y \hookrightarrow Z$, we have $Z^d \hookrightarrow Y^d$ and we can define a representation Z' of Z^d such that $Z' \hookrightarrow Y = Y'$ as follows. Let $(i_Y)\,|_{Z^d}$ denote the restriction of i_Y to Z^d and define Z' as the image of Z^d under this map. Furthermore, define the inner product on Z' by

$$< z_1', z_2' >_{Z'} = < ((i_Y)\,|_{Z^d})^{-1} z_1', ((i_Y)\,|_{Z^d})^{-1} z_2' >_{Z^d} . \tag{2.55}$$

Then the pair $\{Z', ((i_Y)\,|_{Z^d})^{-1}\}$ is a representation of Z^d and the duality pairing is given by (compare with (2.47))

$$< z', z >_{<Z',Z>} = ((i_Y)\,|_{Z^d})^{-1} z'(z). \tag{2.56}$$

Hence for all $(z', z) \in Z' \times Y$ we have

$$< z', z >_{<Z',Z>} = ((i_Y)\,|_{Z^d})^{-1} z'(z) = (i_Y)^{-1} z'(z) = < z', z >_Y \tag{2.57}$$

(in the last step, apply (2.45) with X replaced by Y) and $Z' \hookrightarrow Y' = Y \hookrightarrow Z$ so that Y is also a pivot space for Z. Schematically we now have

$$\begin{array}{ccccc} Z^d & \hookrightarrow & Y^d & \hookrightarrow & X^d \\ \downarrow (i_Y)\,|_{Z^d} & & \downarrow i_Y & & \downarrow \overline{i_Y} \\ Z' & \hookrightarrow & Y' = Y & \hookrightarrow & X'. \end{array} \tag{2.58}$$

A consequence of (2.54) and (2.57) is that for all $(x', x) \in Z' \times X$

$$< x', x >_{<X',X>} = < x', x >_Y = < x', x >_{<Z',Z>} . \tag{2.59}$$

In Chapter 6 an example is given of how these procedures work in practice.

Next we recall the concept of adjoint operators. Suppose that X and Y are real Hilbert spaces and suppose that $\{X', j\}$ and $\{Y', k\}$ are representations of X^d and Y^d, with the pairings denoted by $< \cdot, \cdot >_{<X',X>}$ and $< \cdot, \cdot >_{<Y',Y>}$, respectively. Note that now $i_X j$ is an isometry from X' to X and that $i_Y k$ is an isometry from Y' to Y (see also (2.46)). The *adjoint* of a densely defined linear operator A from $D(A) \subset X$ to Y is an operator A' from $D(A') \subset Y'$ to X', defined by

$$\left.\begin{aligned} &D(A') = \{y' \in Y' \mid \text{ there exists an } x' \in X' \text{ such that for} \\ &\text{all } x \in D(A) \text{ we have } < y', Ax >_{<Y',Y>} = < x', x >_{<X',X>}\} \\ &< A'y', x >_{<X',X>} = < y', Ax >_{<Y',Y>} \\ &\text{for all } y' \in D(A') \text{ and all } x \in D(A). \end{aligned}\right\} \tag{2.60}$$

We note that the *Hilbert adjoint* of A (denoted by A^*) from $D(A^*) \subset Y$ to X is related to A' by

$$\left.\begin{aligned} &D(A^*) = (i_Y k) D(A'), \\ &A^* y = (i_X j) A' (i_Y k)^{-1} y \quad \text{for all } y \in D(A^*), \end{aligned}\right\} \tag{2.61}$$

so that for all $y \in D(A^*)$ and $x \in D(A)$ we have

$$\begin{aligned} < A^* y, x >_X &= < (i_X j)^{-1} (i_X j) A' (i_Y k)^{-1} y, x >_{<X',X>} \\ &\quad \text{(using (2.47) and the formula for } A^* \text{ in (2.61))} \\ &= < (i_Y k)^{-1} y, Ax >_{<Y',Y>} = < y, Ax >_Y \end{aligned} \tag{2.62}$$

using the formula for A' in (2.60) and formula (2.47) applied to Y.
For a *bounded* operator $T \in \mathcal{L}(X,Y)$ we have $T' \in \mathcal{L}(Y',X')$, $T^* = (i_X j)T'(i_Y k)^{-1} \in \mathcal{L}(Y,X)$ and $\|T'\|_{\mathcal{L}(Y',X')} = \|T\|_{\mathcal{L}(X,Y)}$. We note that, because of the identification of the bidual of a Hilbert space with itself, an operator $T \in \mathcal{L}(X,Y)$ satisfies $(T')' = T$.

The adjoint of a linear densely defined *closed* operator $A : X \supset D(A) \to Y$, denoted by $A' : Y' \supset D(A') \to X'$ is closed and densely defined. Hence, the adjoint of A', denoted by $(A')' : X = X'' \supset D((A')') \to Y = Y''$ is well defined and moreover, $D((A')') = D(A)$ and $(A')'x = Ax$ for all $x \in D(A)$ (cf. [53, section 10.3]).

Since we always identify the bidual of a Hilbert space with itself, i.e. we have $X = X''$ and (2.49), it follows that the adjoint of an operator $T \in \mathcal{L}(X,X')$ satisfies $T' \in \mathcal{L}(X,X')$. We call an operator $T \in \mathcal{L}(X,X')$ *nonnegative definite* (or $T \geq 0$), if $T = T'$ and

$$< Tx, x >_{<X',X>} \geq 0 \qquad \text{for all } x \in X. \tag{2.63}$$

In this case the operator $S := (i_X j)T \in \mathcal{L}(X)$ satisfies $< Sx, y >_X = < Tx, y >_{<X',X>}$ (using (2.47)) so that $S = S^*$ and so T is nonnegative definite if and only S is.

Next, we give a lemma that gives some relations between a C_0-semigroup and its generator with the adjoint of the C_0-semigroup and its generator. The proof follows immediately from [69, section 1.10]. In this lemma we need the notion of the complexification of a Hilbert space X and we recall that we denote this complexification by the same symbol. We note that the duality theory given above is (mutatis mutandi) also valid for complex Hilbert spaces.

Lemma 2.15

Suppose that we have a C_0-semigroup $S(\cdot)$ on a real separable Hilbert space X, with infinitesimal generator A from $D(A) \subset X$ to X and let $\{X', j\}$ be a representation of X^d. Then $S'(\cdot)$ is a C_0-semigroup on X' and its infinitesimal generator is given by the adjoint of A, denoted by $A' : D(A') \subset X' \to X'$. Furthermore, we have $\omega^X_{S(\cdot)} = \omega^{X'}_{S'(\cdot)}$, $\rho(A) = \rho(A')$ and for all $s \in \rho(A)$ we have

$$((sI - A)^{-1})' = ((sI)' - A')^{-1}.$$

We note that either $(sI)' = sI'$ or $(sI)' = \bar{s}I'$, depending on the choice of the representation of X^d. For instance, the first situation occurs if we choose $\{X', j\} = \{X^d, I\mid_{X^d}\}$ and the latter situation occurs if $\{X', j\} = \{X, (i_X)^{-1}\}$ (then A' corresponds to the Hilbert adjoint A^*).

Finally, we are ready to derive some duality results for Pritchard-Salamon systems. Suppose that we have a Pritchard-Salamon system $\Sigma(S(\cdot), B, C, D)$ of the form (2.7). For the dual spaces of $\mathcal{W}$ and $\mathcal{V}$ (recall that $\mathcal{W} \hookrightarrow \mathcal{V}$) we use the notion of representations and duality pairings as explained above. We suppose that $\{\mathcal{W}', j\}$ is a representation of $\mathcal{W}^d$, so that the pairing $< \cdot, \cdot >_{<\mathcal{W}',\mathcal{W}>}$ is given by

$$< w', w >_{<\mathcal{W}',\mathcal{W}>} = < jw', w >_{<\mathcal{W}^d,\mathcal{W}>} \quad \text{(cf. (2.47))}.$$

Furthermore, we let $\mathcal{V}^d$ be represented by $\{\mathcal{V}', k\}$, where

$$\mathcal{V}' = j^{-1}\mathcal{V}^d \quad \text{and} \quad k = j\mid_{\mathcal{V}'} \tag{2.64}$$

and the inner product on $\mathcal{V}'$ is given by

$$< x', y' >_{\mathcal{V}'} = < kx', ky' >_{\mathcal{V}^d} = < jx', jy' >_{\mathcal{V}^d}$$

(compare this with the construction of the representation $\{Z', ((i_Y)\mid_{Z^d})^{-1}\}$ of Z^d, formulas (2.55)-(2.57)). The pairing for $\mathcal{V}'$ is given by

$$< v', v >_{<\mathcal{V}',\mathcal{V}>} = < kv', v >_{<\mathcal{V}^d,\mathcal{V}>} \quad \text{(cf. (2.47))}$$

and for all $(v', w) \in \mathcal{V}' \times \mathcal{W}$ we have

$$< v', w >_{<\mathcal{V}',\mathcal{V}>} = < kv', w >_{<\mathcal{V}^d,\mathcal{V}>} = < jv', w >_{<\mathcal{V}^d,\mathcal{V}>} =$$

$$< jv', w >_{<\mathcal{W}^d,\mathcal{W}>} = < v', w >_{<\mathcal{W}',\mathcal{W}>}, \tag{2.65}$$

where we have used (2.47) and (2.50). Furthermore,

$$\mathcal{V}' \hookrightarrow \mathcal{W}' \tag{2.66}$$

(properties (2.65) and (2.66) were in fact the reason for the choice in (2.64)). Schematically we have

$$\begin{array}{ccc} \mathcal{V}^d & \hookrightarrow & \mathcal{W}^d \\ \uparrow k = j\mid_{\mathcal{V}'} & & \uparrow j \\ \mathcal{V}' & \hookrightarrow & \mathcal{W}'. \end{array} \tag{2.67}$$

As explained above, we identify the bidual of a Hilbert space with itself so that the representation of $(\mathcal{W}')^d$ is given by $\{\mathcal{W}, c_\mathcal{W}\}$ with $c_\mathcal{W} = (i_{\mathcal{W}'})^{-1}j^{-1}(i_\mathcal{W})^{-1}$ and the representation of $(\mathcal{V}')^d$ is given by $\{\mathcal{V}, c_\mathcal{V}\}$ with $c_\mathcal{V} = (i_{\mathcal{V}'})^{-1}k^{-1}(i_\mathcal{V})^{-1}$ (cf. (2.48)). Schematically we have

$$\begin{array}{ccc} (\mathcal{W}')^d & \hookrightarrow & (\mathcal{V}')^d \\ \uparrow c_\mathcal{W} = c_\mathcal{V}\mid_\mathcal{W} & & \uparrow c_\mathcal{V} \\ \mathcal{W} & \hookrightarrow & \mathcal{V}. \end{array} \tag{2.68}$$

We note that in the set up of [72], the representations of $\mathcal{W}^d$ and $\mathcal{V}^d$ are determined by a pivot space formulation with respect to a third real separable Hilbert space $\mathcal{H}$ such that $\mathcal{W} \hookrightarrow \mathcal{H} \hookrightarrow \mathcal{V}$ (see also Section 2.2). Identifying $\mathcal{H}$ with its dual it follows that

$$\begin{array}{ccccc} \mathcal{V}^d & \hookrightarrow & \mathcal{H}^d & \hookrightarrow & \mathcal{W}^d \\ \downarrow i_\mathcal{H}\mid_{\mathcal{V}^d} & & \downarrow i_\mathcal{H} & & \downarrow \overline{i_\mathcal{H}} \\ \mathcal{V}' & \hookrightarrow & \mathcal{H}' = \mathcal{H} & \hookrightarrow & \mathcal{W}' \end{array} \tag{2.69}$$

(compare this with (2.58)). Later we shall also choose this formulation, but for the moment we prefer to use the more general formulation corresponding to (2.67).

We proceed with a lemma which shows the relationship between the extension of an operator and the restriction of its adjoint.

Lemma 2.16

Suppose that we have three real Hilbert spaces X, Y and Z such that $X \hookrightarrow Y$. Let $\{X', j\}$ be a representation of X^d and assume that the representation of Y^d is given by $\{Y', k\}$, where $Y' = j^{-1}Y^d$ with inner product given by $< x', y' >_{Y'} = < jx', jy' >_{Y^d}$ and $k = j \mid_{Y'}$ (compare with (2.64)-(2.66) and note that $Y' \hookrightarrow X'$). Let $\{Z', l\}$ be a representation of Z.

(i) *Suppose that $T \in \mathcal{L}(X, Z)$. Then there exists a continuous extension $\overline{T} \in \mathcal{L}(Y, Z)$ of T if and only if the adjoint of T, denoted by $T' \in \mathcal{L}(Z', X')$ satisfies $T' \in \mathcal{L}(Z', Y')$.*

(ii) *Suppose that $S \in \mathcal{L}(Z, Y)$. Then there exist a continuous extension $\overline{S'} \in \mathcal{L}(X', Z')$ of $S' \in \mathcal{L}(Y', Z')$ if and only if $S \in \mathcal{L}(Z, X)$.*

Proof

Proof of (i):
First of all, we note that for all $(y', x) \in Y' \times X$ we have

$$< y', x >_{<Y',Y>} = < y', x >_{<X',X>} \tag{2.70}$$

(see also (2.65)).
Necessity:
We know that for all $z' \in Z'$ we have $T'z' \in X'$ and since for all $x \in X$

$$< T'z', x >_{<X',X>} = < z', Tx >_{<Z',Z>} \quad \text{(cf. (2.60))}$$

$$= < z', \overline{T}x >_{<Z',Z>} = < (\overline{T})'z', x >_{<Y',Y>} \quad \text{(using } \overline{T} \in \mathcal{L}(Y, Z) \text{ and (2.60))}$$

$$= < (\overline{T})'z', x >_{<X',X>} \quad \text{(using (2.70))}$$

we infer that $T'z' = (\overline{T})'z' \in Y'$. Since this holds for all $z' \in Z'$ it follows that $T' = (\overline{T})' \in \mathcal{L}(Z', Y')$.
Sufficiency:
We have for all $x \in X$

$$\|T(x)\|_Z^2 = < Tx, Tx >_Z = < (i_Z l)^{-1}Tx, Tx >_{<Z',Z>}$$

$$\text{(using (2.47) applied to } Z)$$

$$= < T'(i_Z l)^{-1}Tx, x >_{<X',X>} = < T'(i_Z l)^{-1}Tx, x >_{<Y',Y>}$$

$$\text{(using (2.70))}$$

$$\le \|T'(i_Z l)^{-1}Tx\|_{Y'}\|x\|_Y \le const\|(i_Z l)^{-1}Tx\|_{Z'}\|x\|_Y$$

$$\text{(using the fact that } T' \in \mathcal{L}(Z', Y'))$$

$$= const\|Tx\|_Z\|x\|_Y$$

so that $\|T(x)\|_Z \leq const\|x\|_Y$ and the result follows.
Proof of (ii):
This follows from part (i), noting that the bidual of a Hilbert space is identified with itself so that $(S')' = S \in \mathcal{L}(Z, Y)$. ■

The next theorem is the main result of this section on duality.

Theorem 2.17
Suppose that we have a Pritchard-Salamon system $\Sigma(S(\cdot), B, C, D)$ of the form (2.7) and suppose that the representations of the dual spaces $\mathcal{W}^d$ and $\mathcal{V}^d$ are given as in (2.67). Then $S'(\cdot)$ is a C_0-semigroup on $\mathcal{V}' \hookrightarrow \mathcal{W}'$ and the following hold:

- (i) *Let Z be any real separable Hilbert space and let $\{Z', l\}$ be a representation of Z^d. Then, $H \in \mathcal{L}(Z, \mathcal{V})$ is an admissible input operator with respect to $(\mathcal{W}, \mathcal{V})$ for $S(\cdot)$ if and only if $H' \in \mathcal{L}(\mathcal{V}', Z')$ is an admissible output operator with respect to $(\mathcal{V}', \mathcal{W}')$ for $S'(\cdot)$,*
  $F \in \mathcal{L}(\mathcal{W}, Z)$ is an admissible output operator with respect to $(\mathcal{W}, \mathcal{V})$ for $S(\cdot)$ if and only if $F' \in \mathcal{L}(Z', \mathcal{W}')$ is an admissible input operator with respect to $(\mathcal{V}', \mathcal{W}')$ for $S'(\cdot)$,
  $\Sigma(S'(\cdot), C', B', D')$ is a Pritchard-Salamon system.
- (ii) *Let Z be any real separable Hilbert space and let $\{Z', l\}$ be a representation of Z^d. For any admissible output operator $F \in \mathcal{L}(\mathcal{W}, Z)$ and any admissible input operator $H \in \mathcal{L}(Z, \mathcal{V})$ we have $(S_{HF})'(\cdot) = S'_{F'H'}(\cdot)$.*
- (iii) *If $D(A^{\mathcal{V}}) \hookrightarrow \mathcal{W}$ is satisfied (i.e. assumption (2.8)), then $D((A^{\mathcal{W}})') \hookrightarrow \mathcal{V}'$.*

Proof
According to (2.67), we have $\mathcal{V}' \hookrightarrow \mathcal{W}'$ and it follows from Lemma 2.15 that $S'(t)$ defines a C_0-semigroup on both $\mathcal{W}'$ and $\mathcal{V}'$, with infinitesimal generators $(A^{\mathcal{W}})'$ and $(A^{\mathcal{V}})'$ respectively.
Proof of (i):
The first part of this result is also given in [72], but for completeness we give a detailed proof here. Let $H \in \mathcal{L}(Z, \mathcal{V})$ be given and let $\tau > 0$. Define the operators $\mathcal{B}^\tau$ from $L_2(0, \tau; Z)$ to $\mathcal{V}$ and $\mathcal{C}^\tau$ from $\mathcal{V}'$ to $L_2(0, \tau; Z')$ by

$$\mathcal{B}^\tau z(\cdot) := \int_0^\tau S(s)Hz(s)ds \text{ for } z(\cdot) \in L_2(0, \tau; Z)$$

and

$$\mathcal{C}^\tau v' = H'S'(\cdot)v' \text{ for } v' \in \mathcal{V}'.$$

We know that $\mathcal{B}^\tau \in \mathcal{L}(L_2(0,\tau;Z),\mathcal{V})$ and $\mathcal{C}^\tau \in \mathcal{L}(\mathcal{V}',L_2(0,\tau;Z'))$. We would like to relate the adjoint of $\mathcal{B}^\tau$ to $\mathcal{C}^\tau$ and in order to do this we show that $L_2(0,\tau;Z')$ can be chosen as a representation of the dual of $L_2(0,\tau;Z)$ (loosely speaking, $L_2(0,\tau;Z)' = L_2(0,\tau;Z')$). We shall show every step of the proof, in order to illustrate the theory given at the beginning of this section. For $f(\cdot) \in L_2(0,\tau;Z')$ and $g(\cdot) \in L_2(0,\tau;Z)$ we define

$$(jf)(g) := \int_0^\tau < f(s), g(s) >_{<Z',Z>} ds. \tag{2.71}$$

It is easy to see that $jf \in \mathcal{L}(L_2(0,\tau;Z),\mathbb{R})$, so that j is a linear map from $L_2(0,\tau;Z')$ to $\mathcal{L}(L_2(0,\tau;Z),\mathbb{R}) = L_2(0,\tau;Z)^d$. Furthermore, j is an isometry because

$$\|jf\|^2_{L_2(0,\tau;Z)^d} = \sup_{\|g\|_{L_2(0,\tau;Z)}=1} |jf(g)|^2 \quad \text{(by definition)}$$

$$= \sup_{\|g\|_{L_2(0,\tau;Z)}=1} \left| \int_0^\tau < f(s), g(s) >_{<Z',Z>} ds \right|^2 \quad \text{(by definition of } j)$$

$$= \sup_{\|g\|_{L_2(0,\tau;Z)}=1} \left| \int_0^\tau < i_Z l f(s), g(s) >_Z ds \right|^2 \quad \text{(using (2.47) applied to } Z)$$

$$= \int_0^\tau < i_Z l f(s), i_Z l f(s) >_Z ds = \int_0^\tau < f(s), f(s) >_{Z'} ds =$$

$$\|f\|^2_{L_2(0,\tau;Z')},$$

where we have used that $i_Z l$ is an isometry from Z' to Z (compare with (2.46)). Furthermore, using Riesz's representation theorem for $\mathcal{L}(L_2(0,\tau;Z),\mathbb{R})$, it easily follows that j maps $L_2(0,\tau;Z')$ *onto* $\mathcal{L}(L_2(0,\tau;Z),\mathbb{R})$. Hence, the pair $\{L_2(0,\tau;Z'), j\}$ is a representation of the dual of $L_2(0,\tau;Z)$ and the duality pairing is given by (see (2.47))

$$< f, g >_{<L_2(0,\tau;Z'),L_2(0,\tau;Z)>} = (jf)(g) =$$

$$\int_0^\tau < f(s), g(s) >_{<Z',Z>} ds. \tag{2.72}$$

Using these facts, we shall prove that

$$(\mathcal{B}^\tau)' = \mathcal{C}^\tau \tag{2.73}$$

(noting that $(\mathcal{B}^\tau)' \in \mathcal{L}(\mathcal{V}',(L_2(0,\tau;Z')))$. Indeed, for arbitrary $z(\cdot) \in L_2(0,\tau;Z)$ and $v' \in \mathcal{V}'$ there holds

$$< v', \mathcal{B}^\tau z(\cdot) >_{<\mathcal{V}',\mathcal{V}>} = < v', \int_0^\tau S(s) H z(s) ds >_{<\mathcal{V}',\mathcal{V}>} =$$

$$\int_0^\tau < v', S(s)Hz(s) >_{<\mathcal{V}',\mathcal{V}>} ds = \int_0^\tau < H'S'(s)v', z(s) >_{<Z',Z>} ds =$$

$$\int_0^\tau < (\mathcal{C}^\tau v')(s), z(s) >_{<Z',Z>} ds =< (\mathcal{C}^\tau v')(\cdot), z(\cdot) >_{<L_2(0,\tau;Z'),L_2(0,\tau;Z)>},$$

where in the last step we have used (2.72). This proves (2.73).
Now we are ready to prove the first part of item (i). Using Definition 2.1, it is easy to see that H is an admissible input operator for $S(\cdot)$ if and only if $\mathcal{B}^\tau \in \mathcal{L}(L_2(0,\tau;U),\mathcal{W})$. Furthermore, $H' \in \mathcal{L}(\mathcal{V}',U')$ is an admissible output operator for $S'(\cdot)$ if and only if $\mathcal{C}^\tau = (\mathcal{B}^\tau)'$ has a continuous extension to $\mathcal{W}'$. Hence, the result follows from Lemma 2.16.

The second part of (i) follows from the fact that we identify the bidual of a Hilbert space with itself (so that $F = (F')'$ and $S(\cdot) = (S')'(\cdot)$) and the third part follows trivially from the first two parts.

Proof of (ii):
It follows from the above that $(S_{HF})'(\cdot)$ is a C_0-semigroup on $\mathcal{V}' \hookrightarrow \mathcal{W}'$ and that $\Sigma((S_{HF})'(\cdot), F', H', 0)$ is a Pritchard-Salamon system. We shall show that for all $v' \in \mathcal{V}'$, $(S_{HF})'(\cdot)$ satisfies

$$(S_{HF})'(t)v' = S'(t)v' + \int_0^t S'(t-s)F'H'(S_{HF})'(s)v'ds. \tag{2.74}$$

Indeed, for all $v' \in \mathcal{V}'$ and $v \in \mathcal{W}$ there holds

$$< (S_{HF})'(t)v', v >_{<\mathcal{V}',\mathcal{V}>} =< v', S_{HF}(t)v >_{<\mathcal{V}',\mathcal{V}>} =< v', S(t)v >_{<\mathcal{V}',\mathcal{V}>} +$$

$$< v', \int_0^t S_{HF}(t-s)HFS(s)vds >_{<\mathcal{V}',\mathcal{V}>} =< v', S(t)v >_{<\mathcal{V}',\mathcal{V}>} +$$

$$\int_0^t < v', S_{HF}(t-s)HFS(s)v >_{<\mathcal{V}',\mathcal{V}>} ds =< S'(t)v', v >_{<\mathcal{V}',\mathcal{V}>} +$$

$$\int_0^t < F'H'(S_{HF})'(t-s)v', S(s)v >_{<\mathcal{W}',\mathcal{W}>} ds$$

(using $F \in \mathcal{L}(\mathcal{W}, Z)$, $H \in \mathcal{L}(Z, \mathcal{V})$ and (2.60))

$$=< S'(t)v', v >_{<\mathcal{V}',\mathcal{V}>} + \int_0^t < S'(s)F'H'(S_{HF})'(t-s)v', v >_{<\mathcal{W}',\mathcal{W}>} ds$$

$$=< S'(t)v', v >_{<\mathcal{V}',\mathcal{V}>} + < \int_0^t S'(s)F'H'(S_{HF})'(t-s)v'ds, v >_{<\mathcal{W}',\mathcal{W}>}$$

$$=< S'(t)v' + \int_0^t S'(s)F'H'(S_{HF})'(t-s)v'ds, v >_{<\mathcal{V}',\mathcal{V}>},$$

where in the last step we have used (2.65) and the fact that F' is an admissible input operator with respect to $(\mathcal{V}', \mathcal{W}')$ for $S'(\cdot)$. Since the above expression is valid for all $v \in \mathcal{W}$ and $\mathcal{W}$ is dense in $\mathcal{V}$, we have obtained (2.74). Now it follows from Lemma 2.13 that $(S_{HF})'(\cdot) = S'_{F'H'}(\cdot)$.

Proof of (*iii*):
The proof consists of three parts: first we prove that $D((A^{\mathcal{W}})') \subset \mathcal{V}'$, then we show that this injection is continuous and finally we will prove that the injection is dense.
First we note that $D((A^{\mathcal{V}})') \subset D((A^{\mathcal{W}})')$ (apply Lemma 2.12 to $\mathcal{V}' \hookrightarrow \mathcal{W}'$). Now let $\sigma > \max(\omega_{S^{\mathcal{W}}(\cdot)}, \omega_{S^{\mathcal{V}}(\cdot)})$ and suppose that $w' \in D((A^{\mathcal{W}})')$. Then for all $w \in \mathcal{W}$,

$$| < w', w >_{<\mathcal{W}',\mathcal{W}>} | = | < w', (\sigma I - A^{\mathcal{W}})(\sigma I - A^{\mathcal{V}})^{-1} w >_{<\mathcal{W}',\mathcal{W}>} | =$$

$$| < (\sigma I - A^{\mathcal{W}})' w', (\sigma I - A^{\mathcal{V}})^{-1} w >_{<\mathcal{W}',\mathcal{W}>} | \leq c_1 \|(\sigma I - A^{\mathcal{V}})^{-1} w)\|_{\mathcal{W}} \leq$$

$$c_2 \|(\sigma I - A^{\mathcal{V}})^{-1} w)\|_{D(A^{\mathcal{V}})} \leq c_3 \|w\|_{\mathcal{V}},$$

where we used $D(A^{\mathcal{V}}) \hookrightarrow \mathcal{W}$ and the fact that $\|(\sigma I - A^{\mathcal{V}}) \cdot \|_{\mathcal{V}}$ is equivalent to the graph norm of $D(A^{\mathcal{V}})$. Hence, $D((A^{\mathcal{W}})') \subset \mathcal{V}'$.
Furthermore, using Lemma 2.12 to obtain $D(A^{\mathcal{W}}) \hookrightarrow D(A^{\mathcal{V}})$, we have for $w' \in D((A^{\mathcal{W}})')$

$$\|w'\|_{\mathcal{V}'} = \sup_{w \in \mathcal{V}} \frac{| < w', w >_{<\mathcal{V}',\mathcal{V}>} |}{\|w\|_{\mathcal{V}}} = \sup_{w \in D(A^{\mathcal{V}})} \frac{| < w', (\sigma I - A^{\mathcal{V}}) w >_{<\mathcal{V}',\mathcal{V}>} |}{\|(\sigma I - A^{\mathcal{V}}) w\|_{\mathcal{V}}}$$

$$= \sup_{w \in D(A^{\mathcal{W}})} \frac{| < w', (\sigma I - A^{\mathcal{W}}) w >_{<\mathcal{W}',\mathcal{W}>} |}{\|(\sigma I - A^{\mathcal{V}}) w\|_{\mathcal{V}}} \quad (\text{since } D(A^{\mathcal{W}}) \hookrightarrow D(A^{\mathcal{V}}))$$

$$\leq c_0 \sup_{w \in D(A^{\mathcal{W}})} \frac{| < (\sigma I - A^{\mathcal{W}})' w', w >_{<\mathcal{W}',\mathcal{W}>} |}{\|w\|_{D(A^{\mathcal{V}})}} \leq$$

$$c_1 \sup_{w \in D(A^{\mathcal{W}})} \frac{| < (\sigma I - A^{\mathcal{W}})' w', w >_{<\mathcal{W}',\mathcal{W}>} |}{\|w\|_{\mathcal{W}}} \quad (\text{using } D(A^{\mathcal{V}}) \hookrightarrow \mathcal{W})$$

$$= c_1 \|(\sigma I - A^{\mathcal{W}})' w'\|_{\mathcal{W}'} \leq c_2 \|w'\|_{D((A^{\mathcal{W}})')},$$

which proves that the injection is continuous.
Finally, to prove that $D((A^{\mathcal{W}})')$ is dense in $\mathcal{V}'$, we note that $D((A^{\mathcal{V}})') \hookrightarrow \mathcal{V}'$ and $D((A^{\mathcal{V}})') \subset D((A^{\mathcal{W}})')$. ∎

2.6 Stability theory

Next, we turn to the issues of stability, stabilizability and detectability. In [15] it is shown that a C_0-semigroup that is exponentially stable on $\mathcal{V}$ (or $\mathcal{W}$), need not be exponentially stable on $\mathcal{W}$ (or $\mathcal{V}$) (see also Remark 2.4). This seems to be rather akward, but using notions of *admissible stabilizability* and *admissible detectability*, a satisfactory stability theory can be given. The following definition is quoted from [15].

Definition 2.18
Let $\Sigma(S(\cdot), B, C, D)$ be a Pritchard-Salamon system of the form (2.7).

(*i*) The pair $(S(\cdot), B)$ is called *admissibly (boundedly) stabilizable* if there exists an admissible output operator $F \in \mathcal{L}(\mathcal{W}, U)$ (an operator $F \in \mathcal{L}(\mathcal{V}, U)$) such that the C_0-semigroup $S_{BF}(\cdot)$ is exponentially stable on $\mathcal{W}$ and $\mathcal{V}$.

(*ii*) The pair $(C, S(\cdot))$ is called *admissibly (boundedly) detectable* if there exists an admissible input operator $H \in \mathcal{L}(Y, \mathcal{V})$ (an operator $H \in \mathcal{L}(Y, \mathcal{W})$) such that the C_0-semigroup $S_{HC}(\cdot)$ is exponentially stable on $\mathcal{W}$ and $\mathcal{V}$.

Remark 2.19
Usually, if $S_0(\cdot)$ is a C_0-semigroup on a Hilbert space X (with infinitesimal generator denoted by A_0) and $B_0 \in \mathcal{L}(U, X)$, where U is another Hilbert space, then the pair (A_0, B_0) is called exponentially stabilizable on X if there exists an $F_0 \in \mathcal{L}(X, U)$ such that $A_0 + B_0 F_0$ generates an exponentially stable C_0-semigroup on X. We note that if $F \in \mathcal{L}(\mathcal{V}, U)$, then F is is an admissible output operator (F is bounded with respect to $\mathcal{V}$) and in this case exponential stability of $S_{BF}(\cdot)$ on $\mathcal{V}$ means that the pair $(A^{\mathcal{V}}, B)$ is exponentially stabilizable. The difference with the notion of 'boundedly stabilizable' above is that in that notion we also demand stability on $\mathcal{W}$ (stability on one space does not imply stability on the other, as explained in [15]). Suitably modified remarks hold for the notions of detectability.

In Theorem 2.20 below, we deduce some interesting results related to Definition 2.18. In particular, we present some useful duality results and we show that if a system is admissibly stabilizable (detectable), then it is also boundedly stabilizable (detectable).

In the proof of Theorem 2.20 we shall use a result of Datko [19] that relates exponential stabilizability to the solvability of the standard LQ-problem for systems with a bounded input operator. His result is summarized in the appendix in Lemma A.2.

Theorem 2.20
Let $\Sigma(S(\cdot), B, C, D)$ be a Pritchard-Salamon system of the form (2.7) and suppose that the representations of the dual spaces $\mathcal{W}^d$ and $\mathcal{V}^d$ are given as in (2.67) and that the representations of the dual spaces U^d and Y^d are given by $\{U', j_U\}$ and $\{Y', j_Y\}$ respectively.

(*i*) *The pair $(S(\cdot), B)$ is admissibly (boundedly) stabilizable if and only if the pair $(B', S'(\cdot))$ is admissibly (boundedly) detectable.*
Similarly, the pair $(C, S(\cdot))$ is admissibly (boundedly) detectable if and only if the pair $(S'(\cdot), C')$ is admissibly (boundedly) stabilizable.

(*ii*) *If the pair $(S(\cdot), B)$ is admissibly stabilizable and $F \in (\mathcal{W}, U)$ is an admissible output operator for $S(\cdot)$ such that $S_{BF}(\cdot)$ is exponentially stable on $\mathcal{V}$, then $S_{BF}(\cdot)$ is also exponentially stable on $\mathcal{W}$.*
Similarly, if the pair $(C, S(\cdot))$ is admissibly detectable and $H \in \mathcal{L}(Y, \mathcal{V})$ is an admissible input operator such that $S_{HC}(\cdot)$ is exponentially stable on $\mathcal{W}$, then $S_{HC}(\cdot)$ is also exponentially stable on $\mathcal{V}$.

(*iii*) *The pair $(S(\cdot), B)$ is admissibly stabilizable if and only if it is boundedly stabilizable.*
Similarly, the pair $(C, S(\cdot))$ is admissibly detectable if and only if it is boundedly detectable.

(*iv*) *Let Z_0 be any real separable Hilbert space and suppose that there exists an admissible input operator $H_0 \in \mathcal{L}(Z_0, \mathcal{V})$ and an admissible output operator $F_0 \in \mathcal{L}(\mathcal{W}, Z_0)$ such that $S_{H_0F_0}(\cdot)$ is exponentially stable on $\mathcal{V}$ and $\mathcal{W}$. Let Z be another real separable Hilbert space. Then, for any admissible input operator $H \in \mathcal{L}(Z, \mathcal{V})$ and any admissible output operator $F \in \mathcal{L}(\mathcal{W}, Z)$ it follows that $S_{HF}(\cdot)$ is exponentially stable on $\mathcal{V}$ if and only if $S_{HF}(\cdot)$ is exponentially stable on $\mathcal{W}$.*

Proof
Proof of (*i*):
Suppose that the pair $(S(\cdot), B)$ is admissibly (boundedly) stabilizable, so that there exists an an admissible output operator $F \in \mathcal{L}(\mathcal{W}, U)$ (an operator $F \in \mathcal{L}(\mathcal{V}, U)$) such that the C_0-semigroup $S_{BF}(\cdot)$ is exponentially stable on $\mathcal{W}$ and $\mathcal{V}$. Using Theorem 2.17 and Lemma 2.15, we infer that $F' \in \mathcal{L}(U', \mathcal{W}')$ is an admissible input operator for $S'(\cdot)$ (or $F' \in \mathcal{L}(U', \mathcal{V}')$ respectively) such that the C_0-semigroup $S'_{F'B'}(\cdot)$ is exponentially stable on $\mathcal{W}'$ and $\mathcal{V}'$. Hence the pair $(B', S'(\cdot))$ is admissibly (boundedly) detectable. The converse of this statement can be proved similarly, using the fact that we identify the bidual of a Hilbert space with itself (if H works for $(B', S'(\cdot))$, then H' does it for $((S')'(\cdot), (B')') = (S(\cdot), B)$).
The proof of the second part of item (*i*) follows from the first part:

$((C')', (S')'(\cdot)) = (C, S(\cdot))$.

Proof of (*ii*):

The first part of this item follows from (the proof of) [15, Lemma 5.3].

The second part follows from the duality results in Theorem 2.17, Lemma 2.15 and item (*i*): if the pair $(C, S(\cdot))$ is admissibly detectable and $H \in \mathcal{L}(Y, \mathcal{V})$ is an admissible input operator such that $S_{HC}(\cdot)$ is exponentially stable on $\mathcal{W}$, then the pair $(S'(\cdot), C')$ is admissibly stabilizable and $H' \in \mathcal{L}(\mathcal{V}', Y')$ is an admissible output operator such that $S'_{C'H'}(\cdot)$ is exponentially stable on $\mathcal{W}'$. Hence we can use the first part of this item to infer the required result.

Proof of (*iii*):

In order to prove the first part we only have to show that admissible stabilizability implies bounded stabilizability. Let $F \in \mathcal{L}(\mathcal{W}, U)$ be an admissible output operator such that $S_{BF}(\cdot)$ is exponentially stable on $\mathcal{W}$ and $\mathcal{V}$. It follows from Lemma 2.13 that for all $x_0 \in \mathcal{V}$

$$S_{BF}(t)x_0 = S(t)x_0 + \int_0^t S(t-s)B\overline{FS}_{BF}(s)x_0 ds.$$

Since $S_{BF}(\cdot)$ is exponentially stable on $\mathcal{V}$, it follows from (2.6) that the control $u(\cdot) := \overline{FS}_{BF}(\cdot)x_0$ is in $L_2(0, \infty; U)$. This means that the infinite-horizon *LQ*-problem for the system on the state-space $\mathcal{V}$

$$x(t) = S(t)x_0 + \int_0^t S(t-s)Bu(s)ds$$

with cost function

$$J := \int_0^\infty (\|x(t)\|_{\mathcal{V}}^2 + \|u(t)\|_U^2)dt$$

is solvable (for any $x_0 \in \mathcal{V}$ there exists a control $u(\cdot)$ such that the cost is finite) and therefore there exists an $F_0 \in \mathcal{L}(\mathcal{V}, U)$ such that $S_{BF_0}(\cdot)$ is exponentially stable on $\mathcal{V}$ (see Lemma A.2). It follows from item (*ii*) that $S_{BF_0}(\cdot)$ is also exponentially stable on $\mathcal{W}$ and so the pair $(S(\cdot), B)$ is boundedly stabilizable.

The second part of this item follows from duality, using item (*i*): the pair $(C, S(\cdot))$ is admissibly detectable if and only if the pair the pair $(S'(\cdot), C')$ is admissibly stabilizable if and only if the pair $(S'(\cdot), C')$ is boundedly stabilizable if and only if the pair $((C')', (S')'(\cdot)) = (C, S(\cdot))$ is boundedly detectable.

Proof of (*iv*):

From the fact that

$$S_{\begin{pmatrix} H_0 & H \end{pmatrix}\begin{pmatrix} F_0 \\ 0 \end{pmatrix}}(\cdot) = S_{H_0F_0}(\cdot)$$

It follows that the pair $(S(\cdot), (\ H_0 \quad H\ ))$ is admissibly stabilizable. Similarly, we infer that the pair

$$\left(\begin{pmatrix} F_0 \\ F \end{pmatrix}, S(\cdot)\right)$$

is admissibly detectable. Now suppose that $S_{HF}(\cdot)$ is exponentially stable on $\mathcal{V}$. Since

$$S_{HF}(\cdot) = S_{\begin{pmatrix} H_0 & H \end{pmatrix}\begin{pmatrix} 0 \\ F \end{pmatrix}}(\cdot),$$

it follows from item (ii) that $S_{HF}(\cdot)$ is exponentially stable on $\mathcal{W}$. Conversely, suppose that $S_{HF}(\cdot)$ is exponentially stable on $\mathcal{W}$. Since

$$S_{HF}(\cdot) = S_{\begin{pmatrix} 0 & H \end{pmatrix}\begin{pmatrix} F_0 \\ F \end{pmatrix}}(\cdot)$$

it follows from item (ii) that $S_{HF}(\cdot)$ is exponentially stable on $\mathcal{V}$.

■

The proof of item (iii) of theorem 2.20 implies the following result:

Corollary 2.21
The following two conditions are equivalent:

(i) *there exists an admissible output operator $F \in \mathcal{L}(\mathcal{W}, U)$ such that $S_{BF}(\cdot)$ is exponentially stable on $\mathcal{V}$*

(ii) *$(A^{\mathcal{V}}, B)$ is exponentially stabilizable on $\mathcal{V}$ (i.e. there exists an operator $F \in \mathcal{L}(\mathcal{V}, U)$ such that $S_{BF}(\cdot)$ is exponentially stable on $\mathcal{V}$).*

Corollary 2.21 will be used in the chapters about LQ-theory and $\mathcal{H}_\infty$-control with state-feedback, where the state of the system is required to be stable only with respect to the topology of $\mathcal{V}$.

Next, we define the notion of *input-output stability* and investigate the relationship with exponential stability, using the concepts of stabilizability and detectability given above.

Let $\Sigma(S(\cdot), B, C, D)$ be a Pritchard-Salamon system of the form (2.7) and let G denote the linear map from $L_2^{loc}(0,\infty;U)$ to $L_2^{loc}(0,\infty;Y)$ defined by

$$(Gu)(\cdot) := C\int_0^{\cdot} S(\cdot - s)Bu(s)ds + Du(\cdot) \tag{2.75}$$

(we recall from Section 2.2 that $\Sigma_G = \Sigma(S(\cdot), B, C, D)$ is called a realization for G). Using Remark 2.2), it is easy to see that $G \in \mathcal{L}(L_2(0,T; U), L_2(0,T;Y))$ for all $T > 0$. We consider G as an unbounded map from $D(G) \subseteq L_2(0,\infty;U)$ to $L_2(0,\infty;Y)$, where $D(G)$ is given by

$$D(G) := \{u \in L_2(0,\infty;U) |\ (Gu)(\cdot) \in L_2(0,\infty;Y)\}.$$

Definition 2.22
Let $G : L_2(0,\infty;U) \supseteq D(G) \to L_2(0,\infty;Y)$ be defined as in (2.75). We call the Pritchard-Salamon system $\Sigma_G = \Sigma(S(\cdot),B,C,D)$ *i/o-stable* (input/output stable) if for all $u(\cdot) \in L_2(0,\infty;U)$, $(Gu)(\cdot) \in L_2(0,\infty;Y)$, i.e. if $D(G) = L_2(0,\infty;U)$.

Since $G \in \mathcal{L}(L_2(0,T;\,U), L_2(0,T;Y))$ for all $T > 0$, it is easy to see that $(G, D(G))$ defines a closed linear map so that $D(G) = L_2(0,\infty;U)$ implies that $G \in \mathcal{L}(L_2(0,\infty;U), L_2(0,\infty;Y))$ (apply the closed graph theorem).

Next, we shall characterize some frequency domain properties of an i/o-stable system (see also Section 2.3). Consider the linear map G defined by (2.75) and suppose that it is i/o-stable. Since G is (right-)shift invariant, it follows from a well known result (see e.g. [33]) that there exists a function $\tilde{G} : \mathbb{C}^+ \to \mathcal{L}(U,Y)$ such that for all $u(\cdot) \in L_2(0,\infty;U)$ and $y(\cdot) = (Gu)(\cdot)$

$$\hat{y}(s) = \tilde{G}(s)\hat{u}(s) \text{ for all } s \in \mathbb{C}^+ \tag{2.76}$$

and

$$\tilde{G}(\cdot) \in \mathcal{H}_\infty(\mathcal{L}(U,Y))$$

(we note that according to Definition 2.11, $\tilde{G}$ is the transfer function of Σ_G). Furthermore,

$$\|G\|_{\mathcal{L}(L_2(0,\infty;U),L_2(0,\infty;Y))} = \|\tilde{G}(\cdot)\|_{\mathcal{H}_\infty(\mathcal{L}(U,Y))} \tag{2.77}$$

(this follows from the Paley-Wiener Theorem (see also (2.17)) and an argument as in [22, Theorem II.6.7]).

The following result is the same as [15, Theorem 5.8], except that in [15] the authors define i/o-stability via the transfer function (it is required to be in $\mathcal{H}_\infty$) rather than in the time-domain, as we have done here. It follows from the above that these notions are equivalent, but the proof of our result is much simpler.

Lemma 2.23
Let $\Sigma(S(\cdot),B,C,D)$ be a Pritchard-Salamon system of the form (2.7) and suppose that the pairs $(S(\cdot),B)$ and $(C,S(\cdot))$ are admissibly stabilizable and admissibly detectable, respectively. Then the system is i/o-stable if and only if $S(\cdot)$ is exponentially stable on $\mathcal{W}$ and $\mathcal{V}$.

Proof
Suppose first that $S(\cdot)$ is exponentially stable on $\mathcal{W}$ and $\mathcal{V}$. It follows from the above that in this case the transfer function $\tilde{G}(s)$ is an element of $\mathcal{H}_\infty(\mathcal{L}(U,Y))$ and that for all $u(\cdot) \in L_2(0,\infty;U)$, $\widehat{Gu}(s) = \hat{y}(s)$ is well-defined on $\mathbb{C}^+$ and

$\hat{y}(s) = \tilde{G}(s)\hat{u}(s)$. We have $\hat{u}(\cdot) \in \mathcal{H}_2(U)$ and so $\hat{y}(\cdot) \in \mathcal{H}_2(Y)$ and using the Paley-Wiener Theorem it follows that $y(\cdot) \in L_2(0, \infty; Y)$. This proves that Σ_G is i/o-stable.

To prove the converse, let $F \in \mathcal{L}(\mathcal{W}, U)$ be an admissible output operator such that $S_{BF}(\cdot)$ is exponentially stable on $\mathcal{W}$ and $\mathcal{V}$, let $H \in \mathcal{L}(Y, \mathcal{V})$ be an admissible input operator such that the C_0-semigroup $S_{HC}(\cdot)$ is exponentially stable on $\mathcal{W}$ and $\mathcal{V}$ and let $x_0 \in \mathcal{V}$. Using the perturbation results in Lemma 2.13 it follows that

$$\overline{CS}_{BF}(t)x_0 = \overline{CS}(t)x_0 + C\int_0^t S(t-s)B\overline{FS}_{BF}(s)x_0 ds.$$

Since $S_{BF}(\cdot)$ is exponentially stable on $\mathcal{V}$ it follows that $\overline{FS}_{BF}(\cdot)x_0 \in L_2(0, \infty; U)$ and $\overline{CS}_{BF}(\cdot)x_0 \in L_2(0, \infty; Y)$ (see (2.6)). Using the fact that the system $\Sigma(S(\cdot), B, C, D)$ is i/o-stable we infer that $\overline{CS}(t)x_0 \in L_2(0, \infty; Y)$ (choose $u(\cdot) = \overline{FS}_{BF}(\cdot)x_0$).
In addition, we have

$$S(t)x_0 = S_{HC}(t)x_0 - \int_0^t S_{HC}(t-s)H\overline{CS}(s)x_0 ds.$$

Now since $S_{HC}(\cdot)$ is exponentially stable on $\mathcal{V}$ and $\overline{CS}(\cdot)x_0 \in L_2(0, \infty; Y)$ it follows that the integral term is in $L_2(0, \infty; \mathcal{V})$ (see Appendix B.1). Hence we may conclude that $S(\cdot)x_0 \in L_2(0, \infty; \mathcal{V})$. Now it follows from Datko's result in Lemma A.1 that $S(\cdot)$ is exponentially stable on $\mathcal{V}$ and because of Theorem 2.20 item *(ii)* it follows that $S(\cdot)$ is also exponentially stable on $\mathcal{W}$. ∎

In the previous lemma we have seen that a Pritchard-Salamon system of the form (2.7) for which $S(\cdot)$ is exponentially stable on both $\mathcal{V}$ and $\mathcal{W}$ is input/output stable. Of course it is also interesting to know if in this case the state function itself is L_2-bounded. Here we have to be careful with the definition of L_2-boundedness; in the next lemma we shall determine under what conditions the state is in $L_2(0, \infty; \mathcal{V})$ or $L_2(0, \infty; \mathcal{W})$ (note that the latter implies the former because $\mathcal{W} \hookrightarrow \mathcal{V}$).

Lemma 2.24
Let $\Sigma(S(\cdot), B, C, D)$ be a Pritchard-Salamon system of the form (2.7) and suppose that $S(\cdot)$ is exponentially stable on $\mathcal{W}$ and $\mathcal{V}$. Consider

$$x(t) = S(t)x_0 + \int_0^t S(t-s)Bu(s)ds$$

and suppose that $u(\cdot) \in L_2(0, \infty; U)$. For $x_0 \in \mathcal{V}$ we have $x(\cdot) \in L_2(0, \infty; \mathcal{V})$ and for $x_0 \in \mathcal{W}$ there holds $x(\cdot) \in L_2(0, \infty; \mathcal{W})$.

Proof

The fact that for $x_0 \in \mathcal{V}$ we have $x(\cdot) \in L_2(0,\infty;\mathcal{V})$ follows immediately from Appendix B.1 and the assumption that $S(\cdot)$ is exponentially stable on $\mathcal{V}$.

Now suppose that $x_0 \in \mathcal{W}$. Since $S(\cdot)$ is exponentially stable on $\mathcal{W}$ it follows that $S(\cdot)x_0 \in L_2(0,\infty;\mathcal{W})$, so that we only have to prove that $w(\cdot)$ defined by

$$w(t) := \int_0^t S(t-\tau)Bu(\tau)d\tau$$

satisfies $w(\cdot) \in L_2(0,\infty;\mathcal{W})$ (Remark 2.2 implies that $w(t) \in \mathcal{W}$ for all $t \geq 0$). It is possible to prove this using only time-domain arguments, but we prefer to do it using some frequency domain results. The idea is to exploit the fact that

$$\sup_{s\in\mathbb{C}^+} \|(sI - A^{\mathcal{V}})^{-1}B\|_{\mathcal{L}(U,\mathcal{W})} < \infty \tag{2.78}$$

(this follows from (2.19), because $S(\cdot)$ is exponentially stable on $\mathcal{W}$ and $\mathcal{V}$). As in the proof of [15, Proposition 2.15] we have $\exp(-s\cdot)w(\cdot) \in L_1(0,\infty;\mathcal{W})$ for all $s \in \mathbb{C}^+$ and

$$\hat{w}(s) = \int_0^\infty \exp(-st)w(t)dt = (sI - A^{\mathcal{V}})^{-1}B\hat{u}(s)$$

for all $s \in \mathbb{C}^+$ (the integral is taken with respect to the topology of $\mathcal{V}$). As in the proof of [15, Proposition 2.15] it can be argued that $(sI - A^{\mathcal{V}})^{-1}B$ is holomorphic with respect to the topology of $\mathcal{L}(U,\mathcal{W})$. Using (2.78) and the fact that $\hat{u}(\cdot) \in \mathcal{H}_2(U)$ we infer that $\hat{w}(\cdot) \in \mathcal{H}_2(\mathcal{W})$. Finally, using the Paley-Wiener Theorem this implies that $w(\cdot) \in L_2(0,\infty;\mathcal{W})$. ■

The last result of this section deals with the 'transpose' of a Pritchard-Salamon system. Let $\Sigma_G = \Sigma(S(\cdot),B,C,D)$ be a Pritchard-Salamon system of the form (2.7) and suppose that the representations of $\mathcal{W}^d$ and $\mathcal{V}^d$ are given as in (2.67) and that the representations of U^d and Y^d are given by $\{U', j_U\}$ and $\{Y', j_Y\}$ respectively. Let G be the corresponding linear input/output map given by (2.75).

We define the *transpose* of Σ_G as the Pritchard-Salamon system

$$\Sigma_{G^\natural} = \Sigma(S'(\cdot),C',B',D'),$$

given by

$$\Sigma_{G^\natural} \left\{ \begin{array}{rcl} x'(t) & = & S'(t)x'_0 + \int_0^t S'(t-s)C'y'(s)ds \\ \\ u'(t) & = & B'x'(t) + D'y'(t). \end{array} \right. \tag{2.79}$$

According to the notational convention in Remark 2.5, now $G^\natural$ denotes the corresponding linear map for $x'_0 = 0$:

$$(G^\natural y')(\cdot) = B' \int_0^{\cdot} S'(\cdot - s)C'y'(s)ds + D'y'(\cdot). \tag{2.80}$$

The following holds:

Lemma 2.25
Suppose that $S(\cdot)$ is exponentially stable on $\mathcal{W}$ and $\mathcal{V}$, so that $G \in \mathcal{L}(L_2(0,\infty; U), L_2(0,\infty;Y))$. Then the C_0-semigroup $S'(\cdot)$ is exponentially stable on $\mathcal{W}'$ and $\mathcal{V}'$ and $G^\natural$ given by (2.80) satisfies $G^\natural \in \mathcal{L}(L_2(0,\infty;Y'), L_2(0,\infty;U'))$ and we have

$$\|G^\natural\| = \|G\|. \tag{2.81}$$

Proof
The fact that $S'(\cdot)$ is exponentially stable on $\mathcal{W}'$ and $\mathcal{V}'$ follows from Lemma 2.15 and it follows from Lemma 2.23 that $G^\natural \in \mathcal{L}(L_2(0,\infty;Y'), L_2(0,\infty;U'))$. To prove (2.81), we first calculate the adjoint of the transfer function corresponding to G. We claim that for all $s \in \mathbb{C}^+$

$$(C(sI - A^{\mathcal{V}})^{-1}B + D)' = B'((sI)' - (A^{\mathcal{W}})')^{-1}C' + D'. \tag{2.82}$$

To prove (2.82) we first show that

$$((sI - A^{\mathcal{V}})^{-1}B)' = \overline{B'((sI)' - (A^{\mathcal{V}})')^{-1}} \in \mathcal{L}(\mathcal{W}', U'). \tag{2.83}$$

Indeed, since $(sI - A^{\mathcal{V}})^{-1}B \in \mathcal{L}(U, \mathcal{V})$, we have that $((sI - A^{\mathcal{V}})^{-1}B)' = B'((sI)' - (A^{\mathcal{V}})')^{-1} \in \mathcal{L}(\mathcal{V}', U')$ (see Lemma 2.15). Now since in fact $(sI - A^{\mathcal{V}})^{-1}B \in \mathcal{L}(U, \mathcal{W})$ (see (2.19)), (2.83) follows from Lemma 2.16.
Finally, we know that $\overline{B'(sI - (A^{\mathcal{V}})')^{-1}}C' = B'(sI - (A^{\mathcal{W}})')^{-1}C'$ (compare with (2.21)), so that

$$(C(sI - A^{\mathcal{V}})^{-1}B)' = \overline{B'((sI)' - (A^{\mathcal{V}})')^{-1}}C' = B'((sI)' - (A^{\mathcal{W}})')^{-1}C',$$

which proves (2.82).
Using the relationship between the norm of the input/ouput operator and the $\mathcal{H}_\infty$-norm of the transfer function (cf. (2.77)), (2.81) easily follows from (2.82). ∎

2.7 Dynamic output-feedback

In this section we shall formalize the notion of dynamic output-feedback (or dynamic measurement-feedback) for Pritchard-Salamon systems and establish some important system theoretic results. In particular, we shall show that the Pritchard-Salamon class is closed under cascade, parallel and (dynamic) feedback interconnections. Furthermore, we shall introduce the feedback configuration that is used for the $\mathcal{H}_\infty$-control problem in Chapters 4-5 (linear fractional transformations), we shall explain the notion of 'stabilizing controllers' and give a version of the small-gain theorem.

Let $\Sigma(S(\cdot), B, C, D)$ be a Pritchard-Salamon system with respect to $(\mathcal{W}, \mathcal{V})$ of the form (2.7) (i.e. with $S(\cdot)$ a C_0-semigroup on $\mathcal{W} \hookrightarrow \mathcal{V}$ etc.). Now suppose that $\mathcal{W}_K$ and $\mathcal{V}_K$ are also real separable Hilbert spaces, satisfying $\mathcal{W}_K \hookrightarrow \mathcal{V}_K$ and suppose that $T(\cdot)$ is a C_0-semigroup on $\mathcal{V}_K$ which restricts to a C_0-semigroup on $\mathcal{W}_K$ (we denote the generators of $T(\cdot)$ on $\mathcal{W}_K$ and $\mathcal{V}_K$ by $M^{\mathcal{W}_K}$ and $M^{\mathcal{V}_K}$ respectively). Furthermore, let $N \in \mathcal{L}(Y, \mathcal{V}_K)$, $L \in \mathcal{L}(\mathcal{W}_K, U)$ and $R \in \mathcal{L}(Y, U)$ be such that $\Sigma(T(\cdot), N, L, R)$ is a Pritchard-Salamon system with respect to $(\mathcal{W}_K, \mathcal{V}_K)$. We want to consider 'well-posedness' of the feedback interconnection in Figure 2.1 (this is also called the *closed-loop system*) The

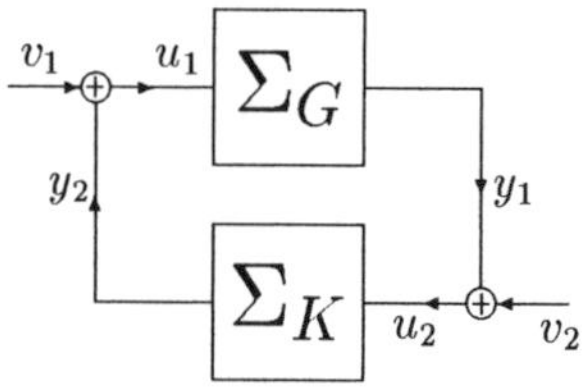

Figure 2.1: Dynamic feedback interconnection.

system equations are given by

$$\Sigma_G \begin{cases} x(t) &= S(t)x_0 + \int_0^t S(t-s)Bu_1(s)ds \\ y_1(t) &= Cx(t) + Du_1(t) \end{cases} \tag{2.84}$$

and

$$\Sigma_K \begin{cases} p(t) &= T(t)p_0 + \int_0^t T(t-s)Nu_2(s)ds \\ y_2(t) &= Lp(t) + Ru_2(t) \end{cases} \tag{2.85}$$

where $u_1(\cdot) = y_2(\cdot) + v_1(\cdot)$ and $u_2(\cdot) = y_1(\cdot) + v_2(\cdot)$. First of all, we define the extended real Hilbert spaces $\tilde{\mathcal{V}} = \mathcal{V} \times \mathcal{V}_K$ and $\tilde{\mathcal{W}} = \mathcal{W} \times \mathcal{W}_K$ and we note that $\tilde{\mathcal{W}} \hookrightarrow \tilde{\mathcal{V}}$. Defining

$$\begin{aligned} \tilde{S}(\cdot) &:= \begin{pmatrix} S(\cdot) & 0 \\ 0 & T(\cdot) \end{pmatrix} \quad ; \quad \tilde{B} := \begin{pmatrix} B & 0 \\ 0 & N \end{pmatrix} \\ \tilde{C} &:= \begin{pmatrix} C & 0 \\ 0 & L \end{pmatrix} \quad ; \quad \tilde{D} := \begin{pmatrix} D & 0 \\ 0 & R \end{pmatrix}, \end{aligned} \tag{2.86}$$

it follows that $\tilde{S}(\cdot)$ is a C_0-semigroup on $\tilde{\mathcal{W}}$ and $\tilde{\mathcal{V}}$ and that $\Sigma(\tilde{S}(\cdot), \tilde{B}, \tilde{C}, \tilde{D})$ is a Pritchard-Salamon system with respect to $(\tilde{\mathcal{W}}, \tilde{\mathcal{V}})$. We can reformulate the combination of (2.84) and (2.85) by introducing the extended state $\begin{pmatrix} x(t) \\ p(t) \end{pmatrix}$:

$$\left\{ \begin{aligned} \begin{pmatrix} x(t) \\ p(t) \end{pmatrix} &= \tilde{S}(t) \begin{pmatrix} x_0 \\ p_0 \end{pmatrix} + \int_0^t \tilde{S}(t-s)\tilde{B} \begin{pmatrix} u_1(s) \\ u_2(s) \end{pmatrix} ds, \\ \begin{pmatrix} y_1(t) \\ y_2(t) \end{pmatrix} &= \tilde{C} \begin{pmatrix} x(t) \\ p(t) \end{pmatrix} + \tilde{D} \begin{pmatrix} u_1(t) \\ u_2(t) \end{pmatrix}. \end{aligned} \right. \tag{2.87}$$

For notational convenience we define

$$\mathcal{J} := \begin{pmatrix} 0 & I \\ I & 0 \end{pmatrix}. \tag{2.88}$$

If we set $u_1(\cdot) = y_2(\cdot) + v_1(\cdot)$ and $u_2(\cdot) = y_1(\cdot) + v_2(\cdot)$ in the output equation of (2.87) (as in Figure 2.1) we obtain

$$\begin{pmatrix} y_2(t) \\ y_1(t) \end{pmatrix} = \mathcal{J}\tilde{C} \begin{pmatrix} x(t) \\ p(t) \end{pmatrix} + \mathcal{J}\tilde{D} \begin{pmatrix} y_2(t) \\ y_1(t) \end{pmatrix} + \mathcal{J}\tilde{D} \begin{pmatrix} v_1(t) \\ v_2(t) \end{pmatrix}$$

and so we call the feedback interconnection *well-posed* if the operator defined by

$$\begin{pmatrix} I & 0 \\ 0 & I \end{pmatrix} - \mathcal{J}\tilde{D} = \begin{pmatrix} I & -R \\ -D & I \end{pmatrix}$$

has a bounded inverse. It is easy to see that this is the case if and only if $I - RD$ and $I - DR$ both have bounded inverses. Moreover, this is equivalent to the statement that either $I - RD$ or $I - DR$ has a bounded inverse (see Appendix D, Lemma D.2). If the feedback interconnection is well posed, we define

$$\bar{D} := \begin{pmatrix} I & -R \\ -D & I \end{pmatrix}^{-1} = \begin{pmatrix} (I-RD)^{-1} & R(I-DR)^{-1} \\ D(I-RD)^{-1} & (I-DR)^{-1} \end{pmatrix}$$

and we have

$$\begin{pmatrix} y_2(t) \\ y_1(t) \end{pmatrix} = \bar{D}\mathcal{J}\tilde{C}\begin{pmatrix} x(t) \\ p(t) \end{pmatrix} + \bar{D}\mathcal{J}\tilde{D}\begin{pmatrix} v_1(t) \\ v_2(t) \end{pmatrix}.$$

Hence,

$$\begin{pmatrix} u_1(t) \\ u_2(t) \end{pmatrix} = \begin{pmatrix} y_2(t) \\ y_1(t) \end{pmatrix} + \begin{pmatrix} v_1(t) \\ v_2(t) \end{pmatrix} = \bar{D}\mathcal{J}\tilde{C}\begin{pmatrix} x(t) \\ p(t) \end{pmatrix} + \bar{D}\begin{pmatrix} v_1(t) \\ v_2(t) \end{pmatrix},$$

(using that $\bar{D}\mathcal{J}\tilde{D} + I = \bar{D}$) so that we can interpret the feedback interconnection as a preliminary state-feedback on the state-space of the system given by (2.87) (cf. Lemma 2.14). The admissible preliminary feedback

$$\begin{pmatrix} u_1(t) \\ u_2(t) \end{pmatrix} = \bar{D}\mathcal{J}\tilde{C}\begin{pmatrix} x(t) \\ p(t) \end{pmatrix} + \bar{D}\begin{pmatrix} v_1(t) \\ v_2(t) \end{pmatrix}, \tag{2.89}$$

applied to (2.87) gives the Pritchard-Salamon system

$$\Sigma(\tilde{S}_{\tilde{B}\bar{D}\mathcal{J}\tilde{C}}(\cdot), \tilde{B}\bar{D}, \mathcal{J}\bar{D}\mathcal{J}\tilde{C}, \tilde{D}\bar{D}):$$

$$\begin{cases} \begin{pmatrix} x(t) \\ p(t) \end{pmatrix} = \tilde{S}_{\tilde{B}\bar{D}\mathcal{J}\tilde{C}}(t)\begin{pmatrix} x_0 \\ p_0 \end{pmatrix} + \int_0^t \tilde{S}_{\tilde{B}\bar{D}\mathcal{J}\tilde{C}}(t-s)\tilde{B}\bar{D}\begin{pmatrix} v_1(s) \\ v_2(s) \end{pmatrix} ds \\ \begin{pmatrix} y_1(t) \\ y_2(t) \end{pmatrix} = \mathcal{J}\bar{D}\mathcal{J}\tilde{C}\begin{pmatrix} x(t) \\ p(t) \end{pmatrix} + \tilde{D}\bar{D}\begin{pmatrix} v_1(t) \\ v_2(t) \end{pmatrix} \end{cases} \tag{2.90}$$

and this is how the feedback interconnection of (2.87) with $u_1(\cdot) = y_2(\cdot) + v_1(\cdot)$ and $u_2(\cdot) = y_1(\cdot) + v_2(\cdot)$ should be interpreted.

The *cascade interconnection* of two Pritchard-Salamon systems Σ_G and Σ_K as in Figure 2.2 can also be described using the formulation of (2.87).

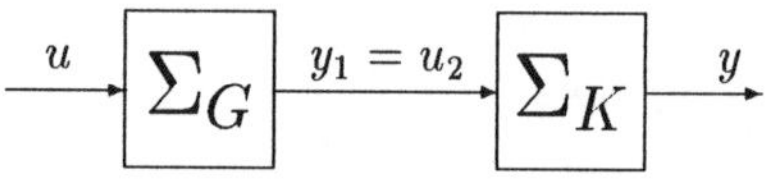

Figure 2.2: Cascade interconnection.

Suppose that we take $u_1(\cdot) = u(\cdot)$ in (2.84) and $u_2(\cdot) = y_1(\cdot)$ and $y(\cdot) = y_2(\cdot)$ in (2.85). We can regard this as a preliminary feedback in (2.87) of the form

$$\begin{pmatrix} u_1(t) \\ u_2(t) \end{pmatrix} = \begin{pmatrix} 0 & 0 \\ I & 0 \end{pmatrix}\tilde{C}\begin{pmatrix} x(t) \\ p(t) \end{pmatrix} + \begin{pmatrix} I \\ D \end{pmatrix}u(t)$$

and we have

$$y(\cdot) = (R \quad I)\tilde{C}\begin{pmatrix} x(t) \\ p(t) \end{pmatrix} + RDu(t).$$

We define

$$\tilde{F} := \begin{pmatrix} 0 & 0 \\ I & 0 \end{pmatrix}\tilde{C}.$$

The cascade system (from u to y) is therefore given by the Pritchard-Salamon system $\Sigma(\tilde{S}_{\tilde{B}\tilde{F}}(\cdot), \tilde{B}\begin{pmatrix} I \\ D \end{pmatrix}, (R \quad I)\tilde{C}, RD)$:

$$\left\{ \begin{array}{rcl} \begin{pmatrix} x(t) \\ p(t) \end{pmatrix} & = & \tilde{S}_{\tilde{B}\tilde{F}}(t)\begin{pmatrix} x_0 \\ p_0 \end{pmatrix} + \int_0^t \tilde{S}_{\tilde{B}\tilde{F}}(t-s)\tilde{B}\begin{pmatrix} I \\ D \end{pmatrix}u(s) \\ y(t) & = & (R \quad I)\tilde{C}\begin{pmatrix} x(t) \\ p(t) \end{pmatrix} + RDu(t). \end{array} \right. \tag{2.91}$$

In the following lemma we characterize the *inverse* of a Pritchard-Salamon system (if it exists).

Lemma 2.26
Let $\Sigma_G = \Sigma(S(\cdot), B, C, D)$ be Pritchard-Salamon system of the form (2.7) such that $U = Y$ and suppose that D has a bounded inverse. Denote the linear map G from $L_2^{loc}(0,\infty;U)$ to $L_2^{loc}(0,\infty;U)$ by

$$(Gu)(\cdot) = C\int_0^{\cdot} S(\cdot - s)Bu(s)ds + Du(\cdot). \tag{2.92}$$

The system $\Sigma(S_{-BD^{-1}C}(\cdot), BD^{-1}, -D^{-1}C, D^{-1})$ is also a Pritchard-Salamon system and defining G^{-1} as the linear map from $L_2^{loc}(0,\infty;U)$ to $L_2^{loc}(0,\infty;U)$ given by

$$(G^{-1}y)(\cdot) := -D^{-1}C\int_0^{\cdot} S_{-BD^{-1}C}(\cdot - s)BD^{-1}y(s)ds + D^{-1}y(\cdot). \tag{2.93}$$

we have for all $u(\cdot) \in L_2^{loc}(0,\infty;U)$

$$G^{-1}(Gu)(\cdot) = G(G^{-1}u)(\cdot) = u(\cdot). \tag{2.94}$$

Proof
We can consider $\Sigma(S_{-BD^{-1}C}(\cdot), BD^{-1}, -D^{-1}C, D^{-1})$ as a system of the form (2.85) and the product $G^{-1}G$ is simply the map from $u(\cdot)$ to $y(\cdot)$ in (2.87), where $x_0 = p_0 = 0$, $u_1(\cdot) = u(\cdot)$, $u_2(\cdot) = y_1(\cdot)$ and $y(\cdot) = y_2(\cdot)$. Proving that

$G^{-1}G = I$ is then equivalent to proving that $y(\cdot) = u(\cdot)$. It is easy to see that for this choice of inputs we get

$$\begin{cases} x(t) &= \int_0^t S(t-s)Bu(s)ds \\ p(t) &= \int_0^t S_{-BD^{-1}C}(t-s)B(D^{-1}Cx(s)+u(s))ds \\ y(t) &= D^{-1}C(x(t)-p(t))+u(t). \end{cases} \tag{2.95}$$

We claim that $x(t) = p(t)$. It follows from Lemma 2.14 that

$$p(t) = \int_0^t S(t-s)B(D^{-1}C(x(s)-p(s))+u(s))ds.$$

Hence $e(t) := x(t) - p(t)$ satisfies

$$e(t) = -\int_0^t S(t-s)BD^{-1}Ce(s)ds.$$

We know that $e(\cdot)$ is continuous with respect to the topology of $\mathcal{W}$ and using Remark 2.2 we infer that for all $T > 0$ there exists a constant c_T such that for all $t \in (0,T)$

$$\|e(t)\|_{\mathcal{W}} \leq c_T \|D^{-1}Ce(\cdot)\|_{L_2(0,t;U)},$$

so that

$$\|e(t)\|_{\mathcal{W}} \leq c_1 \|e(\cdot)\|_{L_2(0,t;\mathcal{W})}.$$

Gronwall's Lemma now implies that $e(\cdot) = x(\cdot) - p(\cdot) = 0$ so that $y(\cdot) = u(\cdot)$ follows from (2.95). Hence, $G^{-1}G = I$. The proof of $GG^{-1} = I$ is similar and is omitted. ∎

Next we shall prove an important result that will be needed to prove the small gain theorem and Redheffer's Lemma for Pritchard-Salamon systems. It relates exponential stability of the closed-loop system to internal stability using admissible stabilizability and detectability. We consider the systems Σ_G and Σ_K (Σ_K is usually regarded as the *controller*) given by (2.84) and (2.85) with the feedback configuration of figure 2.1 (see also (2.87)). We suppose that the interconnection is well-posed and that $x_0 = 0$ and $p_0 = 0$. The closed-loop system from $(v_1(\cdot), v_2(\cdot))$ to $(u_1(\cdot), u_2(\cdot))$ is the Pritchard-Salamon system $\Sigma(\tilde{S}_{\tilde{B}\bar{D}\mathcal{J}\tilde{C}}(\cdot), \tilde{B}\bar{D}, \bar{D}\mathcal{J}\tilde{C}, \bar{D})$ (cf. formula (2.89)):

$$\begin{cases} \begin{pmatrix} x(t) \\ p(t) \end{pmatrix} &= \int_0^t \tilde{S}_{\tilde{B}\bar{D}\mathcal{J}\tilde{C}}(t-s)\tilde{B}\bar{D}\begin{pmatrix} v_1(s) \\ v_2(s) \end{pmatrix} ds \\ \begin{pmatrix} u_1(t) \\ u_2(t) \end{pmatrix} &= \bar{D}\mathcal{J}\tilde{C}\begin{pmatrix} x(t) \\ p(t) \end{pmatrix} + \bar{D}\begin{pmatrix} v_1(t) \\ v_2(t) \end{pmatrix}. \end{cases} \tag{2.96}$$

As before, the linear maps G from $L_2^{loc}(0,\infty;U)$ to $L_2^{loc}(0,\infty\,;Y)$ and K from $L_2^{loc}(0,\infty;Y)$ to $L_2^{loc}(0,\infty;U)$ are given by

$$(Gu)(\cdot) = C\int_0^{\cdot} S(\cdot - s)Bu(s)ds + Du(\cdot) \tag{2.97}$$

and

$$(Ky)(\cdot) = L\int_0^{\cdot} T(\cdot - s)Ny(s)ds + Ru(\cdot). \tag{2.98}$$

Since

$$\begin{pmatrix} u_1(t) \\ u_2(t) \end{pmatrix} = \begin{pmatrix} y_2(t) \\ y_1(t) \end{pmatrix} + \begin{pmatrix} v_1(t) \\ v_2(t) \end{pmatrix} \quad \text{and}$$

$$\begin{pmatrix} y_1(t) \\ y_2(t) \end{pmatrix} = \begin{pmatrix} G & 0 \\ 0 & K \end{pmatrix}\begin{pmatrix} u_1(t) \\ u_2(t) \end{pmatrix},$$

we can reformulate (2.96) as

$$\begin{pmatrix} u_1(t) \\ u_2(t) \end{pmatrix} = G_{cl}\begin{pmatrix} v_1(t) \\ v_2(t) \end{pmatrix},$$

where

$$G_{cl} = \begin{pmatrix} (I-KG)^{-1} & (I-KG)^{-1}K \\ G(I-KG)^{-1} & (I-GK)^{-1} \end{pmatrix}. \tag{2.99}$$

(note that since $(I-DR)^{-1} \in \mathcal{L}(Y)$ and $(I-RD)^{-1} \in \mathcal{L}(U)$, $(I-GK)^{-1}$ and $(I-KG)^{-1}$ are both well defined). In the following theorem we shall relate i/o-stability of (2.96) (which is equivalent to boundedness of G_{cl}) to exponential stability of (2.96). It contains a generalization of one of the main results in [43], where the bounded/finite-rank input/output case is treated.

Theorem 2.27

Consider the systems Σ_G and Σ_K given by (2.84) and (2.85) and the well-posed feedback interconnection determined by (2.96). We have the following equivalence: The C_0-semigroup $\tilde{S}_{\tilde{B}\tilde{D}\mathcal{J}\tilde{C}}(\cdot)$ is exponentially stable on $\tilde{\mathcal{W}}$ and $\tilde{\mathcal{V}}$ if and only if $(S(\cdot),B)$ and $(T(\cdot),N)$ are admissibly stabilizable, $(C,S(\cdot))$ and $(L,T(\cdot))$ are admissibly detectable and the system given by (2.96) is i/o-stable (i.e. G_{cl} given by (2.99) is bounded).

In the case that one of the equivalent statements in Theorem 2.27 holds, we shall call Σ_K a *stabilizing (measurement-feedback) controller* for Σ_G.

Proof

(necessity) Suppose that $\tilde{S}_{\tilde{B}\tilde{D}\mathcal{J}\tilde{C}}(\cdot)$ is exponentially stable on $\tilde{\mathcal{W}}$ and $\tilde{\mathcal{V}}$.

The fact that the system given by (2.96) is i/o-stable follows from Lemma 2.23, so it remains to prove that $(S(\cdot),B)$ and $(T(\cdot),N)$ are admissibly stabilizable and that $(C,S(\cdot))$ and $(L,T(\cdot))$ are admissibly detectable.
It follows directly from our assumption that the pair $(\tilde{S}(\cdot),\tilde{B})$ is admissibly stabilizable. We shall show that this implies that that the pairs $(S(\cdot),B)$ and $(T(\cdot),N)$ are exponentially stabilizable on $\mathcal{V}$ and $\mathcal{V}_K$, respectively (this means that there exists an $F_1 \in \mathcal{L}(U,\mathcal{V})$ and an $F_2 \in \mathcal{L}(U,\mathcal{V}_K)$ such that $S_{BF_1}(\cdot)$ is exponentially stable on $\mathcal{V}$ and $T_{NF_2}(\cdot)$ is exponentially stable on $\mathcal{V}_K$). Indeed, it follows from (2.89) and (2.90) with $v_1(\cdot)=0$ and $v_2(\cdot)=0$ that

$$\begin{pmatrix} x(t) \\ p(t) \end{pmatrix} := \tilde{S}_{\tilde{B}\bar{D}\mathcal{J}\tilde{C}}(t) \begin{pmatrix} x_0 \\ p_0 \end{pmatrix} \tag{2.100}$$

satisfies

$$\begin{pmatrix} x(t) \\ p(t) \end{pmatrix} = \tilde{S}(t)\begin{pmatrix} x_0 \\ p_0 \end{pmatrix} + \int_0^t \tilde{S}(t-s)\tilde{B}\begin{pmatrix} u_1(s) \\ u_2(s) \end{pmatrix} ds,$$

where

$$\begin{pmatrix} u_1(t) \\ u_2(t) \end{pmatrix} = \bar{D}\mathcal{J}\tilde{C}\begin{pmatrix} x(t) \\ p(t) \end{pmatrix}. \tag{2.101}$$

Using the structure of $\tilde{S}(\cdot)$ and $\tilde{B}$ in (2.86), we infer that

$$x(t) = S(t)x_0 + \int_0^t S(t-s)Bu_1(s)ds. \tag{2.102}$$

Since $\tilde{S}_{\tilde{B}\bar{D}\mathcal{J}\tilde{C}}(\cdot)$ is exponentially stable on $\tilde{\mathcal{W}}$ and $\tilde{\mathcal{V}}$ it follows from (2.100) that $x(\cdot) \in L_2(0,\infty;\mathcal{V})$ and so (2.101) and (2.6) imply that $u_1(\cdot) \in L_2(0,\infty;U)$. Since x_0 was arbitrary, we can apply Lemma A.2 to the system given by (2.102) and infer the existence of an $F_1 \in \mathcal{L}(\mathcal{V},U)$ such that $S_{BF_1}(\cdot)$ is exponentially stable on $\mathcal{V}$. The proof of the existence of $F_2 \in \mathcal{L}(\mathcal{V}_K,U)$ such that $T_{NF_2}(\cdot)$ is exponentially stable on $\mathcal{V}_K$ is of course similar. It follows that,

$$\tilde{S}_{\tilde{B}\begin{pmatrix} F_1 & 0 \\ 0 & F_2 \end{pmatrix}}(\cdot) = \tilde{S}_{\begin{pmatrix} B & 0 \\ 0 & N \end{pmatrix}\begin{pmatrix} F_1 & 0 \\ 0 & F_2 \end{pmatrix}}(\cdot) =$$

$$\begin{pmatrix} S_{BF_1}(\cdot) & 0 \\ 0 & S_{NF_2}(\cdot) \end{pmatrix} \tag{2.103}$$

is exponentially stable on $\tilde{\mathcal{V}} = \mathcal{V}\times\mathcal{V}_K$. We have shown above that the pair $(\tilde{S}(\cdot),\tilde{B})$ is admissibly stabilizable and so, using Theorem 2.20 item (ii), we infer that $S_{BF_1}(\cdot)$ is also exponentially stable on $\mathcal{W}$ and $T_{NF_2}(\cdot)$ is also exponentially stable on $\mathcal{W}_K$. Hence we have proved that $(S(\cdot),B)$ and $(T(\cdot),N)$ are admissibly stabilizable.

To prove that $(C, S(\cdot))$ and $(L, T(\cdot))$ are admissibly detectable, we first note that the pair $(\tilde{C}, \tilde{S}(\cdot))$ is admissibly detectable. The result now follows from the result that we have just proved and some duality arguments using Theorem 2.20, item (i).

(sufficiency) Suppose that the system given by (2.96) *is i/o-stable, that* $(S(\cdot), B)$ *and* $(T(\cdot), N)$ *are admissibly stabilizable and that* $(C, S(\cdot))$ *and* $(L, T(\cdot))$ *are admissibly detectable.*

Since the pairs $(S(\cdot), B)$ and $(T(\cdot), N)$ are admissibly stabilizable it is easy to see from the definitions of $\tilde{S}(\cdot)$ and $\tilde{B}$ in (2.86) and formula (2.103) that the pair $(\tilde{S}(\cdot), \tilde{B})$ is admissibly stabilizable which, in turn, implies that the pair $(\tilde{S}_{\tilde{B}\bar{D}\mathcal{J}\tilde{C}}(\cdot), \tilde{B}\bar{D})$ is admissibly stabilizable. Similarly, since the pairs $(C, S(\cdot))$ and $(L, T(\cdot))$ are admissibly detectable, we see that the pair $(\bar{D}\mathcal{J}\tilde{C}, \tilde{S}_{\tilde{B}\bar{D}\mathcal{J}\tilde{C}}(\cdot))$ is admissibly detectable. The result now follows from Lemma 2.23. ∎

Now we are ready to give the small gain theorem with internal stability for Pritchard-Salamon systems:

Theorem 2.28

Suppose that we have two Pritchard-Salamon systems as in (2.84) *and* (2.85), *where* $S(\cdot)$ *is exponentially stable on* $\mathcal{V}$ *and* $\mathcal{W}$ *and* $T(\cdot)$ *is exponentially stable on* $\mathcal{V}_K$ *and* $\mathcal{W}_K$. *Furthermore, suppose that the corresponding bounded linear maps* G *and* K *given in* (2.97) *and* (2.98) *satisfy* $\|G\| < 1$ *and* $\|K\| \leq 1$. *Then the closed-loop system produced by letting* $u_2(\cdot) = y_1(\cdot)$ *and* $u_1(\cdot) = y_2(\cdot)$ *is well-posed and the corresponding closed-loop* C_0*-semigroup* $\tilde{S}_{\tilde{B}\bar{D}\mathcal{J}\tilde{C}}(\cdot)$ *(given as in Theorem* 2.27) *is exponentially stable on* $\mathcal{W}$ *and* $\mathcal{V}$ *(hence* Σ_K *is a stabilizing controller for* Σ_G).

Proof

We denote the transfer functions of Σ_G and Σ_K by $\tilde{G}$ and $\tilde{K}$. Then for $s \in \mathbb{C}^+$

$$\tilde{G}(s) = C(sI - A^{\mathcal{V}})^{-1}B + D \quad \text{and} \quad \tilde{K}(s) = L(sI - M^{\mathcal{V}_K})^{-1}N + R.$$

According to (2.77), we have $\|G\| = \|\tilde{G}\|_\infty$ and $\|K\| = \|\tilde{K}\|_\infty$. Hence, $\|\tilde{G}\|_\infty < 1$ and $\|\tilde{K}\|_\infty \leq 1$ and so it follows from (2.23) that $\|D\|_{\mathcal{L}(U,Y)} < 1$ and $\|R\|_{\mathcal{L}(Y,U)} \leq 1$. Hence, $I - RD$ and $I - DR$ both have bounded inverses and the feedback interconnection is well-posed. Furthermore, it is easy to see that $\|GK\| < 1$ and $\|KG\| < 1$, so that $(I - GK)^{-1}$ and $(I - KG)^{-1}$ are linear and bounded. Hence, G_{cl} given by (2.99) is linear and bounded and the system given by (2.96) is i/o-stable. The result now follows from Theorem 2.27. ∎

Remark 2.29

It is easy to see that Theorem 2.28 can be stated in a more general way, for instance the condition that $\|G\| < 1$ and $\|K\| \leq 1$ can be replaced by the condition that $\|GK\| < 1$ and $\|KG\| < 1$. In fact, the most general version is given precisely by Theorem 2.27, because the essential element is the boundedness of G_{cl}. Theorem 2.28 will be used to generalize an important result in $\mathcal{H}_\infty$-theory, called Redheffer's Lemma, to the Pritchard-Salamon class (see Section 5.2).

Next, we give a theorem about existence of stabilizing measurement-feedback controllers for stabilizable/detectable Pritchard-Salamon systems.

Theorem 2.30
Let $\Sigma(S(\cdot), B, C, D)$ be a Pritchard-Salamon system of the form (2.7) and suppose that $F \in \mathcal{L}(\mathcal{W}, U)$ and $H \in \mathcal{L}(Y, \mathcal{V})$ are admissible ouput and input operators, such that $S_{BF}(\cdot)$ and $S_{HC}(\cdot)$ are exponentially stable on $\mathcal{W}$ and $\mathcal{V}$. Then the controller Σ_K given by

$$\Sigma_K \left\{ \begin{array}{rcl} p(t) &=& T(t)p_0 + \int_0^t T(t-s)Ny(s)ds \\ \\ u(t) &=& Lp(t) + Ru(t), \end{array} \right. \tag{2.104}$$

where

$$\left. \begin{array}{rcl} T(\cdot) &:=& S_{BF+H(DF+C)}(\cdot) \\ N &:=& -H \\ L &:=& F \\ R &:=& 0, \end{array} \right\} \tag{2.105}$$

is a Pritchard-Salamon system such that the closed-loop system is well-posed and the corresponding closed-loop semigroup is exponentially stable on $\mathcal{V} \times \mathcal{V}$ and $\mathcal{W} \times \mathcal{W}$.

Proof
Since $R = 0$, it follows from the above that the closed-loop system is well-posed and that it is given by (2.90) with $v_1(\cdot) = 0$ and $v_2(\cdot) = 0$, so that

$$\begin{pmatrix} x(t) \\ p(t) \end{pmatrix} = \tilde{S}_{\tilde{B}\tilde{D}\mathcal{J}\tilde{C}}(t) \begin{pmatrix} x_0 \\ p_0 \end{pmatrix}, \tag{2.106}$$

(of course using the definition of the controller parameters in (2.105)). We have to show that the C_0-semigroup $\tilde{S}_{\tilde{B}\tilde{D}\mathcal{J}\tilde{C}}(\cdot)$ is exponentially stable on $\mathcal{W} \times \mathcal{W}$ and $\mathcal{V} \times \mathcal{V}$. We claim that $x(\cdot)$ and $x(\cdot) - p(\cdot)$ can be expressed as

$$x(t) = S_{BF}(t)x_0 + \int_0^t S_{BF}(t-s)BF(p(s) - x(s))ds \tag{2.107}$$

and

$$x(t) - p(t) = S_{HC}(t)(x_0 - p_0). \tag{2.108}$$

The idea is to make use of the result about preliminary feedbacks in Lemma 2.14. First of all, (2.104)-(2.105) imply that $u(\cdot) = Fp(\cdot)$ and so (2.7) can be reformulated as

$$x(t) = S(t)x_0 + \int_0^t S(t-s)B(Fx(s) + F(p(s) - x(s))ds,$$

which, using Lemma 2.14, implies (2.107). Furthermore,

$$x(t) = S(t)x_0 + \int_0^t S(t-s)BFp(s)ds =$$

$$S_{HC}(t)x_0 + \int_0^t S_{HC}(t-s)(BFp(s) - HCx(s))ds, \qquad \text{using Lemma 2.14}$$

and using (2.104)-(2.105) and (2.7) we have

$$p(t) = T(t)p_0 - \int_0^t T(t-s)(HCx(s) + HDFp(s))ds =$$

$$S_{HC}(t)p_0 + \int_0^t S_{HC}(t-s)(BFp(s) - HCx(s))ds,$$

again using Lemma 2.14. Subtraction of these last equations for p and x gives (2.108).

Now suppose that $(x_0, p_0) \in \mathcal{V} \times \mathcal{V}$. Since $S_{HC}(\cdot)$ is exponentially stable on $\mathcal{V}$, it follows from (2.108) that $x(\cdot) - p(\cdot) \in L_2(0,\infty;\mathcal{V})$ and (2.6) implies that $F(x(\cdot) - p(\cdot)) \in L_2(0,\infty;U)$. Since $S_{BF}(\cdot)$ is exponentially stable on $\mathcal{V}$, we conclude from (2.107) that $x(\cdot) \in L_2(0,\infty;\mathcal{V})$, so that in fact

$$\begin{pmatrix} x(\cdot) \\ p(\cdot) \end{pmatrix} \in L_2(0,\infty;\mathcal{V}\times\mathcal{V}).$$

Combination with (2.106) and Datko's result in Lemma A.1 implies that $\tilde{S}_{\tilde{B}\tilde{D}\mathcal{J}\tilde{C}}(\cdot)$ is exponentially stable on $\mathcal{V} \times \mathcal{V}$. In order to prove that $\tilde{S}_{\tilde{B}\tilde{D}\mathcal{J}\tilde{C}}(\cdot)$ is also exponentially stable on $\mathcal{W} \times \mathcal{W}$, we want to use Theorem 2.20, item (*ii*). In order to apply that result we shall show that the pair

$$(\tilde{S}(\cdot), \tilde{B}) = (\begin{pmatrix} S(\cdot) & 0 \\ 0 & T(\cdot) \end{pmatrix}, \begin{pmatrix} B & 0 \\ 0 & N \end{pmatrix})$$

is admissibly stabilizable. Indeed, using the perturbation results of Lemma 2.13, it follows that

$$\begin{pmatrix} S(\cdot) & 0 \\ 0 & T(\cdot) \end{pmatrix}_{\begin{pmatrix} B & 0 \\ 0 & -H \end{pmatrix}\begin{pmatrix} F & 0 \\ 0 & DF+C \end{pmatrix}} = \begin{pmatrix} S_{BF}(\cdot) & 0 \\ 0 & S_{BF}(\cdot) \end{pmatrix},$$

and since $S_{BF}(\cdot)$ is exponentially stable on $\mathcal{W}$ and $\mathcal{V}$ the result follows. Now Theorem 2.20, item (*ii*) implies that $\tilde{S}_{\tilde{B}\tilde{D}\mathcal{J}\tilde{C}}(\cdot)$ is also exponentially stable on $\mathcal{W} \times \mathcal{W}$ and this completes the proof of the theorem. ■

We conclude this section with a description of the feedback configuration that will be used in the $\mathcal{H}_\infty$-control context. Let $\mathcal{W}$ and $\mathcal{V}$ be real separable Hilbert spaces, satisfying $\mathcal{W} \hookrightarrow \mathcal{V}$ and let $S(\cdot)$ be a C_0-semigroup on $\mathcal{V}$ which restricts to a C_0-semigroup on $\mathcal{W}$. Furthermore, suppose that U, W, Y and Z are also real separable Hilbert spaces and let $B_1 \in \mathcal{L}(W, \mathcal{V}), B_2 \in \mathcal{L}(U, \mathcal{V}), C_1 \in \mathcal{L}(\mathcal{W}, Z), C_2 \in \mathcal{L}(\mathcal{W}, Y), D_{11} \in \mathcal{L}(W, Z), D_{12} \in \mathcal{L}(U, Z), D_{21} \in \mathcal{L}(W, Y)$ and $D_{22} \in \mathcal{L}(U, Y)$. We assume that B_1 and B_2 are both admissible input operators and that C_1 and C_2 are admissible output operators. We consider the Pritchard-Salamon system (see also Figure 1.1)

$$\Sigma_G = \Sigma(S(\cdot), \begin{pmatrix} B_1 & B_2 \end{pmatrix}, \begin{pmatrix} C_1 \\ C_2 \end{pmatrix}, \begin{pmatrix} D_{11} & D_{12} \\ D_{21} & D_{22} \end{pmatrix}):$$

$$\Sigma_G \begin{cases} x(t) &= S(t)x_0 + \int_0^t S(t-s)(B_1 w(s) + B_2 u(s))ds \\ z(t) &= C_1 x(t) + D_{11} w(t) + D_{12} u(t) \\ y(t) &= C_2 x(t) + D_{21} w(t) + D_{22} u(t), \end{cases} \tag{2.109}$$

where $x_0 \in \mathcal{V}, t \geq 0, w(\cdot) \in L_2^{loc}(0, \infty; W)$ and $u(\cdot) \in L_2^{loc}(0, \infty; U)$.
If $x_0 = 0$, we can express (2.109) in the following way:

$$\begin{pmatrix} z(\cdot) \\ y(\cdot) \end{pmatrix} = \begin{pmatrix} G_{11} & G_{12} \\ G_{21} & G_{22} \end{pmatrix} \begin{pmatrix} w(\cdot) \\ u(\cdot) \end{pmatrix}, \tag{2.110}$$

where G_{ij} represent the corresponding linear maps denoted below:

$$\begin{aligned} (G_{11}w)(t) &= C_1 \textstyle\int_0^t S(t-s)B_1 w(s)ds &+& \; D_{11}w(t) \\ (G_{12}u)(t) &= C_1 \textstyle\int_0^t S(t-s)B_2 u(s)ds &+& \; D_{12}u(t) \\ (G_{21}w)(t) &= C_2 \textstyle\int_0^t S(t-s)B_1 w(s)ds &+& \; D_{21}w(t) \\ (G_{22}u)(t) &= C_2 \textstyle\int_0^t S(t-s)B_2 u(s)ds &+& \; D_{22}u(t). \end{aligned} \tag{2.111}$$

Now suppose that we have a second Pritchard-Salamon system $\Sigma_K = \Sigma(T(\cdot), N, L, R)$ of the form (2.85) (Σ_K will be the controller). The linear map K from $L_2^{loc}(0, \infty; Y)$ to $u(\cdot) \in L_2^{loc}(0, \infty; U)$ is given by

$$(Ku_2)(t) := L \int_0^t T(t-s)Nu_2(s)ds + Ru_2(t). \tag{2.112}$$

Using the procedure at the beginning of this section, we can make sense of the feedback interconnection of (2.109) and (2.85) with $u_2(\cdot) = y(\cdot)$ and $u(\cdot) =$

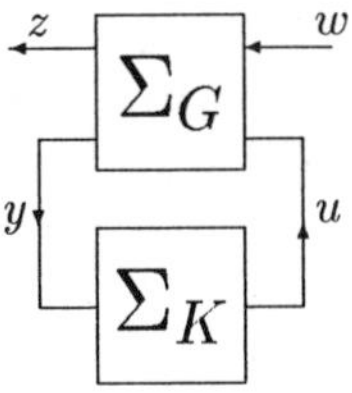

Figure 2.3: $\Sigma_{G_{zw}} = \Sigma_{\mathcal{F}(G,K)}$.

$y_2(\cdot)$ as in Figure 2.3 (this is also called the *closed-loop system*). We define the real Hilbert spaces $\tilde{\mathcal{V}} = \mathcal{V} \times \mathcal{V}_K$ and $\tilde{\mathcal{W}} = \mathcal{W} \times \mathcal{W}_K$ and we note that $\tilde{\mathcal{W}} \hookrightarrow \tilde{\mathcal{V}}$. Furthermore, we define

$$\begin{aligned} \tilde{S}(\cdot) &:= \begin{pmatrix} S(\cdot) & 0 \\ 0 & T(\cdot) \end{pmatrix} \quad ; \ \tilde{B} := \begin{pmatrix} B_2 & 0 \\ 0 & N \end{pmatrix} \\ \tilde{C} &:= \begin{pmatrix} C_2 & 0 \\ 0 & L \end{pmatrix} \quad ; \ \tilde{D} := \begin{pmatrix} D_{22} & 0 \\ 0 & R \end{pmatrix} \end{aligned} \tag{2.113}$$

(note the relationship with (2.86)). We can consider the combination of (2.109) and (2.85) in Figure 2.3 as the Pritchard-Salamon system with state-space equations

$$\begin{cases} \begin{pmatrix} x(t) \\ p(t) \end{pmatrix} = \tilde{S}(t)\begin{pmatrix} x_0 \\ p_0 \end{pmatrix} + \int_0^t \tilde{S}(t-s) \\ \qquad \left\{ \tilde{B}\begin{pmatrix} u(s) \\ u_2(s) \end{pmatrix} + \begin{pmatrix} B_1 \\ 0 \end{pmatrix} w(s) \right\} ds \\ z(t) = (C_1 \ \ 0)\begin{pmatrix} x(t) \\ p(t) \end{pmatrix} + (D_{12} \ \ 0)\begin{pmatrix} u(t) \\ u_2(t) \end{pmatrix} + D_{11}w(t) \\ \begin{pmatrix} y(t) \\ y_2(t) \end{pmatrix} = \tilde{C}\begin{pmatrix} x(t) \\ p(t) \end{pmatrix} + \tilde{D}\begin{pmatrix} u(t) \\ u_2(t) \end{pmatrix} + \begin{pmatrix} D_{21} \\ 0 \end{pmatrix} w(t). \end{cases} \tag{2.114}$$

Recall that

$$J = \begin{pmatrix} 0 & I \\ I & 0 \end{pmatrix}.$$

We assume that $I - D_{22}R$ (or, equivalently, $I - RD_{22}$) has a bounded inverse so that the feedback system will be well-posed (see also the discussion at the beginning of this section). We define

$$\bar{D} := (I - \mathcal{J}\tilde{D})^{-1} = \begin{pmatrix} I & -R \\ -D_{22} & I \end{pmatrix}^{-1}. \tag{2.115}$$

The feedback interconnection of Figure 2.3 can be formalized as follows. According to the procedure at the beginning of this section we can consider the choice of $u_2(\cdot) = y(\cdot)$ and $u(\cdot) = y_2(\cdot)$ as a preliminary state-feedback in (2.114):

$$\begin{pmatrix} u(t) \\ u_2(t) \end{pmatrix} = \bar{D}\mathcal{J}\tilde{C}\begin{pmatrix} x(t) \\ p(t) \end{pmatrix} + \bar{D}\mathcal{J}\begin{pmatrix} D_{21} \\ 0 \end{pmatrix} w(t).$$

This admissible preliminary state-feedback applied to (2.114) gives the Pritchard-Salamon system $\Sigma(\mathcal{S}(\cdot), \mathcal{B}, \mathcal{C}, \mathcal{D})$

$$\left\{ \begin{array}{rcl} \begin{pmatrix} x(t) \\ p(t) \end{pmatrix} & = & \mathcal{S}(\cdot)\begin{pmatrix} x_0 \\ p_0 \end{pmatrix} + \int_0^t \mathcal{S}(t-s)\mathcal{B}w(s)ds \\ z(t) & = & \mathcal{C}\begin{pmatrix} x(t) \\ p(t) \end{pmatrix} + \mathcal{D}w(t), \end{array} \right. \tag{2.116}$$

where

$$\begin{aligned} \mathcal{S}(\cdot) &= \tilde{S}_{\tilde{B}\bar{D}\mathcal{J}\tilde{C}}(\cdot) \\ \mathcal{B} &= \tilde{B}\bar{D}\mathcal{J}\begin{pmatrix} D_{21} \\ 0 \end{pmatrix} + \begin{pmatrix} B_1 \\ 0 \end{pmatrix} \\ \mathcal{C} &= (C_1 \;\; 0) + (D_{12} \;\; 0)\bar{D}\mathcal{J}\tilde{C} \\ \mathcal{D} &= (D_{12} \;\; 0)\bar{D}\mathcal{J}\begin{pmatrix} D_{21} \\ 0 \end{pmatrix} + D_{11}. \end{aligned} \tag{2.117}$$

In Chapter 5 we shall at certain places assume that $D_{11} = 0$ and $D_{22} = 0$. In this case the closed-loop system is automatically well-posed and we have

$$\begin{aligned} \tilde{B}\bar{D}\mathcal{J}\tilde{C} &= \begin{pmatrix} B_2RC_2 & B_2L \\ NC_2 & 0 \end{pmatrix} \\ \mathcal{B} &= \begin{pmatrix} B_2RD_{21} + B_1 \\ ND_{21} \end{pmatrix} \\ \mathcal{C} &= (C_1 + D_{12}RC_2 \;\; D_{12}L) \\ \mathcal{D} &= D_{12}RD_{21}. \end{aligned} \tag{2.118}$$

In the case that $x_0 = 0$ and $p_0 = 0$, we can formulate (2.116)-(2.117) differently, using (2.110) with $u(\cdot) = Ky(\cdot)$. Since $I - RD_{22}$ and $I - D_{22}R$ are both boundedly invertible, it follows that $I - KG_{22}$ and $I - G_{22}K$ are both invertible (as maps from L_2^{loc} to L_2^{loc}). Hence we can express $z(\cdot)$ as

$$z(\cdot) = (G_{zw}w)(\cdot) = (G_{11} + G_{12}K(I - G_{22}K)^{-1}G_{21})w(\cdot).$$

For any G of the form (2.110) and K of the form (2.112) such that $I - RD_{22}$ (or, equivalently, $I - D_{22}R$) is boundedly invertible we define $\mathcal{F}$ (a *linear fractional transformation*) as

$$\mathcal{F}(G, K) := G_{11} + G_{12}K(I - G_{22}K)^{-1}G_{21}. \tag{2.119}$$

Remark 2.31
It follows from the above that $\mathcal{F}$ is well defined and that $\mathcal{F}(G, K)$ has a realization of the form (2.116) so that $\Sigma_{\mathcal{F}(G,K)} = \Sigma(\mathcal{S}(\cdot), \mathcal{B}, \mathcal{C}, \mathcal{D})$ (the notion of realization was introduced in Remark 2.5). Unless stated otherwise, we shall use this realization of $\mathcal{F}(G, K)$ throughout. We note that if $\mathcal{S}(\cdot)$ is exponentially stable on $\mathcal{V}$ and $\mathcal{W}$, then $\mathcal{F}(G, K)$ is a bounded linear map from $\mathcal{L}(L_2(0, \infty; W)$ to $L_2(0, \infty; Z)$ (see Lemma 2.23). Furthermore, if $(S(\cdot), B_2)$ and $(T(\cdot), N)$ are admissibly stabilizable and $(C_2, S(\cdot))$ and $(L, T(\cdot))$ are admissibly detectable, then $\mathcal{S}(\cdot)$ is exponentially stable on $\mathcal{V}$ and $\mathcal{W}$ if and only if

$$\begin{pmatrix} (I - KG_{22})^{-1} & (I - KG_{22})^{-1}K \\ G_{22}(I - KG_{22})^{-1} & (I - G_{22}K)^{-1} \end{pmatrix} \tag{2.120}$$

is i/o-stable (see Theorem 2.27). As before, we shall in this case call Σ_K a *stabilizing controller* for Σ_G (compare with the remark after Theorem 2.27).

2.8 Riccati equations

In this section we derive several general results regarding Riccati equations for Pritchard-Salamon systems (the relationship with Hamiltonians, uniqueness of stabilizing solutions, etc.). The first result is a useful technical lemma.

Lemma 2.32
Let X be a real separable Hilbert space and suppose that $S_1(\cdot), S_2(\cdot)$ are C_0-semigroups on X, with infinitesimal generators denoted by A_1 and A_2 respectively. Let $\{X', j\}$ be a representation of X^d with the pairing given by $< x', x >_{<X',X>} = < jx', x >_{<X^d,X>}$. The following statements hold.

(*i*) *Suppose that $T \in \mathcal{L}(X, X')$ is such that $TD(A_1) \subset D((A_2)')$ and*

$$((A_2)'T + TA_1)x = 0 \quad \textit{for all } x \in D(A_1). \tag{2.121}$$

If $S_1(\cdot)$ and $S_2(\cdot)$ are exponentially stable on X, then it follows that $T = 0$.

(*ii*) *Suppose that $T \in \mathcal{L}(X)$ is such that $TD(A_1) \subset D(A_2)$ and*

$$A_2Tx = TA_1x \quad \text{for all } x \in D(A_1). \tag{2.122}$$

If $T^{-1} \in \mathcal{L}(X)$, then it follows that $TD(A_1) = D(A_2)$.

Proof

Proof of (*i*):
Suppose that $x \in D(A_1)$ and $y \in D(A_2)$. Since $T \in \mathcal{L}(X, X')$ it follows that $TS_1(\cdot)x$ is differentiable with respect to the topology of X' and we have

$$\frac{d}{dt} < TS_1(t)x, S_2(t)y >_{<X',X>}=$$

$$< TA_1S_1(t)x, S_2(t)y >_{<X',X>} + < TS_1(t)x, A_2S_2(t)y >_{<X',X>}=$$

$$< TA_1S_1(t)x, S_2(t)y >_{<X',X>} + < A_2'TS_1(t)x, S_2(t)y >_{<X',X>}= 0,$$

for all $t \geq 0$, since $TD(A_1) \subset D((A_2)')$ and (2.121) holds. Hence,

$$< TS_1(t)x, S_2(t)y >_{<X',X>}= const$$

and since $S_1(\cdot)$ and $S_2(\cdot)$ are exponentially stable on X, $const = 0$. Therefore, $< Tx, y >_{<X',X>}= 0$ for all $x \in D(A_1)$ and $y \in D(A_2)$ and since $D(A_1)$ and $D(A_2)$ are dense in X we may conclude that T =0.

Proof of (*ii*):
We only have to show that $T^{-1}D(A_2) \subset D(A_1)$, because that implies that $D(A_2) \subset TD(A_1)$.

Let $0 < \lambda \in \mathbb{R}$ be larger than $\omega_{S_1(\cdot)}$ and $\omega_{S_2(\cdot)}$. Using (2.122) we infer that for all $x \in D(A_1)$ we have $(\lambda I - A_2)Tx = T(\lambda I - A_1)x$ so that

$$x = T^{-1}(\lambda I - A_2)^{-1}T(\lambda I - A_1)x. \tag{2.123}$$

Now let $y \in D(A_2)$ and define

$$x_y := (\lambda I - A_1)^{-1}T^{-1}(\lambda I - A_2)y \in D(A_1).$$

It follows that

$$y = (\lambda I - A_2)^{-1}T(\lambda I - A_1)x_y$$

and so we conclude from (2.123) that $T^{-1}y = x_y \in D(A_1)$, which completes the proof. ∎

Using Lemma 2.32 item (*i*), we can show that stabilizing solutions of Riccati equations are unique:

Lemma 2.33
Let $\mathcal{W}$ and $\mathcal{V}$ be real separable Hilbert spaces such that $\mathcal{W} \hookrightarrow \mathcal{V}$ and let $S(\cdot)$ be a C_0-semigroup on these spaces with infinitesimal generators $A^{\mathcal{W}}$ and $A^{\mathcal{V}}$ respectively. We assume that the duals of $\mathcal{W}$ and $\mathcal{V}$ are represented as in (2.67), so that for all $(v', w) \in \mathcal{V}' \times \mathcal{W}$ we have (see (2.65))

$$< v', w >_{<\mathcal{V}',\mathcal{V}>} = < v', w >_{<\mathcal{W}',\mathcal{W}>}$$

and $\mathcal{V}' \hookrightarrow \mathcal{W}'$. We assume that $D(A^{\mathcal{V}}) \hookrightarrow \mathcal{W}$. Furthermore, let $R_1 = R_1' \in \mathcal{L}(\mathcal{V}', \mathcal{V})$ and $R_2 = R_2' \in \mathcal{L}(\mathcal{W}, \mathcal{W}')$. Then there exists at most one operator P with the following properties:

- *$P \in \mathcal{L}(\mathcal{V}, \mathcal{V}')$ with $P = P'$ satisfies*

$$< Px, A^{\mathcal{V}}y >_{<\mathcal{V}',\mathcal{V}>} + < PA^{\mathcal{V}}x, y >_{<\mathcal{V}',\mathcal{V}>} +$$
$$< PR_1Px, y >_{<\mathcal{V}',\mathcal{V}>} = < R_2x, y >_{<\mathcal{W}',\mathcal{W}>} \tag{2.124}$$

  for all $x, y \in D(A^{\mathcal{V}})$.
- *$(A^{\mathcal{V}})_{R_1P}$ defined by $D((A^{\mathcal{V}})_{R_1P}) := D(A^{\mathcal{V}})$ and $(A^{\mathcal{V}})_{R_1P} := A^{\mathcal{V}} + R_1P$, is the infinitesimal generator of an exponentially stable C_0-semigroup on $\mathcal{V}$.*

Proof
Suppose that we have two operators P and $\bar{P}$ satisfying the assumptions of the theorem and denote the corresponding exponentially stable semigroups generated by $A^{\mathcal{V}} + R_1P$ and $A^{\mathcal{V}} + R_1\bar{P}$, by $T_1(\cdot)$ and $T_2(\cdot)$ respectively. It takes an easy calculation to show that

$$< (P - \bar{P})(A^{\mathcal{V}} + R_1P)x, y >_{<\mathcal{V}',\mathcal{V}>} + < (P - \bar{P})x, (A^{\mathcal{V}} + R_1\bar{P})y >_{<\mathcal{V}',\mathcal{V}>} = 0$$

for all $x, y \in D(A^{\mathcal{V}})$. Defining $T := P - \bar{P} \in \mathcal{L}(\mathcal{V}, \mathcal{V}')$, it is now easy to see that $T = T'$ and that for all $x \in D(A^{\mathcal{V}})$ we have $Tx \in D((A^{\mathcal{V}} + R_1\bar{P})')$ and

$$\left((A^{\mathcal{V}} + R_1\bar{P})'T + T(A^{\mathcal{V}} + R_1P)\right)x = 0.$$

Hence Lemma 2.32 implies that $T = 0$ and so we have obtained the result. ∎

Remark 2.34
The form of the Riccati equation above is very general. It contains the Riccati equations corresponding to linear quadratic control problems, but also $\mathcal{H}_\infty$-type Riccati equations. We note that for the standard linear quadratic control problem for Pritchard-Salamon systems as treated in [72], R_1 is given by $BR^{-1}B' \in \mathcal{L}(\mathcal{V}', \mathcal{V})$ and R_2 is given by $C'C \in \mathcal{L}(\mathcal{W}, \mathcal{W}')$. In the case that P satisfies the properties of Lemma 2.33, we call P the *unique stabilizing solution* of the Riccati equation (2.124).

The next lemma is concerned with the connection between Riccati equations and Hamiltonians.

Lemma 2.35

Let $\mathcal{W}, \mathcal{V}, \mathcal{W}', \mathcal{V}'$ be as in the above lemma and suppose again that $D(A^{\mathcal{V}}) \hookrightarrow \mathcal{W}$ so that $D((A^{\mathcal{W}})') \hookrightarrow \mathcal{V}'$ (see Theorem 2.17 item (iii)). Furthermore, let $R_1 = R_1' \in \mathcal{L}(\mathcal{V}', \mathcal{V})$ and $R_2 = R_2' \in \mathcal{L}(\mathcal{W}, \mathcal{W}')$. Define the (Hamiltonian) operator H_P by

$$H_P : D(A^{\mathcal{V}}) \times D((A^{\mathcal{W}})') \subset \mathcal{V} \times \mathcal{W}' \to \mathcal{V} \times \mathcal{W}',$$

$$H_P := \begin{pmatrix} A^{\mathcal{V}} & R_1 \\ R_2 & -(A^{\mathcal{W}})' \end{pmatrix}. \tag{2.125}$$

Then for any operator $P \in \mathcal{L}(\mathcal{V}, \mathcal{V}')$ with $P = P'$, the following are equivalent:

(*i*)
$$< Px, A^{\mathcal{V}}y >_{<\mathcal{V}',\mathcal{V}>} + < PA^{\mathcal{V}}x, y >_{<\mathcal{V}',\mathcal{V}>} + < PR_1Px, y >_{<\mathcal{V}',\mathcal{V}>}$$
$$=< R_2x, y >_{<\mathcal{W}',\mathcal{W}>} \text{ for all } x, y \in D(A^{\mathcal{V}}), \tag{2.126}$$

(*ii*) $PD(A^{\mathcal{V}}) \subset D((A^{\mathcal{W}})')$ *and*
$$\left((A^{\mathcal{W}})'P + PA^{\mathcal{V}} + PR_1P - R_2\right)x = 0 \text{ for all } x \in D(A^{\mathcal{V}}), \tag{2.127}$$

(*iii*) $P \in \mathcal{L}(D(A^{\mathcal{V}}), D((A^{\mathcal{W}})'))$ *and*
$$\left((A^{\mathcal{W}})'P + PA^{\mathcal{V}} + PR_1P - R_2\right)x = 0 \text{ for all } x \in D(A^{\mathcal{V}}), \tag{2.128}$$

(*iv*) $P \in \mathcal{L}(D(A^{\mathcal{V}}), D((A^{\mathcal{W}})'))$ *and* $H_P \begin{pmatrix} I & 0 \\ P & I \end{pmatrix} \begin{pmatrix} x \\ y \end{pmatrix} =$
$$\begin{pmatrix} I & 0 \\ P & I \end{pmatrix} \begin{pmatrix} A^{\mathcal{V}} + R_1P & R_1 \\ 0 & -(A^{\mathcal{W}})' - PR_1 \end{pmatrix} \begin{pmatrix} x \\ y \end{pmatrix} \tag{2.129}$$
for all $(x, y) \in D(A^{\mathcal{V}}) \times D((A^{\mathcal{W}})')$.

Proof

We have $D(A^{\mathcal{V}}) \hookrightarrow \mathcal{W}$ and we recall from Lemma 2.12 that $D(A^{\mathcal{W}}) \hookrightarrow D(A^{\mathcal{V}})$.
Proof of $(i) \Rightarrow (ii)$:
We know that for all $x \in D(A^{\mathcal{V}})$ and $y \in D(A^{\mathcal{W}})$

$$< Px, A^{\mathcal{W}}y >_{<\mathcal{W}',\mathcal{W}>}=< Px, A^{\mathcal{V}}y >_{<\mathcal{V}',\mathcal{V}>} \quad \text{(using (2.65))}$$

$$= - < PA^{\mathcal{V}}x, y >_{<\mathcal{V}',\mathcal{V}>} - < PR_1Px, y >_{<\mathcal{V}',\mathcal{V}>} + < R_2x, y >_{<\mathcal{W}',\mathcal{W}>}=$$

$$- < PA^{\mathcal{V}}x, y >_{<\mathcal{W}',\mathcal{W}>} - < PR_1Px, y >_{<\mathcal{W}',\mathcal{W}>} + < R_2x, y >_{<\mathcal{W}',\mathcal{W}>}$$

(again using (2.65)). Hence $Px \in D((A^{\mathcal{W}})')$ for all $x \in D(A^{\mathcal{V}})$ and (2.127) is satisfied.

Proof of $(ii) \Leftrightarrow (iii)$:
The only thing that we have to show is that $PD(A^{\mathcal{V}}) \subset D((A^{\mathcal{W}})')$ implies that $P \in \mathcal{L}(D(A^{\mathcal{V}}), D((A^{\mathcal{W}})'))$ and so we assume the former. We shall show that P is closed as a map from $D(A^{\mathcal{V}})$ to $D((A^{\mathcal{W}})')$: suppose that we have a sequence $x_n \in D(A^{\mathcal{V}})$ and $x \in D(A^{\mathcal{V}})$ such that $\|x_n - x\|_{D(A^{\mathcal{V}})} \to 0$ and $\|Px_n - y\|_{D((A^{\mathcal{W}})')} \to 0$ as $n \to \infty$. Since $D((A^{\mathcal{W}})') \hookrightarrow \mathcal{V}'$ it follows that $\|Px_n - y\|_{\mathcal{V}'} \to 0$ as $n \to \infty$. On the other hand, $P \in \mathcal{L}(\mathcal{V}, \mathcal{V}')$ implies

$$\|Px_n - Px\|_{\mathcal{V}'} \leq \|P\|_{\mathcal{L}(\mathcal{V},\mathcal{V}')} \|x - x_n\|_{\mathcal{V}} \leq const \|x - x_n\|_{D(A^{\mathcal{V}})},$$

and so it follows that $\|Px_n - Px\|_{\mathcal{V}'} \to 0$ as $n \to \infty$. Hence we infer that $y = Px$ and that P is closed. Finally, the result follows from the Closed Graph Theorem.

Proof of $(ii) \Rightarrow (i)$:
For all $x \in D(A^{\mathcal{V}})$ and $y \in D(A^{\mathcal{W}})$ we have

$$< \left((A^{\mathcal{W}})'P + PA^{\mathcal{V}} + PR_1P - R_2\right) x, y >_{<\mathcal{W}',\mathcal{W}>} = 0.$$

As in the proof of $(i) \Rightarrow (ii)$, this equation can be reformulated as the Riccati equation (2.126), in the sense that (2.126) is satisfied for all $x \in D(A^{\mathcal{V}})$ and $y \in D(A^{\mathcal{W}})$. Using the fact that $D(A^{\mathcal{W}}) \hookrightarrow D(A^{\mathcal{V}})$ it follows that for any $y \in D(A^{\mathcal{V}})$ there exists a sequence $y_n \in D(A^{\mathcal{W}})$ such that $\|y_n - y\|_{D(A^{\mathcal{V}})} \to 0$ as $n \to \infty$. Since (2.126) is satisfied for $x \in D(A^{\mathcal{V}})$ and $y_n \in D(A^{\mathcal{W}})$ and $D(A^{\mathcal{V}}) \hookrightarrow \mathcal{W} \hookrightarrow \mathcal{V}$, we can obtain (i) by letting n tend to ∞.

Proof of $(iii) \Leftrightarrow (iv)$:
This follows from a straightforward calculation. ∎

We have a similar result for the 'dual' Riccati equation:

Corollary 2.36
Let $\mathcal{W}, \mathcal{V}, \mathcal{W}', \mathcal{V}'$ be as in the above lemmas and suppose again that $D(A^{\mathcal{V}}) \hookrightarrow \mathcal{W}$ so that $D((A^{\mathcal{W}})') \hookrightarrow \mathcal{V}'$. Furthermore, let $R_1 = R_1' \in \mathcal{L}(\mathcal{W}, \mathcal{W}')$ and $R_2 = R_2' \in \mathcal{L}(\mathcal{V}', \mathcal{V})$. Define the (Hamiltonian) operator H_Q by

$$H_Q : D((A^{\mathcal{W}})') \times D(A^{\mathcal{V}}) \subset \mathcal{W}' \times \mathcal{V} \to \mathcal{W}' \times \mathcal{V},$$

$$H_Q = \begin{pmatrix} (A^{\mathcal{W}})' & R_1 \\ R_2 & -A^{\mathcal{V}} \end{pmatrix}. \tag{2.130}$$

Then for any operator $Q \in \mathcal{L}(\mathcal{W}', \mathcal{W})$ with $Q = Q'$, the following are equivalent:

(i) $$< Qx, (A^{\mathcal{W}})'y >_{<\mathcal{W},\mathcal{W}'>} + < Q(A^{\mathcal{W}})'x, y >_{<\mathcal{W},\mathcal{W}'>} +$$

$$< QR_1Qx, y >_{<\mathcal{W},\mathcal{W}'>} = < R_2x, y >_{<\mathcal{V},\mathcal{V}'>} \tag{2.131}$$

for all $x, y \in D((A^{\mathcal{W}})')$,

(ii) $QD((A^{\mathcal{W}})') \subset D(A^{\mathcal{V}})$ *and for all* $x \in D((A^{\mathcal{W}})')$

$$\left(A^{\mathcal{V}}Q + Q(A^{\mathcal{W}})' + QR_1Q - R_2\right)x = 0, \tag{2.132}$$

(iii) $Q \in \mathcal{L}(D((A^{\mathcal{W}})'), D(A^{\mathcal{V}}))$ *and for all* $x \in D((A^{\mathcal{W}})')$

$$\left(A^{\mathcal{V}}Q + Q(A^{\mathcal{W}})' + QR_1Q - R_2\right)x = 0, \tag{2.133}$$

(iv) $Q \in \mathcal{L}(D((A^{\mathcal{W}})'), D(A^{\mathcal{V}}))$ *and* $H_Q \begin{pmatrix} I & 0 \\ Q & I \end{pmatrix} \begin{pmatrix} x \\ y \end{pmatrix} =$

$$\begin{pmatrix} I & 0 \\ Q & I \end{pmatrix} \begin{pmatrix} (A^{\mathcal{W}})' + R_1Q & R_1 \\ 0 & -A^{\mathcal{V}} - QR_1 \end{pmatrix} \begin{pmatrix} x \\ y \end{pmatrix} \tag{2.134}$$

for all $(x, y) \in D((A^{\mathcal{W}})') \times D(A^{\mathcal{V}})$.

Proof

We could give a proof of this result similar to the proof of Lemma 2.35, but in fact the result *follows* from Lemma 2.35, if we replace $\mathcal{V}$ by $\mathcal{W}'$, $\mathcal{W}$ by $\mathcal{V}'$, $S(\cdot)$ by $S'(\cdot)$, $A^{\mathcal{V}}$ by $(A^{\mathcal{W}})'$ and $(A^{\mathcal{W}})'$ by $((A^{\mathcal{V}})')' = A^{\mathcal{V}}$ (we identify the duals of $\mathcal{W}'$ and $\mathcal{V}'$ with $\mathcal{W}$ and $\mathcal{V}$, see section 2.5). ■

Chapter 3

Linear quadratic control and frequency domain inequalities

3.1 Introduction

In the state-space approach to the $\mathcal{H}_\infty$-control problem the linear quadratic control problem (LQ-problem) plays an important role. In Chapter 4 we relate the $\mathcal{H}_\infty$-control problem with state-feedback to a certain sup-inf-problem, which can be solved using the LQ-theory of this chapter. In some sense the Riccati equations corresponding to the solution of the $\mathcal{H}_\infty$-control problem are reminiscent of the LQ-Riccati equations. The results in this chapter also have applications to other system theoretical problems, such as questions arising in optimal control, identification theory, and robust stability theory (positive real lemma, bounded real lemma and so on). Therefore this chapter can be considered as an entity, interesting in its own right.

The first solutions for the (finite-dimensional) LQ-problem were obtained for the case that the cost criterion is positive definite (sometimes called the *standard case*). However, for many applications a more general formulation is needed. In Willems [100], the LQ-problem is treated for finite-dimensional systems in its most general form, i.e. the cost criterion is not necessarily positive definite (the *nonstandard LQ-problem*). In particular, we refer to Theorem 5 of [100] which shows the equivalence of the solvability of the LQ-problem with stability, to the existence of a solution to a (nonstandard) Riccati equation and a frequency domain inequality. This theorem is very similar to the well-known Kalman-Yakubovich-Popov Lemma (see Popov [71, Theorem 1 pp.98 and Lemma 1 pp.106]). One difference is that in [100] the *strict* frequency domain inequality is related to *stabilizing* solutions of the Riccati equation. Furthermore, in [71] Popov considers the (generalized) Lur'e equations, rather than the less general Riccati equation.

In the infinite-dimensional systems theory literature the attention was almost exclusively focussed on the standard LQ-problem. In Curtain and Pritchard [16] and Balakrishnan [3] (and the many references therein) the case is treated where the input and output operators are bounded and in Lions [59], Pritchard and Salamon [72], Lasiecka and Triggiani [57], Bensoussan et al. [8] (and references therein) the unbounded case is considered. The 'non-standard' problem was considered by Yakubovich [101] for bounded generators and by Wexler [99] and Louis and Wexler [61, 62] for the case of unbounded generators, but bounded input and output operators. All of these authors have obtained a kind of generalization of the afore-mentioned result in [100] for their class of systems. It is the purpose of this chapter to do the same for the Pritchard-Salamon class. We note that in Pritchard and Townley [74] some partial results in this direction were obtained (bounded real lemma), in connection with the stability radius optimization problem (cf. Section 1.1).

In Section 3.2 we give some preliminary results. In particular, we show what happens to the stability theory of Section 2.6 if we only demand that the semigroup of a Pritchard-Salamon system is stable on the 'larger space' $\mathcal{V}$. Furthermore, we give a new (short) proof of the generalization of [100, Theorem 5] to the class of infinite-dimensional systems with bounded input and output operators (this is the result obtained by Louis and Wexler [61], [62, Theorem 2]). The short proof is partly based on the result in [99], where the stable case is considered. Some of the arguments in this proof shall be used to derive the generalization of [62, Theorem 2] to the Pritchard-Salamon class, which is in fact the main result of this chapter (see Section 3.3). In order to deal with the unboundedness of B and C, we follow the approach of Pritchard and Salamon [72], in the sense that we consider the system with initial conditions on the 'larger space' $\mathcal{V}$ and with stability on $\mathcal{V}$. Hence, we only have to deal with the unboundedness of the output operators with respect to $\mathcal{V}$. The idea of the proof, given in Section 3.4, is to reformulate the 'unbounded part' of the cost function to something bounded and to use the result for the bounded input/output case.

3.2 Preliminary results

Suppose that we have a smooth Pritchard-Salamon $\Sigma_G = \Sigma(S(\cdot), B, C, D)$ of the form

$$\Sigma_G \begin{cases} x(t) &= S(t)x_0 + \int_0^t S(t-s)Bu(s)ds \\ \\ y(t) &= Cx(t) + Du(t), \end{cases} \tag{3.1}$$

where $x_0 \in \mathcal{V}, t \geq 0$ and $u(\cdot) \in L_2^{loc}(0, \infty; U)$ (recall that smoothness means that $D(A^{\mathcal{V}}) \hookrightarrow \mathcal{W}$). The first lemma is a Lyapunov type result for Pritchard-

Salamon systems whose C_0-semigroup is exponentially stable on the 'larger' space $\mathcal{V}$ (this does not imply stability on $\mathcal{W}$ in general, as explained in [15]). We recall that $*$ is used to denote Hilbert adjoints.

Lemma 3.1

Let Σ_G be a smooth Pritchard-Salamon system of the form (3.1) such that $S(\cdot)$ is exponentially stable on $\mathcal{V}$. Then the operator $\mathcal{C}^\infty$ defined on $\mathcal{W}$ by

$$\mathcal{C}^\infty x := CS(\cdot)x, \text{ for } x \in \mathcal{W}, \tag{3.2}$$

satisfies $\mathcal{C}^\infty \in \mathcal{L}(\mathcal{W}, L_2(0,\infty;Y))$ and it has a unique bounded extension on $\mathcal{V}$ (denoted by the same symbol), so that

$$\mathcal{C}^\infty \in \mathcal{L}(\mathcal{V}, L_2(0,\infty;Y)). \tag{3.3}$$

Furthermore, the nonnegative definite operator $P \in \mathcal{L}(\mathcal{V})$ defined by

$$P := (\mathcal{C}^\infty)^* \mathcal{C}^\infty, \tag{3.4}$$

is the unique self-adjoint solution to the Lyapunov equation

$$< A^{\mathcal{V}}x, Py >_{\mathcal{V}} + < x, PA^{\mathcal{V}}y >_{\mathcal{V}} + < Cx, Cy >_Y = 0 \tag{3.5}$$

for all $x, y \in D(A^{\mathcal{V}})$.

Proof

It follows from (2.6), that there exists some $c > 0$ such that for all $x \in \mathcal{W}$,

$$\|CS(\cdot)x\|_{L_2(0,\infty;Y)} \leq c\|x\|_{\mathcal{V}} \tag{3.6}$$

Therefore, $\mathcal{C}^\infty \in \mathcal{L}(\mathcal{V}, L_2(0, \infty; Y)$ (cf. also Remark 2.2).
Now let $x, y \in D(A^{\mathcal{W}})$ and define the function

$$f(t) := < CS(t)x, CS(t)y >_Y .$$

Then $f(\cdot)$ is defined for all t and it is differentiable, so

$$\dot{f}(t) = < CS(t)A^{\mathcal{W}}x, CS(t)y >_Y + < CS(t)x, CS(t)A^{\mathcal{W}}y >_Y . \tag{3.7}$$

For all $x \in D(A^{\mathcal{W}})$ we have

$$\|CS(t)x\|_Y \leq \|C\|_{\mathcal{L}(\mathcal{W},Y)} \|S(t)x\|_{\mathcal{W}} \leq const \|S(t)x\|_{D(A^{\mathcal{V}})},$$

where we have used the fact that $D(A^{\mathcal{V}}) \hookrightarrow \mathcal{W}$. Furthermore, for all $x \in D(A^{\mathcal{W}})$

$$S(t)x \stackrel{D(A^{\mathcal{V}})}{\rightarrow} 0 \text{ as } t \rightarrow \infty,$$

because $S(\cdot)$ is exponentially stable on $\mathcal{V}$ and therefore also on $D(A^{\mathcal{V}})$ (use Lemma D.1 and the fact that for large enough $\sigma \in \mathbb{R}$, $(\sigma I - A^{\mathcal{V}})^{-1}$ is an isomorphism from $\mathcal{V}$ to $D(A^{\mathcal{V}})$). Therefore, we conclude that $f(t) \to 0$ as $t \to \infty$. Now integrating (3.7) from 0 to T gives

$$f(T) - f(0) = \int_0^T (< CS(t)A^{\mathcal{W}}x, CS(t)y >_Y + < CS(t)x, CS(t)A^{\mathcal{W}}y >_Y)dt.$$

Since $CS(\cdot)x \in L_2(0,\infty;Y)$ for all $x \in \mathcal{W}$ (see (3.6)), we can take the limit as $T \to \infty$ to obtain

$$\int_0^\infty (< CS(t)A^{\mathcal{W}}x, CS(t)y >_Y + < CS(t)x, CS(t)A^{\mathcal{W}}y >_Y)dt$$
$$= - < Cx, Cy >_Y . \tag{3.8}$$

Since $x, y \in D(A^{\mathcal{W}})$ were arbitrary, we can use (3.2) and the fact that $A^{\mathcal{W}}x = A^{\mathcal{V}}x$ for all $x \in D(A^{\mathcal{W}})$ to conclude that

$$< \mathcal{C}^\infty A^{\mathcal{V}}x, \mathcal{C}^\infty y >_{L_2(0,\infty;Y)} + < \mathcal{C}^\infty x, \mathcal{C}^\infty A^{\mathcal{V}}y >_{L_2(0,\infty;Y)}$$
$$= - < Cx, Cy >_Y, \text{ for all } x, y \in D(A^{\mathcal{W}}). \tag{3.9}$$

Using the fact that $D(A^{\mathcal{W}}) \hookrightarrow D(A^{\mathcal{V}})$ (see Lemma 2.12) and the assumption that $D(A^{\mathcal{V}}) \hookrightarrow \mathcal{W}$, we can extend (3.9) to all $x, y \in D(A^{\mathcal{V}})$, by choosing sequences $x_n, y_n \in D(A^{\mathcal{W}})$ converging to $x, y \in D(A^{\mathcal{V}})$.

Finally, it is clear that then P defined by (3.4) satisfies (3.5). The uniqueness follows from Lemma 2.33. ∎

We are interested in linear quadratic control problems with stability. This means that we are going to look for controls $u(\cdot)$ in our system given by (3.1) that minimize a certain quadratic cost function with the additional constraint that the corresponding state $x(\cdot)$ in (3.1) is 'stable'. Here we shall demand L_2-stability with respect to $\mathcal{V}$, i.e. $x(\cdot)$ is required to be an element of $L_2(0,\infty;\mathcal{V})$. In the next lemma we shall show that, if the pair $(A^{\mathcal{V}}, B)$ is exponentially stabilizable on $\mathcal{V}$ (see Remark 2.19 for the definition), then $y(\cdot) \in L_2(0,\infty;Y)$. This is not true in general, because the growth constant of $S(\cdot)$ on $\mathcal{W}$ need not equal the growth constant on $\mathcal{V}$ (see [15]), and so we cannot conclude that $x(\cdot) \in L_2(0,\infty;\mathcal{W})$ in general.

We note that if the pair $(A^{\mathcal{V}}, B)$ is exponentially stabilizable on $\mathcal{V}$, the perturbation results of Lemma 2.13 imply that the set U_{adm} defined by

$$U_{adm} := \{u(\cdot) \in L_2(0,\infty,U) \text{ such that } x(\cdot) \text{ given by}$$
$$(3.1) \text{ satisfies } x(\cdot) \in L_2(0,\infty;\mathcal{V})\},$$

is nonempty.

Lemma 3.2
Let $\Sigma_G = \Sigma(S(\cdot), B, C, D)$ be a smooth Pritchard-Salamon system of the form (3.1) and suppose that $(A^{\mathcal{V}}, B)$ is exponentially stabilizable on $\mathcal{V}$. For all $u(\cdot) \in U_{adm}$ it follows that $y(\cdot)$ given by (3.1) is in $L_2(0,\infty;Y)$ and if $x_0 = 0$ there holds

$$\|y(\cdot)\|_{L_2(0,\infty;Y)} \leq const(\|u(\cdot)\|_{L_2(0,\infty;U)} + \|x(\cdot)\|_{L_2(0,\infty;\mathcal{V})}). \tag{3.10}$$

If, in addition, $S(\cdot)$ is exponentially stable on $\mathcal{V}$, it follows that the linear map G defined by

$$y(t) = (Gu)(t) = C\int_0^t S(t-s)Bu(s)ds, \tag{3.11}$$

an element of $\mathcal{L}(L_2(0,\infty;U), L_2(0,\infty;Y))$.

Proof
If C were an element of $\mathcal{L}(\mathcal{V}, Y)$, the fact that $x(\cdot) \in L_2(0,\infty;\mathcal{V})$ would immediately imply that the lemma is true. However, we only know that $C \in \mathcal{L}(\mathcal{W}, Y)$ is admissible and so we have to do some work to establish the result.
Since $(A^{\mathcal{V}}, B)$ is exponentially stabilizable on $\mathcal{V}$, there exists an $F \in \mathcal{L}(\mathcal{V}, U)$ such that $S_{BF}(\cdot)$ is exponentially stable on $\mathcal{V}$. Furthermore, Lemma 2.13 shows that $D(A^{\mathcal{V}}) = D(A^{\mathcal{V}}_{BF}) \hookrightarrow \mathcal{W}$ and $A^{\mathcal{V}}_{BF} = A^{\mathcal{V}} + BF$. Using Lemma 3.1 it follows that there exists a nonnegative definite $P_F \in \mathcal{L}(\mathcal{V})$ that satisfies

$$< A^{\mathcal{V}}_{BF}x, P_F y >_{\mathcal{V}} + < x, P_F A^{\mathcal{V}}_{BF} y >_{\mathcal{V}} + < Cx, Cy >_Y = 0$$

$$\text{for all } x, y \in D(A^{\mathcal{V}}) = D(A^{\mathcal{V}}_{BF}).$$

For reasons that will become clear later on we reformulate this as

$$< A^{\mathcal{V}}x, P_F y >_{\mathcal{V}} + < x, P_F A^{\mathcal{V}} y >_{\mathcal{V}} = - < Cx, Cy >_Y$$

$$- < (P_F BF + F^* B^* P_F)x, y >_{\mathcal{V}} \text{ for all } x, y \in D(A^{\mathcal{V}}) = D(A^{\mathcal{V}}_{BF}). \tag{3.12}$$

Now consider the system (3.1) and suppose that $u(\cdot) \in U_{adm}$. Let $u_n(\cdot) \in CD(0,\infty;U)$ be such that $u_n(\cdot) \to u(\cdot)$ (in the $L_2(0,\infty;U)$-norm) and $x_{0n} \in D(A^{\mathcal{V}})$ be such that $x_{0n} \overset{\mathcal{V}}{\to} x_0$. Let $x_n(\cdot)$ be given by

$$x_n(t) = S(t)x_{0n} + \int_0^t S(t-s)Bu_n(s)ds.$$

It follows from Appendices B.3 and B.6, that $x_n(\cdot)$ is continuously differentiable, $x_n(t) \in D(A^{\mathcal{V}})$ for all $t \geq 0$,

$$\dot{x}_n(t) = A^{\mathcal{V}}x_n(t) + Bu_n(t), \text{ for all } t \geq 0, \tag{3.13}$$

and for all $T > 0$

$$\|x_n(\cdot) - x(\cdot)\|_{L_2(0,T;\mathcal{V})} \to 0 \text{ as } n \to \infty, \tag{3.14}$$

$$x_n(T) \xrightarrow{\mathcal{V}} x(T) \text{ as } n \to \infty \tag{3.15}$$

and

$$\|Cx_n(\cdot) - Cx(\cdot)\|_{L_2(0,T;Y)} \to 0 \text{ as } n \to \infty. \tag{3.16}$$

Now using (3.12) and (3.13), it is straightforward to show that

$$\frac{d}{dt} < x_n(t), P_F x_n(t) >_{\mathcal{V}} = - < Cx_n(t), Cx_n(t) >_Y -$$

$$< (P_F BF + F^* B^* P_F) x_n(t), x_n(t) >_{\mathcal{V}} + 2 < Bu_n(t), P_F x_n(t) >_{\mathcal{V}} . \tag{3.17}$$

Integrating (3.17) from 0 to T gives

$$< x_n(T), P_F x_n(T) >_{\mathcal{V}} - < x_{0n}, P_F x_{0n} >_{\mathcal{V}} + \int_0^T \|Cx_n(t)\|_Y^2 dt =$$

$$\int_0^T (< -(P_F BF + F^* B^* P_F) x_n(t), x_n(t) >_{\mathcal{V}} + 2 < Bu_n(t), P_F x_n(t) >_{\mathcal{V}}) dt.$$

Now we wish to take the limit as $n \to \infty$. Since $P_F \in \mathcal{L}(\mathcal{V})$, $B \in \mathcal{L}(U, \mathcal{V})$ and $F \in \mathcal{L}(\mathcal{V}, U)$, the only difficult part is the term with $\|Cx_n(t)\|_Y^2$ because C is not bounded w.r.t. $\mathcal{V}$ (whence the need for a result of the form (3.16)). Using (3.14)-(3.16) we obtain

$$< x(T), P_F x(T) >_{\mathcal{V}} - < x_0, P_F x_0 >_{\mathcal{V}} + \int_0^T \|Cx(t)\|_Y^2 dt =$$

$$\int_0^T (< -(P_F BF + F^* B^* P_F) x(t), x(t) >_{\mathcal{V}} + 2 < Bu(t), P_F x(t) >_{\mathcal{V}}) dt.$$

Since P_F is nonnegative definite, $u(\cdot) \in L_2(0, \infty; U)$ and $x(\cdot) \in L_2(0, \infty; \mathcal{V})$, the last equation implies that $Cx(\cdot) \in L_2(0, \infty; Y)$ and therefore that $y(\cdot) \in L_2(0, \infty; Y)$. Furthermore, we have that $x(T) \xrightarrow{\mathcal{V}} 0$ as $T \to \infty$ (see Appendix B.2) and so, using the fact that $u(\cdot) \in L_2(0, \infty; U)$ and $x(\cdot) \in L_2(0, \infty; \mathcal{V})$, we may let $T \to \infty$ to obtain

$$\int_0^\infty \|Cx(t)\|_Y^2 dt = < x_0, P_F x_0 >_{\mathcal{V}} +$$

$$\int_0^\infty (< -(P_F BF + F^* B^* P_F) x(t), x(t) >_{\mathcal{V}} + 2 < Bu(t), P_F x(t) >_{\mathcal{V}}) dt. \tag{3.18}$$

It is clear that if $x_0 = 0$, then (3.10) follows from (3.18).
Finally, if $S(\cdot)$ is exponentially stable on $\mathcal{V}$, it follows that $U_{adm} = L_2(0,\infty;U)$ (see Appendix B.1) and for all $u(\cdot) \in L_2(0,\infty;\ U)$ we have $x(\cdot) \in L_2(0,\infty;\mathcal{V})$ and

$$\|x(\cdot)\|_{L_2(0,\infty;\mathcal{V})} \leq const\|u(\cdot)\|_{L_2(0,\infty;U)}, \tag{3.19}$$

for zero initial conditions (see Appendix B.1, formula B.5). Hence, combination of (3.18) and (3.19) shows that the map G from $u(\cdot) \in L_2(0,\infty;U)$ to $y(\cdot) \in L_2(0,\infty;Y)$ is linear and bounded. ∎

Remark 3.3
If $S(\cdot)$ is exponentially stable on $\mathcal{V}$, then the fact that $y(\cdot) \in L_2(0,\infty;Y)$ in Lemma 3.2, also follows from very general results in [98]. An important aspect of the proof given here is that it provides us with formula (3.18) that will play an important role in our treatment of the LQ-problem. In fact it shows how we can transform the 'unbounded part' with $Cx(\cdot)$ into 'something bounded'.

In the next two lemmas we calculate a frequency domain representation of a Pritchard-Salamon system. If $S(\cdot)$ is exponentially stable on $\mathcal{V}$, we find an expression for its transfer function on $\mathbb{C}^+$, and prove that it is in $\mathcal{H}_\infty$. A particular difficulty here is that $S(\cdot)$ need not be stable on $\mathcal{W}$ and that $\widehat{Cx}(i\omega) = C\hat{x}(i\omega)$ is not immediate because in general $C \notin \mathcal{L}(\mathcal{V},Y)$ ($\hat{}$ denotes Fourier/Laplace transform). We note that in Section 2.3 the transfer function is given for the case that $S(\cdot)$ is exponentially stable on $\mathcal{V}$ *and* $\mathcal{W}$. Some of the arguments in the proofs that are given here, can also be found in [98].

Lemma 3.4
Let $\Sigma(S(\cdot),B,C,D)$ be a smooth Pritchard-Salamon system of the form (3.1) *and suppose that $S(\cdot)$ is exponentially stable on $\mathcal{V}$. Denote the corresponding bounded linear map from $u(\cdot) \in L_2(0,\infty;U)$ to $y(\cdot) \in L_2(0,\infty;Y)$ by G, just as in* (3.11) *(cf. Lemma* 3.2*). Then $\tilde{G}(s)$ defined by*

$$\tilde{G}(s) := C(sI - A^{\mathcal{V}})^{-1}B + D, \tag{3.20}$$

is well-defined for all $Re(s) \geq 0$ and

$$\tilde{G}(s) \in \mathcal{H}_\infty(\mathcal{L}(U,Y)). \tag{3.21}$$

Furthermore, if $u(\cdot) \in L_2(0,\infty;U)$, then $y(\cdot) = (Gu)(\cdot) \in L_2(0,\infty;Y)$ satisfies

$$\hat{y}(s) = \tilde{G}(s)\hat{u}(s) \text{ for all } s \in \mathbb{C}^+ \tag{3.22}$$

and

$$\hat{y}(i\omega) = \tilde{G}(i\omega)\hat{u}(i\omega) \text{ for almost all } \omega \in \mathbb{R}. \tag{3.23}$$

Finally,

$$\|G\| = \|\tilde{G}\|_{\infty}. \tag{3.24}$$

Proof

First of all we show that $\tilde{G}(s)$ is well-defined for all s with $\mathrm{Re}(s) \geq 0$ and holomorphic on $\mathrm{Re}(s) \geq 0$, w.r.t. the topology of $\mathcal{L}(U,Y)$. Using the fact that $S(\cdot)$ is exponentially stable on $\mathcal{V}$, the resolvent identity for $(sI - A^{\mathcal{V}})^{-1}$ gives

$$(sI - A^{\mathcal{V}})^{-1} = (\alpha I - A^{\mathcal{V}})^{-1} + (\alpha - s)(\alpha I - A^{\mathcal{V}})^{-1}(sI - A^{\mathcal{V}})^{-1} \tag{3.25}$$

for $\alpha, s \in \mathrm{Re}(s) \geq 0$. We know that $(sI - A^{\mathcal{V}})^{-1}$ is holomorphic on $\mathrm{Re}(s) \geq 0$ (w.r.t. the topology of $\mathcal{L}(\mathcal{V})$), and that $(\alpha I - A^{\mathcal{V}})^{-1} \in \mathcal{L}(\mathcal{V}, \mathcal{W})$ for $\mathrm{Re}(\alpha) \geq 0$ (use the facts that $(\alpha I - A^{\mathcal{V}})^{-1} \in \mathcal{L}(\mathcal{V}, D(A^{\mathcal{V}}))$ and $D(A^{\mathcal{V}}) \hookrightarrow \mathcal{W}$). Hence we can conclude from (3.25) that $(sI - A^{\mathcal{V}})^{-1} \in \mathcal{L}(\mathcal{V}, \mathcal{W})$ on $\mathrm{Re}(s) \geq 0$ and $(sI - A^{\mathcal{V}})^{-1}$ is holomorphic on $\mathrm{Re}(s) \geq 0$ w.r.t. the topology of $\mathcal{L}(\mathcal{V}, \mathcal{W})$. Therefore $\tilde{G}(s) = C(sI - A^{\mathcal{V}})^{-1}B + D$ is well-defined and holomorphic on $\mathrm{Re}(s) \geq 0$ w.r.t. the topology of $\mathcal{L}(U,Y)$.

Next we show (3.21),(3.22) and (3.24): Consider the linear map G defined by (3.11). Since $S(\cdot)$ is exponentially stable on $\mathcal{V}$ it follows from Lemma 3.2 that $G \in \mathcal{L}(L_2(0,\infty;\, U), L_2(0,\infty;Y))$. It is easy to see that G is also shift invariant, so that a well-known result (see e.g. [33]) implies that there exists a (transfer) function $\bar{G} : \mathbb{C}^+ \to \mathcal{L}(U,Y)$ such that for $u(\cdot) \in L_2(0,\infty;U)$ and $y(\cdot) = (Gu)(\cdot) \in L_2(0,\infty;Y)$ there holds

$$\hat{y}(s) = \bar{G}(s)\hat{u}(s) \text{ for all } s \in \mathbb{C}^+.$$

Furthermore,

$$\bar{G}(\cdot) \in \mathcal{H}_{\infty}(\mathbb{C}^+; \mathcal{L}(U,Y))$$

and

$$\|G\| = \|\bar{G}\|_{\infty}$$

(see sections 2.3 and 2.6). The next step is to show that $\bar{G}(s) = \tilde{G}(s) = C(sI - A^{\mathcal{V}})^{-1}B + D$ on $\mathbb{C}^+$. We know from Section 2.3 (formula (2.21) and further) that for $\mathrm{Re}(s)$ large enough there holds

$$\bar{G}(s) = C(sI - A^{\mathcal{V}})^{-1}B + D. \tag{3.26}$$

We have seen above that $\tilde{G}(s) = C(sI - A^{\mathcal{V}})^{-1}B + D$ is holomorphic on $\mathbb{C}^+$, w.r.t. the topology of $\mathcal{L}(U,Y)$. Since also $\bar{G}(s)$ is holomorphic on $\mathbb{C}^+$ w.r.t. the topology of $\mathcal{L}(U,Y)$, we conclude (3.21), (3.22) and (3.24).

Finally we show (3.23).

Since $\hat{u}(\cdot) \in \mathcal{H}_2(U)$ and $\hat{y}(\cdot) \in \mathcal{H}_2(Y)$, it follows from Fatou's Theorem (see e.g. [78, section 4.6]) that for almost all $\omega \in \mathbb{R}$,

$$\lim_{\epsilon \downarrow 0} \|\hat{u}(i\omega + \epsilon) - \hat{u}(i\omega)\|_U = 0, \text{ and } \lim_{\epsilon \downarrow 0} \|\hat{y}(i\omega + \epsilon) - \hat{y}(i\omega)\|_Y = 0. \quad (3.27)$$

Formula (3.23) now follows from (3.22), (3.27), and the fact that $\tilde{G}(s)$ is holomorphic on $\text{Re}(s) \geq 0$ w.r.t. the topology on $\mathcal{L}(U, Y)$. ■

Now we derive a frequency domain representation of (3.1), using Lemma 3.4.

Lemma 3.5

Let $\Sigma(S(\cdot), B, C, D)$ be a smooth Pritchard-Salamon system of the form (3.1) *and suppose that $(A^{\mathcal{V}}, B)$ is exponentially stabilizable on $\mathcal{V}$. If $u(\cdot) \in U_{adm}$ and $x_0 = 0$, then*

$$\hat{x}(i\omega) \in D(A^{\mathcal{V}}) \textit{ for almost all } \omega \in \mathbb{R} \quad (3.28)$$

$$i\omega\hat{x}(i\omega) = A^{\mathcal{V}}\hat{x}(i\omega) + B\hat{u}(i\omega) \textit{ for almost all } \omega \in \mathbb{R} \quad (3.29)$$

and

$$\hat{y}(i\omega) = C\hat{x}(i\omega) + D\hat{u}(i\omega) \textit{ for almost all } \omega \in \mathbb{R}. \quad (3.30)$$

Proof

Since $u(\cdot) \in U_{adm}$ and $(A^{\mathcal{V}}, B)$ is exponentially stabilizable on $\mathcal{V}$, we can use the perturbation results of Lemma 2.14 to infer that

$$x(t) = \int_0^t S_{BF}(t-s)B(u(s) - Fx(s))ds,$$

where $F \in \mathcal{L}(\mathcal{V}, U)$ is such that $S_{BF}(\cdot)$ is exponentially stable on $\mathcal{V}$. We know that $x(\cdot) \in L_2(0, \infty; \mathcal{V})$ so that an application of Lemma 3.4 (with $\mathcal{W} = \mathcal{V}$ and $C = I$) implies

$$\hat{x}(i\omega) = (i\omega I - A^{\mathcal{V}}_{BF})^{-1}B(\hat{u}(i\omega) - F\hat{x}(i\omega)) \in D(A^{\mathcal{V}}_{BF}) \quad (3.31)$$

for almost all $\omega \in \mathbb{R}$. Lemma 2.13 shows that $D(A^{\mathcal{V}}_{BF}) = D(A^{\mathcal{V}})$ and $A^{\mathcal{V}}_{BF}x = (A^{\mathcal{V}} + BF)x$ for $x \in D(A^{\mathcal{V}}_{BF})$ and so (3.28) and (3.29) follow from (3.31). In Lemma 3.2 we have seen that $y(\cdot) \in L_2(0, \infty, Y)$. We have $y(t) = Cx(t) + Du(t)$ and so Lemma 3.4 tells us that

$$\hat{y}(i\omega) = C(i\omega I - A^{\mathcal{V}}_{BF})^{-1}B(\hat{u}(i\omega) - F\hat{x}(i\omega)) +$$

$$D\hat{u}(i\omega) \text{ for almost all } \omega \in \mathbb{R}. \quad (3.32)$$

Combining (3.31) and (3.32) gives (3.30). ■

The next lemma is a crucial result for the theory of optimal control. The result follows immediately from [59, Theorem 1.1].

Lemma 3.6

Let H be a real Hilbert space, suppose that $T \in \mathcal{L}(H)$ is coercive and let y be an arbitrary element of H. Then there exists a unique $x^* \in H$ such that

$$< Tx, x >_H + < x, y >_H \geq$$

$$< Tx^*, x^* >_H + < x^*, y >_H, \text{ for all } x \in H. \tag{3.33}$$

The last lemma of this section deals with the linear quadratic control problem for infinite-dimensional systems with *bounded* input and output operators. It relates the solvability of the LQ-problem to the solvability of a Riccati equation and a frequency domain inequality, just as in [100], where the finite-dimensional case is treated. The lemma is proved in full detail in [61, 62], but here we prefer to give a new, relatively short proof. In [99], a part of this lemma is proved, under the extra assumption that the system is exponentially stable. This shall be used in our proof below and it is for this reason that our proof is quite short. Some of the arguments in the proof shall be extended in the next section in order to generalize the result to the Pritchard-Salamon class.

We note that the result holds for both complex valued systems and for real valued systems, as explained in [61]. As mentioned before, we shall use complexifications of Hilbert spaces wherever necessary. In particular, we are obliged to consider complex spaces whenever we are dealing with the frequency domain inequality.

Lemma 3.7

Let A be the infinitesimal generator of a semigroup $S(\cdot)$ on a real Hilbert separable space H, let $B \in \mathcal{L}(U, H)$, where U is another real separable Hilbert space, and suppose that (A, B) is exponentially stabilizable. Furthermore, let $Q = Q^* \in \mathcal{L}(H)$, $L \in \mathcal{L}(H, U)$ and $R = R^* \in \mathcal{L}(U)$, with R coercive. For all $(x, u) \in H \times U$ we define the quadratic form

$$\mathcal{F}(x, u) := < Qx, x >_H + 2\mathrm{Re} < Lx, u >_U + < Ru, u >_U \tag{3.34}$$

(in the case that $< Lx, u >_U$ is not real, we take the real part of it). Consider the system

$$x(t) = S(t)x_0 + \int_0^t S(t-s)Bu(s)ds, \ x_0 \in H, \ t \geq 0. \tag{3.35}$$

Define

$$U_{adm} := \{u(\cdot) \in L_2(0, \infty; U) \text{ such that } x(\cdot) \text{ given by } (3.35)$$

$$\text{satisfies } x(\cdot) \in L_2(0, \infty; H)\}, \tag{3.36}$$

For all $u(\cdot) \in U_{adm}$ *we define the cost function*

$$J(x_0, u(\cdot)) := \int_0^\infty \mathcal{F}(x(t), u(t))dt. \tag{3.37}$$

Now the following are equivalent:

(*i*) *For all* $x_0 \in H$ *there exists a unique* $\bar{u}(\cdot) \in L_2(0, \infty, U)$ *such that*

$$\inf_{u(\cdot)\in U_{adm}} J(x_0, u(\cdot)) = \min_{u(\cdot)\in U_{adm}} J(x_0, u(\cdot)) = J(x_0, \bar{u}(\cdot)), \tag{3.38}$$

(*ii*) *There exists a self-adjoint* $P \in \mathcal{L}(H)$ *such that for all* $x, y \in D(A)$

$$< Ax, Py >_H + < Px, Ay >_H -$$

$$< (B^*P + L)^* R^{-1}(B^*P + L)x, y >_H + < Qx, y >_H = 0 \tag{3.39}$$

and $A - BR^{-1}(B^*P + L)$ *is the infinitesimal generator of an exponentially stable semigroup,*

(*iii*) *There exists an* $\epsilon > 0$ *such that for all* $(\omega, x, u) \in \mathbb{R} \times D(A) \times U$ *satisfying* $i\omega x = Ax + Bu$ *there holds*

$$\mathcal{F}(x, u) \geq \epsilon \|x\|_H^2. \tag{3.40}$$

Furthermore, if one of these conditions holds, the minimizing $\bar{u}(\cdot)$ *in* (3.38) *can be given in feedback form:*

$$\bar{u}(\cdot) = -R^{-1}(B^*P + L)x(\cdot), \tag{3.41}$$

and

$$\inf_{u(\cdot)\in U_{adm}} J(x_0, u(\cdot)) = J(x_0, \bar{u}(\cdot)) =< x_0, Px_0 >_H . \tag{3.42}$$

Remark 3.8

We note that in the formulation of Louis and Wexler in [61, 62], the frequency domain inequality is given by

$$\mathcal{F}(x, u) \geq \epsilon_1(\|x\|_H^2 + \|u\|_U^2),$$

rather than by (3.40) above. Our condition appears to be weaker, but in the proof we shall see that both formulations are in fact equivalent.

In [100], where the finite-dimensional case is treated, the frequency domain inequality is given as in (3.40), with x replaced by $(i\omega I - A)^{-1}Bu$. In the next section we give the generalization of Lemma 3.7 to the Pritchard-Salamon class, using also a formula like (3.40) for the frequency domain equality.

Proof

First of all, we prove that condition (*iii*) is equivalent to the existence of an $\epsilon_1 > 0$ such that for all $(\omega, x, u) \in \mathbb{R} \times D(A) \times U$ satisfying $i\omega x = Ax + Bu$ there holds

$$\mathcal{F}(x,u) \geq \epsilon_1(\|x\|_H^2 + \|u\|_U^2), \tag{3.43}$$

as mentioned in Remark 3.8. It is obvious that (3.43) implies (3.40). Hence we only have to show that (*iii*) implies (3.43). First we shall prove that for all $(\omega, x, u) \in \mathbb{R} \times D(A) \times U$ satisfying $i\omega x = Ax + Bu$ and $\|u\|_U = 1$ we have $\mathcal{F}(x,u) \geq \epsilon_2$, for some $\epsilon_2 > 0$.
We argue by contradiction, i.e. suppose that we had a sequence $(\omega_n, x_n, u_n) \in \mathbb{R} \times D(A) \times U$ satisfying $i\omega_n x_n = Ax_n + Bu_n$ with $\|u_n\| = 1$, such that $\mathcal{F}(x_n, u_n) < 1/n$. Then it follows from (3.40) that $x_n \overset{X}{\to} 0$ as $n \to \infty$. However, considering the definition of $\mathcal{F}$ in (3.34), this would imply that $< Ru_n, u_n >_U \to 0$. But this contradicts the facts that R is coercive and that $\|u_n\| = 1$. Hence we see that for all $(\omega, x, u) \in \mathbb{R} \times D(A) \times U$ satisfying $i\omega x = Ax + Bu$

$$\mathcal{F}(x,u) \geq \epsilon_2 \|u\|_U^2, \text{ for some } \epsilon_2 \geq 0. \tag{3.44}$$

Now the condition with (3.43) follows from a suitable combination of (*iii*) and (3.44). In the rest of the proof we can therefore use (3.43) instead of (3.40), wherever convenient.

Next, we shall prove the implication (*ii*) $\Rightarrow$ (*iii*) (which is not treated in [99] and not very elegantly in [61, 62]). From a straightforward calculation, using the Riccati equation in (*ii*), we have that for all $(x, u) \in D(A) \times U$

$$2\text{Re} < Ax + Bu, Px >_H + \mathcal{F}(x,u) = \|R^{\frac{1}{2}}(u + R^{-1}(B^*P + L)x)\|_U^2.$$

Hence, for all $(\omega, x, u) \in \mathbb{R} \times D(A) \times U$ satisfying $i\omega x = Ax + Bu$, there holds

$$\mathcal{F}(x,u) = \|R^{\frac{1}{2}}(u + R^{-1}(B^*P + L)x)\|_U^2. \tag{3.45}$$

For all $(\omega, x, u) \in \mathbb{R} \times D(A) \times U$ satisfying $i\omega x = Ax + Bu$ we have

$$(i\omega I - A + BR^{-1}(B^*P + L))x = B(u + R^{-1}(B^*P + L)x)$$

and since $A - BR^{-1}(B^*P + L)$ generates an exponentially stable C_0-semigroup we infer that

$$x = (i\omega I - A + BR^{-1}(B^*P + L))^{-1} B(u + R^{-1}(B^*P + L)x).$$

Hence, again using the stability of $A - BR^{-1}(B^*P + L)$, we obtain

$$\|x\|_H \leq const \|R^{\frac{1}{2}}(u + R^{-1}(B^*P + L)x)\|_U. \tag{3.46}$$

Combining (3.45) and (3.46) implies the existence of some $\epsilon > 0$ such that for all $(\omega, x, u) \in \mathbb{R} \times D(A) \times U$ satisfying $i\omega x = Ax + Bu$,

$$\mathcal{F}(x,u) \geq \epsilon \|x\|_H^2 \tag{3.47}$$

and this is precisely (iii).

In [99, (proof of) Theorem 3], Wexler proves the implications $(iii) \Rightarrow (i)$ and $(i) \Rightarrow (ii)$ under the extra assumption that $S(\cdot)$ is exponentially stable. Therefore, we reduce the general case to the 'stable' case by introducing a preliminary feedback

$$u(\cdot) = Fx(\cdot) + v(\cdot), \tag{3.48}$$

where $F \in \mathcal{L}(H,U)$ is such that the C_0-semigoup $S_{BF}(\cdot)$ generated by $A_F := A + BF$ is exponentially stable. The precise meaning of preliminary feedback is explained in Lemma 2.14. We define the transformed system

$$x_F(t) = S_{BF}(t)x_0 + \int_0^t S_{BF}(t-s)Bv(s)ds, \ x_0 \in H, t \geq 0,$$

and for all $v(\cdot) \in L_2(0,\infty;U)$ we define the transformed cost function

$$J_F(x_0, v(\cdot)) := \int_0^\infty \mathcal{F}_F(x_F(t), v(t))dt,$$

where

$$\mathcal{F}_F(x,v) = \mathcal{F}(x, Fx+v) = <(Q + F^*L + LF + F^*RF)x, x>_H +$$

$$2\mathrm{Re} < (L+RF)x, v>_U + < Rv, v>_U . \tag{3.49}$$

First of all, we note that because of Lemma 2.14, we have $x(\cdot) = x_F(\cdot)$ (actually, this follows from known perturbation results, whereas Lemma 2.14 treats a more general case; here we have $\mathcal{V} = \mathcal{W} = H$). Therefore, since $S_{BF}(\cdot)$ is exponentially stable, we have $u(\cdot) \in U_{adm}$ if and only if $v(\cdot) \in L_2(0,\infty,U)$ and in this case $J(x_0, u(\cdot)) = J_F(x_0, v(\cdot))$. Hence, (i) is equivalent to the existence of a unique $\bar{v}(\cdot) \in L_2(0,\infty,U)$ such that

$$\inf_{v(\cdot)\in L_2(0,\infty;U)} J_F(x_0, v(\cdot)) = \min_{v(\cdot)\in L_2(0,\infty;U)} J_F(x_0, v(\cdot)) = J_F(x_0, \bar{v}(\cdot)). \tag{3.50}$$

Furthermore, because of (3.49) and the result at the beginning of the proof, it is easy to see that (iii) is equivalent to the existence of some ϵ_F such that for all $(\omega, x, v) \in \mathbb{R} \times D(A_F) \times U$ satisfying $i\omega x = A_F x + Bv$ there holds

$$\mathcal{F}_F(x,v) \geq \epsilon_F(\|x\|_H^2 + \|v\|_U^2). \tag{3.51}$$

Using the equivalence between (i) and (3.50) and the equivalence between (iii) and (3.51), we prove the implications $(iii) \Rightarrow (i)$ and $(i) \Rightarrow (ii)$:

Suppose that (iii) is satisfied. We have seen above that this implies (3.51). Since $S_{BF}(\cdot)$ is exponentially stable, we can use Wexler's result [99] applied to the transformed system to derive (3.50). It follows from the above that (3.50) implies (i).

Next, suppose that (i) is satisfied. It follows from the above that this implies that (3.50) is satisfied. Since $S_{BF}(\cdot)$ is exponentially stable, we can use Wexler's result applied to the transformed system. Comparing (3.34) with (3.49), we see that Q is replaced by $Q + F^*L + LF + F^*RF$ and L by $L + RF$. Hence, there exists a self-adjoint $P \in \mathcal{L}(H)$ such that

$$< A_F x, Py >_H + < Px, A_F y >_H -$$

$$< (B^*P + L + RF)^* R^{-1}(B^*P + L + RF)x, y >_H +$$

$$< (Q + F^*L + LF + F^*RF)x, y >_H = 0 \text{ for all } x, y \in D(A_F),$$

and $A_F - BR^{-1}(B^*P + L + RF)$ is the infinitesimal generator of an exponentially stable C_0-semigroup. Using the fact that $A_F = A + BF$, (ii) easily follows from this.

Using Wexler's result in [99], the proof of the last few statements of the lemma is now straightforward and left to the reader. ∎

Remark 3.9

Suppose that $S(\cdot)$ is exponentially stable, as in [99]. Then, if $i\omega x = Ax + Bu$, we have $x = (i\omega I - A)^{-1}Bu$ and $\|x\| \leq const\|u\|$. Since according to the proof given above (iii) still holds with (3.40) replaced by (3.43), it follows that (iii) is equivalent to the existence of some ϵ_1 such that for all $u \in U$

$$\mathcal{F}((i\omega I - A)^{-1}Bu, u) \geq \epsilon_1 \|u\|_U^2. \tag{3.52}$$

3.3 Problem formulation and main result

The purpose of this section is to find a generalization of Lemma 3.7 to the Pritchard-Salamon class. So let $\mathcal{W}$ and $\mathcal{V}$ be two real separable Hilbert spaces satisfying $\mathcal{W} \hookrightarrow \mathcal{V}$ and suppose that we have a C_0-semigroup $S(\cdot)$ on both spaces. Furthermore, let U, Y_1, and Y_2 be real separable Hilbert spaces (as before U will play the role of the input space, Y_1 and Y_2 will be output spaces). Let $B \in \mathcal{L}(U, \mathcal{V})$ be an admissible input operator and $C_1 \in \mathcal{L}(\mathcal{W}, Y_1), C_2 \in \mathcal{L}(\mathcal{W}, Y_2)$ and $L \in \mathcal{L}(\mathcal{W}, U)$ be admissible output operators. Furthermore, let

$R = R^* \in \mathcal{L}(U)$ be a coercive operator (i.e. R is positive definite and has a bounded inverse). For all $(x, u) \in \mathcal{W} \times U$ we define the quadratic form

$$\mathcal{F}(x,u) := < C_1 x, C_1 x >_{Y_1} - < C_2 x, C_2 x >_{Y_2} +$$

$$2\text{Re} < Lx, u >_U + < Ru, u >_U . \tag{3.53}$$

In (3.34), Q could have been expressed as $Q = C_1^* C_1 - C_2^* C_2$ for some $C_1 \in \mathcal{L}(H, Y)$ and $C_2 \in \mathcal{L}(H, Y)$ (use $Q = cI - (cI - Q)$ with c large enough), so that (3.53) represents an appropriate generalization of (3.34). Our system is given by

$$x(t) = S(t)x_0 + \int_0^t S(t-s)Bu(s)ds, \tag{3.54}$$

where $x_0 \in \mathcal{V}$ and $t \geq 0$, and we assume that

$$(A^{\mathcal{V}}, B) \text{ is exponentially stabilizable on } \mathcal{V} . \tag{3.55}$$

We define the class of admissible inputs as

$$U_{adm} := \{u(\cdot) \in L_2(0,\infty;U) \text{ such that } x(\cdot) \text{ given by (3.54)}$$

$$\text{satisfies } x(\cdot) \in L_2(0,\infty;\mathcal{V})\}. \tag{3.56}$$

From Lemma 3.2 it follows that for all $u(\cdot) \in U_{adm}$, the cost functional $J(x_0, u(\cdot))$ defined by

$$J(x_0, u(\cdot)) := \int_0^\infty \mathcal{F}(x(t), u(t))dt \tag{3.57}$$

is finite. Note that $\mathcal{F}(x(t), u(t))$ should not be interpreted pointwise, but as explained in Section 2.2:

$$\int_0^\infty \mathcal{F}(x(t), u(t))dt = \|C_1 x(\cdot)\|^2_{L_2(0,\infty;Y_1)} - \|C_2 x(\cdot)\|^2_{L_2(0,\infty;Y_2)} +$$

$$+ 2 < Lx(\cdot), u(\cdot) >_{L_2(0,\infty;U)} + < Ru(\cdot), u(\cdot) >_{L_2(0,\infty;U)} . \tag{3.58}$$

The following result is a complete generalization of Lemma 3.7:

Theorem 3.10
Suppose that we have a smooth Pritchard-Salamon system of the form (3.54) *such that* $(A^{\mathcal{V}}, B)$ *is exponentially stabilizable on* $\mathcal{V}$. *Furthermore, let* $\mathcal{F}$ *and* J *be defined as in* (3.53) *and* (3.57). *The following three conditions are equivalent:*

(*i*) *for all* $x_0 \in \mathcal{V}$ *there exists a unique* $\bar{u}(\cdot) \in L_2(0,\infty,U)$ *such that*

$$\inf_{u(\cdot)\in U_{adm}} J(x_0,u(\cdot)) = \min_{u(\cdot)\in U_{adm}} J(x_0,u(\cdot)) = J(x_0,\bar{u}(\cdot)), \tag{3.59}$$

(*ii*) *there exists a self-adjoint* $P \in \mathcal{L}(\mathcal{V})$ *such that for all* $x, y \in D(A^{\mathcal{V}})$

$$< (A^{\mathcal{V}} - BR^{-1}L)x, Py >_{\mathcal{V}} + < Px, (A^{\mathcal{V}} - BR^{-1}L)y >_{\mathcal{V}} -$$

$$< PBR^{-1}B^*Px, y >_{\mathcal{V}} - < R^{-1}Lx, Ly >_U +$$

$$< C_1x, C_1y >_{Y_1} - < C_2x, C_2y >_{Y_2} = 0 \tag{3.60}$$

and $A^{\mathcal{V}} - BR^{-1}(B^*P + L)$ *is the generator of an exponentially stable* C_0*-semigroup on* $\mathcal{V}$,

(*iii*) *there exists an* $\epsilon > 0$ *such that for all* $(\omega, x, u) \in \mathbb{R}\times D(A^{\mathcal{V}})\times U$ *satisfying* $i\omega x = A^{\mathcal{V}}x + Bu$ *there holds*

$$\mathcal{F}(x,u) \geq \epsilon \|x\|^2_{\mathcal{V}}. \tag{3.61}$$

Furthermore, if one of these conditions holds, the minimizing $\bar{u}(\cdot)$ *in* (3.59) *can be given in feedback form:*

$$\bar{u}(\cdot) = -R^{-1}(B^*P + L)x(\cdot), \tag{3.62}$$

and

$$\inf_{u(\cdot)\in U_{adm}} J(x_0,u(\cdot)) = J(x_0,\bar{u}(\cdot)) = < x_0, Px_0 >_{\mathcal{V}} . \tag{3.63}$$

Finally, there exists at most one operator $P \in \mathcal{L}(\mathcal{V})$ *that satisfies item* (*ii*).

Remark 3.11

As in [72, 74], the Riccati equation (3.60) can be expressed in a 'stronger' form: we use the same formulation as in [72, 74] and section 2.5, i.e. we suppose that $\mathcal{H}$ is a pivot space such that $D(A^{\mathcal{V}}) \hookrightarrow \mathcal{W} \hookrightarrow \mathcal{H} \hookrightarrow \mathcal{V}$ and $\mathcal{V}' \hookrightarrow \mathcal{H} \hookrightarrow \mathcal{W}' \hookrightarrow (D(A^{\mathcal{V}}))'$. The isometry from $\mathcal{V}$ to $\mathcal{V}'$ is given by $c := i_{\mathcal{H}} \mid_{\mathcal{V}'} (i_{\mathcal{V}})^{-1}$. Furthermore, we identify the duals of U, Y_1 and Y_2 with themselves. Then P satisfies (3.60) if and only if $\tilde{P} = cP \in \mathcal{L}(\mathcal{V}, \mathcal{V}')$ satisfies the equation

$$< (A^{\mathcal{V}} - BR^{-1}L)x, \tilde{P}y >_{<\mathcal{V},\mathcal{V}'>} + < x, \tilde{P}(A^{\mathcal{V}} - BR^{-1}L)y >_{<\mathcal{V},\mathcal{V}'>} -$$

$$< x, \tilde{P}BR^{-1}B'\tilde{P}y >_{<\mathcal{V},\mathcal{V}'>} - < R^{-1}Lx, Ly >_U +$$

$$< C_1x, C_1y >_{Y_1} - < C_2x, C_2y >_{Y_2} = 0 \text{ for all } x, y \in D(A^{\mathcal{V}}). \tag{3.64}$$

Denoting the infinitesimal generators of $S_{-BR^{-1}L}(\cdot)$ on $\mathcal{W}$ and $\mathcal{V}$ by $A^{\mathcal{W}}_{-BR^{-1}L}$ and $A^{\mathcal{V}}_{-BR^{-1}L}$ respectively, equation (3.64) can in turn be reformulated as

$$\left((A^{\mathcal{W}}_{-BR^{-1}L})'\tilde{P} + \tilde{P}A^{\mathcal{V}}_{-BR^{-1}L} - \tilde{P}BR^{-1}B'\tilde{P} - L'R^{-1}L + C_1'C_1 - C_2'C_2\right)x = 0$$

for all $x \in D(A^{\mathcal{V}})$ (see Section 2.8). We note that $A^{\mathcal{V}}_{-BR^{-1}L}$ can also be considered as a bounded map from $D(A^{\mathcal{V}})$ to $\mathcal{V}$ ($D(A^{\mathcal{V}}) = D(A^{\mathcal{V}}_{-BR^{-1}L})$ should then be considered as the Hilbert space with inner product given by the graph inner product). The adjoint of this bounded map is given by $(A^{\mathcal{V}}_{-BR^{-1}L})' \in \mathcal{L}(\mathcal{V}', (D(A^{\mathcal{V}}))')$ (in fact, this is the *extension* of the map $(A^{\mathcal{V}}_{-BR^{-1}L})'$ from $D((A^{\mathcal{V}})') \subset \mathcal{V}'$ to $\mathcal{V}'$). The point is that $(A^{\mathcal{V}}_{-BR^{-1}L})' = (A^{\mathcal{W}}_{-BR^{-1}L})'$ on $D((A^{\mathcal{W}}_{-BR^{-1}L})')$, because $D(A^{\mathcal{V}}_{-BR^{-1}L}) \hookrightarrow \mathcal{W}$ so that $D((A^{\mathcal{W}}_{-BR^{-1}L})') \hookrightarrow \mathcal{V}'$ (see Lemma 2.13 and Theorem 2.17 item (*iii*)). This clarifies the relationship with the formulations that are used in [72, 74].
Finally, we note that because of $\mathcal{H} \hookrightarrow \mathcal{V}$, $\mathcal{V}' \hookrightarrow \mathcal{H}$ and the fact that $\tilde{P} \in \mathcal{L}(\mathcal{V}, \mathcal{V}')$, there holds $\tilde{P} \in \mathcal{L}(\mathcal{H})$ and that for all $x_0 \in \mathcal{H}$ we have

$$\inf_{u(\cdot)\in U_{adm}} J(x_0, u(\cdot)) = < x_0, Px_0 >_{\mathcal{V}} = < x_0, \tilde{P}x_0 >_{<\mathcal{V},\mathcal{V}'>} = < x_0, \tilde{P}x_0 >_{\mathcal{H}} .$$

Remark 3.12

As explained before, we do not demand that the state should be in $L_2(0, \infty; \mathcal{W})$ (this is not possible for $x_0 \in \mathcal{V}$ in general). However, if $x_0 \in \mathcal{W}$ and if, in addition, the pair $(S(\cdot), B)$ is admissibly stabilizable, it follows from Theorem 2.20 item (*ii*) that the C_0-semigroup $S_{BF_{optimal}}(\cdot)$ is stable on $\mathcal{V}$ and $\mathcal{W}$ so that the optimal state trajectory will be in $L_2(0, \infty; \mathcal{W})$. Of course a similar remark holds for $\mathcal{H}$.

Remark 3.13

The first results for the infinite-dimensional LQ-problem were obtained for the case that $L = 0, C_2 = 0$, and (C_1, A) exponentially detectable (see e.g. [16, 3, 59, 72] and references therein). It is not difficult to show that for the Pritchard-Salamon case with $L = 0$ and $C_2 = 0$ the frequency domain inequality is *implied* by the assumption that there exists an admissible input operator $H \in \mathcal{L}(Y_1, \mathcal{V})$ such that $S_{HC_1}(\cdot)$ is exponentially stable on $\mathcal{V}$ (see Appendix C.1, item (*ii*)). In fact, if the system is finite-dimensional, then the frequency domain inequality is in this case equivalent to the statement that the poles of A on the imaginary axis are (C_1, A)-detectable:

$$\begin{pmatrix} i\omega I - A \\ C_1 \end{pmatrix} \text{ has full rank for all } \omega \in \mathbb{R}.$$

We note that Theorem 3.10 is a considerable improvement on existing results for systems in the Pritchard-Salamon class:
In [72], the equivalence between (i) and (ii) is proved for the special case that $C_2 = 0, L = 0$ and $(C_1, A^{\mathcal{V}})$ is exponentially detectable (this means that there exists some $H \in \mathcal{L}(Y_1, \mathcal{V})$ such that $S_{HC_1}(\cdot)$ is exponentially stable on $\mathcal{V}$). In [74], the equivalence between (i),(ii) and (iii) is proved for the special case that $C_1 = 0, L = 0$ and $S(\cdot)$ is stable on $\mathcal{V}$ and $\mathcal{W}$.

Remark 3.14
As in Remark 3.9, if $S(\cdot)$ is exponentially stable on $\mathcal{V}$, then the frequency domain inequality in $((iii))$ is equivalent to to the existence of some ϵ_1 such that for all $u \in U$

$$\mathcal{F}((i\omega I - A^{\mathcal{V}})^{-1}Bu, u) \geq \epsilon_1 \|u\|_H^2 \tag{3.65}$$

In the case that $L = 0, C_1 = 0, R = I$ and $S(\cdot)$ is exponentially stable on $\mathcal{V}$, Theorem 3.10 corresponds to the *bounded real lemma* (cf. [1, Chapter 7]) and if $C_1 = C_2 = 0, R = D + D^*$ and $S(\cdot)$ is exponentially stable on $\mathcal{V}$, it corresponds to the *positive real lemma* (cf. [1, Chapter 5]).

Finally, we note that in the next two chapters the frequency domain inequality will turn up again with $\mathcal{F}$ given by

$$\mathcal{F}(x, u) = \|C_1 x + D_{12} u\|_Z^2,$$

where $D_{12} \in \mathcal{L}(U, Z)$ and $C_1 \in \mathcal{L}(\mathcal{W}, Z)$ is an admissible output operator. In this case it is easy to see that the solution P to the coresponding Riccati equation is nonnegative definite, because then the cost function is given by

$$J(x_0, u(\cdot)) := \int_0^\infty \|C_1 x(t) + D_{12} u(t)\|_Z^2 dt$$

and we have for all $x_0 \in \mathcal{V}$

$$J(x_0, \bar{u}(\cdot)) = < x_0, P x_0 >_{\mathcal{V}} \geq 0. \tag{3.66}$$

3.4 Proof of the main result

Proof
We shall prove the implications $(i) \Rightarrow (ii)$, $(ii) \Rightarrow (iii)$ and $(iii) \Rightarrow (i)$.

Proof of $(i) \Rightarrow (ii)$.
We first assume that $L = 0$ (later we remove this extra assumption by applying

some preliminary feedback). The idea is to use Lemma 3.2 and in particular, formula (3.18), to transform the criterion (3.57) in such a way that the new criterion can be treated using Lemma 3.7. Since $(A^{\mathcal{V}}, B)$ is exponentially stabilizable on $\mathcal{V}$, there exists some $\bar{F} \in \mathcal{L}(U, \mathcal{V})$ such that $S_{B\bar{F}}(\cdot)$ is exponentially stable on $\mathcal{V}$. Using Lemma 3.1 it follows that there exist nonnegative definite operators $P_1, P_2 \in \mathcal{L}(\mathcal{V})$ such that for all $x, y \in D(A^{\mathcal{V}}) = D(A^{\mathcal{V}}_{B\bar{F}})$

$$< A^{\mathcal{V}}x, P_1 y >_{\mathcal{V}} + < x, P_1 A^{\mathcal{V}} y >_{\mathcal{V}} =$$

$$- < C_1 x, C_1 y >_{Y_1} - < (P_1 B\bar{F} + \bar{F}^* B^* P_1)x, y >_{\mathcal{V}} \tag{3.67}$$

and

$$< A^{\mathcal{V}}x, P_2 y >_{\mathcal{V}} + < x, P_2 A^{\mathcal{V}} y >_{\mathcal{V}} =$$

$$- < C_2 x, C_2 y >_{Y_2} - < (P_2 B\bar{F} + \bar{F}^* B^* P_2)x, y >_{\mathcal{V}} . \tag{3.68}$$

Since we assume that $L = 0$ we have

$$J(x_0, u(\cdot)) = \int_0^\infty \left(\|C_1 x(t)\|^2_{Y_1} - \|C_2 x(t)\|^2_{Y_2} + \|R^{\frac{1}{2}} u(t)\|^2_U \right) dt.$$

Hence, we can use formula (3.18) in the proof of Lemma 3.2, to infer that

$$J(x_0, u(\cdot)) =< x_0, (P_1 - P_2)x_0 >_{\mathcal{V}} +$$

$$\int_0^\infty (< ((P_2 - P_1)B\bar{F} + \bar{F}^* B^*(P_2 - P_1))x(t), x(t) >_{\mathcal{V}} +$$

$$2 < B^*(P_1 - P_2)x(t), u(t) >_U + \|R^{\frac{1}{2}} u(t)\|^2_U) dt. \tag{3.69}$$

Note that the integral term is as in Lemma 3.7 (formulas (3.34) and (3.37)), with Q replaced by $(P_2 - P_1)B\bar{F} + \bar{F}^* B^*(P_2 - P_1)$ and L replaced by $B^*(P_1 - P_2)$. Hence, because of our assumption (i), we can apply this lemma on $\mathcal{V}$ and infer the existence of a self-adjoint $P_3 \in \mathcal{L}(\mathcal{V})$ such that for all $x, y \in D(A^{\mathcal{V}})$

$$< A^{\mathcal{V}}x, P_3 y >_{\mathcal{V}} + < P_3 x, A^{\mathcal{V}} y >_{\mathcal{V}} -$$

$$((P_3 + P_1 - P_2)BR^{-1}B^*(P_3 + P_1 - P_2)x, y >_{\mathcal{V}} +$$

$$< ((P_2 - P_1)B\bar{F} + \bar{F}^* B^*(P_2 - P_1))x, y >_{\mathcal{V}} = 0 \tag{3.70}$$

and $A^{\mathcal{V}} - BR^{-1}B^*(P_3 + P_1 - P_2)$ is the infinitesimal generator of an exponentially stable semigroup on $\mathcal{V}$. Combining (3.67),(3.68) and (3.70), with $P := P_3 + P_1 - P_2$ implies that $P \in \mathcal{L}(\mathcal{V})$ is self-adjoint and satisfies

$$< A^{\mathcal{V}}x, P y >_{\mathcal{V}} + < Px, A^{\mathcal{V}} y >_{\mathcal{V}} - < PBR^{-1}B^* Px, y >_{\mathcal{V}} +$$

$$< C_1 x, C_1 y >_{Y_1} - < C_2 x, C_2 y >_{Y_2} = 0 \tag{3.71}$$

for all $x, y \in D(A^{\mathcal{V}})$ and $A^{\mathcal{V}} - BR^{-1}B^*P$ is the generator of an exponentially stable semigroup on $\mathcal{V}$. We note that (3.71) is just (3.60) with $L = 0$ and that the optimal control is given by

$$\bar{u}(\cdot) = -R^{-1}B^*Px(\cdot) \tag{3.72}$$

and that the optimal cost is given by

$$J(x_0, \bar{u}(\cdot)) =< x_0, (P_1 - P_2)x_0 >_{\mathcal{V}} + < x_0, P_3 x_0 >_{\mathcal{V}} =< x_0, Px_0 >_{\mathcal{V}} \tag{3.73}$$

(compare with (3.69)).

Next we show how to reduce the general case to the case where $L = 0$: We remove the 'cross term' of the cost criterion by applying the preliminary feedback

$$u(\cdot) = Fx(\cdot) + v(\cdot), \tag{3.74}$$

where $F \in \mathcal{L}(\mathcal{W}, U)$ is given by

$$F = -R^{-1}L, \tag{3.75}$$

To make this more precise we note that F given by (3.75) is an admissible output operator because L is, and we define the transformed system

$$x_F(t) = S_{BF}(t)x_0 + \int_0^t S_{BF}(t-s)Bv(s)ds, \tag{3.76}$$

just as in Lemma 2.14. It follows from Lemma 2.13 that C_1, C_2 and L are admissible output operators for this transformed system as well. Furthermore, it follows from Corollary 2.21 and Lemma 2.13, that $(A_{BF}^{\mathcal{V}}, B)$ is exponentially stabilizable on $\mathcal{V}$. We define the class of admissible inputs for this system as

$$\bar{U}_{adm} := \{v(\cdot) \in L_2(0, \infty; U) \text{ such that } x_F(\cdot) \text{ given by (3.76)}$$

$$\text{satisfies } x_F(\cdot) \in L_2(0, \infty; \mathcal{V})\}. \tag{3.77}$$

Using the perturbation results in Lemma 2.14, we see that if $u(\cdot)$ and $v(\cdot)$ are related by (3.74) (or, more specifically, by (2.31) and (2.34)), we have $x(\cdot) = x_F(\cdot)$, $u(\cdot) = Fx_F(\cdot) + v(\cdot)$ and $v(\cdot) = -Fx(\cdot) + u(\cdot)$. Hence, we conclude that $u(\cdot) \in U_{adm}$ if and only if $v(\cdot) \in \bar{U}_{adm}$. Next, for all $v(\cdot) \in \bar{U}_{adm}$, we define the transformed cost function:

$$J_F(x_0, v(\cdot)) := \int_0^\infty \mathcal{F}_F(x_F(t), v(t))dt,$$

where for $x \in \mathcal{W}$ and $v \in U$ we have

$$\mathcal{F}_F(x,v) = \mathcal{F}(x, Fx+v) =< C_1x, C_1x >_{Y_1} - < C_2x, C_2x >_{Y_2} +$$

$$2\text{Re} < Lx, Fx+v >_U + < R(Fx+v), (Fx+v) >_U,$$

so that because of $F = -R^{-1}L$, we have

$$\mathcal{F}_F(x,v) =< C_1x, C_1x >_{Y_1} - < C_2x, C_2x >_{Y_2} -$$

$$< R^{-\frac{1}{2}}Lx, R^{-\frac{1}{2}}Lx >_U + < Rv, v >_U .$$

Note that the transformed cost function has the same form as in (3.57), with $L = 0$ and $< C_2x, C_2x >_{Y_2}$ replaced by $< C_2x, C_2x >_{Y_2} + < R^{-\frac{1}{2}}Lx, R^{-\frac{1}{2}}Lx >_U$. Using Lemma 2.14 and the above, we have

$$J(x_0, u(\cdot)) = J_F(x_0, v(\cdot)). \tag{3.78}$$

Now (i) implies that there exists a unique $\bar{v}(\cdot) \in L_2(0,\infty,U)$ such that

$$\inf_{v(\cdot)\in\bar{U}_{adm}} J_F(x_0, v(\cdot)) = \min_{v(\cdot)\in\bar{U}_{adm}} J_F(x_0, v(\cdot)) = J_F(x_0, \bar{v}(\cdot)). \tag{3.79}$$

We have proved the implication $(i) \Rightarrow (ii)$ under the assumption that there is no 'cross term' (i.e. $L = 0$), so we conclude from (3.79) that there exists a self-adjoint $P \in \mathcal{L}(\mathcal{V})$ such that for all $x, y \in D(A^{\mathcal{V}}_{BF}) = D(A^{\mathcal{V}})$, we have

$$< A^{\mathcal{V}}_{BF}x, Py >_{\mathcal{V}} + < Px, A^{\mathcal{V}}_{BF}y >_{\mathcal{V}} - < PBR^{-1}B^*Px, y >_{\mathcal{V}} +$$

$$< C_1x, C_1y >_{Y_1} - < C_2x, C_2y >_{Y_2} - < R^{-\frac{1}{2}}Lx, R^{-\frac{1}{2}}Lx >_U = 0 \tag{3.80}$$

and $A^{\mathcal{V}}_{BF} - BR^{-1}B^*P$ is the generator of an exponentially stable semigroup on $\mathcal{V}$.
Since $A^{\mathcal{V}}_{BF} = A^{\mathcal{V}} - BR^{-1}L$, we have proved (ii).
It follows from (3.72) (applied to the transformed system) that the optimal control $\bar{v}(\cdot)$ for the *transformed* system (3.76) is given by

$$\bar{v}(\cdot) = -R^{-1}B^*Px_F(\cdot),$$

so that the optimal control $\bar{u}(\cdot)$ for the *original* system (3.54) is given by

$$\bar{u}(\cdot) = (-R^{-1}L - R^{-1}B^*P)x(\cdot) \tag{3.81}$$

(compare with (3.74) and (3.75)). Furthermore, it follows from (3.73) that the optimal cost for the transformed system is given by

$$J_F(x_0, \bar{v}(\cdot)) =< x_0, Px_0 >_{\mathcal{V}}$$

and because of (3.78) we have

$$J(x_0, \bar{u}(\cdot)) = J_F(x_0, \bar{v}(\cdot)) = < x_0, Px_0 >_{\mathcal{V}} . \tag{3.82}$$

Proof of $(ii) \Rightarrow (iii)$.
The proof of this part is similar to the proof of the implication $(ii) \Rightarrow (iii)$ in the proof of Lemma 3.7. Using (ii), it is straightforward to show that for all $(\omega, x, u) \in \mathbb{R} \times D(A^{\mathcal{V}}) \times U$ satisfying $i\omega x = A^{\mathcal{V}}x + Bu$ there holds

$$\mathcal{F}(x,u) = \|R^{\frac{1}{2}}(u + R^{-1}(B^*P + L)x)\|_U^2. \tag{3.83}$$

Now since $A^{\mathcal{V}} - BR^{-1}(B^*P + L)$ is the generator of an exponentially stable semigroup on $\mathcal{V}$, we see that for all $(\omega, x, u) \in \mathbb{R} \times D(A^{\mathcal{V}}) \times U$ satisfying $i\omega x = A^{\mathcal{V}}x + Bu$, there holds

$$x = (i\omega I - A^{\mathcal{V}} + BR^{-1}(B^*P + L))^{-1}B(u + R^{-1}(B^*P + L)x), \tag{3.84}$$

and so

$$\|x\|_{\mathcal{V}} \leq c_1\|u + R^{-1}(B^*P + L)x\|_U, \tag{3.85}$$

for some $c_1 > 0$. Combining (3.83) and (3.85) (using the fact that R is coercive) implies the existence of some $\epsilon_1 > 0$ such that for all $(\omega, x, u) \in \mathbb{R} \times D(A^{\mathcal{V}}) \times U$ satisfying $i\omega x = A^{\mathcal{V}}x + Bu$, $\mathcal{F}(x,u) \geq \epsilon_1\|x\|_{\mathcal{V}}^2$ and this proves (iii).

Proof of $(iii) \Rightarrow (i)$.
The idea is to apply Lemma 3.6 (a similar procedure is used in [61, 62] and [99] for the bounded case). In order to apply this lemma we need to show that the frequency domain inequality of (iii) implies that there exists an ϵ_1 such that for all $(\omega, x, u) \in \mathbb{R} \times D(A^{\mathcal{V}}) \times U$ satisfying $i\omega x = A^{\mathcal{V}}x + Bu$,

$$\mathcal{F}(x,u) \geq \epsilon_1(\|x\|_{\mathcal{V}}^2 + \|u\|_U^2). \tag{3.86}$$

We note the analogy with Remark 3.8 and the beginning of the proof of Lemma 3.7 for the bounded case. However, the proof here is more involved, due to the unboundedness of C_1, C_2 and L with respect to $\mathcal{V}$. First of all, we claim that there exists an $\epsilon_2 > 0$ such that for all $(\omega, x, u) \in \mathbb{R} \times D(A^{\mathcal{V}}) \times U$ satisfying $i\omega x = A^{\mathcal{V}}x + Bu$ and $\|u\|_U = 1$, there holds

$$\mathcal{F}(x,u) \geq \epsilon_2 \tag{3.87}$$

(we know from (iii) that for such x, u there holds $\mathcal{F}(x,u) \geq \epsilon\|x\|_{\mathcal{V}}^2$). If there were no such ϵ_2, we would have a sequence $(\omega_n, x_n, u_n) \in \mathbb{R} \times D(A^{\mathcal{V}}) \times U$ such that $i\omega_n x_n = A^{\mathcal{V}}x_n + Bu_n$ and $\|u_n\|_U = 1$ and

$$\mathcal{F}(x_n, u_n) \rightarrow 0 \text{ as } n \rightarrow \infty. \tag{3.88}$$

It follows from (iii) that then $x_n \stackrel{\mathcal{V}}{\to} 0$ (so far, we have followed the 'bounded' proof). Now let $\tilde{C} \in \mathcal{L}(\mathcal{W}, Z)$ be an arbitrary admissible output operator, where Z is some separable Hilbert space. We shall prove that $\tilde{C}x_n \stackrel{Z}{\to} 0$, despite the fact that $\tilde{C} \notin \mathcal{L}(\mathcal{V}, Z)$ in general.

Let η be larger than the growth bounds of $S(\cdot)$ on $\mathcal{W}$ and $\mathcal{V}$. We know that for $\sigma > \eta$ there exists a $c > 0$ such that

$$\|\tilde{C}((\sigma+i\omega)I - A^{\mathcal{W}})^{-1}x\|_Z = \|\tilde{C}((\sigma+i\omega)I - A^{\mathcal{V}})^{-1}x\|_Z \leq \frac{c\|x\|_{\mathcal{V}}}{(\sigma-\eta)^{\frac{1}{2}}} \tag{3.89}$$

for all $x \in \mathcal{W}$ and all $\omega \in \mathbb{R}$ (see (2.20)).
If $i\omega x = A^{\mathcal{V}}x + Bu$, then we have $(\sigma + i\omega)x = A^{\mathcal{V}}x + \sigma x + Bu$ so that $x = ((\sigma+i\omega)I - A^{\mathcal{V}})^{-1}\sigma x + ((\sigma+i\omega)I - A^{\mathcal{V}})^{-1}Bu$. Thus it follows from (3.89) and the fact that $x \in D(A^{\mathcal{V}}) \subset \mathcal{W}$ that

$$\|\tilde{C}x\|_Z \leq \frac{c\sigma\|x\|_{\mathcal{V}}}{(\sigma-\eta)^{\frac{1}{2}}} + \frac{c_1\|u\|_U}{(\sigma-\eta)^{\frac{1}{2}}},$$

for some $c_1 > 0$. Therefore, using $\|u_n\|_U = 1$, we infer that

$$\|\tilde{C}x_n\|_Z \leq \frac{c\sigma\|x_n\|_{\mathcal{V}}}{(\sigma-\eta)^{\frac{1}{2}}} + \frac{c_1}{(\sigma-\eta)^{\frac{1}{2}}}. \tag{3.90}$$

This shall be used to prove that $\tilde{C}x_n \stackrel{Z}{\to} 0$. Let $\delta > 0$ be arbitrary. For σ sufficiently large we have $\frac{c_1}{(\sigma-\eta)^{\frac{1}{2}}} < \frac{1}{2}\delta$. Since $x_n \stackrel{\mathcal{V}}{\to} 0$ as $n \to \infty$ we infer that for n sufficiently large, we have $\frac{c\sigma\|x_n\|_{\mathcal{V}}}{(\sigma-\eta)^{\frac{1}{2}}} < \frac{1}{2}\delta$. Combining this fact with (3.90) implies that indeed $\tilde{C}x_n \stackrel{Z}{\to} 0$ as $n \to \infty$.
Hence, for n tending to infinity there holds

$$< C_1x_n, C_1x_n >_{Y_1} - < C_2x_n, C_2x_n >_{Y_2} + 2\text{Re} < Lx_n, u_n >_U \to 0.$$

Combining this with (3.88) implies that $< Ru_n, u_n >_U \to 0$ for $n \to \infty$. But this contradicts the facts that $\|u_n\|_U = 1$ and that R is coercive, so that we have proved (3.87). This, in turn, implies that for all $(\omega, x, u) \in \mathbb{R} \times D(A^{\mathcal{V}}) \times U$ satisfying $i\omega x = A^{\mathcal{V}}x + Bu$, there holds

$$\mathcal{F}(x, u) \geq \epsilon_2\|u\|_U^2.$$

Combining this with (iii) implies (3.86).
Next, let $x_0 \in \mathcal{V}$ be given and define

$$H := \{(x(\cdot), u(\cdot)) \in L_2(0, \infty; \mathcal{V}) \times L_2(0, \infty; U) \text{ such that}$$

$$u(\cdot) \in U_{adm} \text{ and } x(t) = S(t)x_0 + \int_0^t S(t-s)Bu(s)ds\}, \tag{3.91}$$

and

$$H_0 := \{(x(\cdot), u(\cdot)) \in L_2(0,\infty;\mathcal{V}) \times L_2(0,\infty;U) \text{ such that}$$

$$u(\cdot) \in U_{adm} \text{ and } x(t) = \int_0^t S(t-s)Bu(s)ds\}. \tag{3.92}$$

It is straightforward to show that H_0 is a closed subspace of $L_2(0,\infty;\mathcal{V}) \times L_2(0,\infty;U)$ and so H_0 is a Hilbert space (with the obvious inner product determined by the inner products of $L_2(0,\infty;\mathcal{V})$ and $L_2(0,\infty;U)$). Since $(A^{\mathcal{V}}, B)$ is exponentially stabilizable on $\mathcal{V}$, there exists an $F \in \mathcal{L}(\mathcal{V}, U)$ such that $S_{BF}(\cdot)$ is exponentially stable on $\mathcal{V}$. We define

$$(x_0(\cdot), u_0(\cdot)) := (S_{BF}(\cdot)x_0, FS_{BF}(\cdot)x_0).$$

Using the perturbation results of Lemma 2.13, it is easy to see that $(x_0(\cdot), u_0(\cdot)) \in H$ and that $H = H_0 + (x_0(\cdot), u_0(\cdot))$ and we have

$$\inf_{u(\cdot)\in U_{adm}} J(x_0, u(\cdot)) = \inf_{(x(\cdot),u(\cdot))\in H} \int_0^\infty \mathcal{F}(x(t), u(t))dt =$$

$$\inf_{(x(\cdot),u(\cdot))\in H_0} \int_0^\infty \mathcal{F}(x(t)+x_0(t), u(t)+u_0(t))dt. \tag{3.93}$$

Now for $(x(\cdot), u(\cdot)) \in H_0$ we have

$$\int_0^\infty \mathcal{F}(x(t)+x_0(t), u(t)+u_0(t))dt = \int_0^\infty (\|C_1x(t) + C_1x_0(t)\|^2_{Y_1} -$$

$$\|C_2x(t) + C_2x_0(t)\|^2_{Y_2} + 2 < L(x(t)+x_0(t)),$$

$$u(t) + u_0(t) >_U + < R(u(t)+u_0(t)), u(t)+u_0(t) >_U)dt =$$

$$\int_0^\infty \mathcal{F}(x(t), u(t))dt + 2 < C_1x(\cdot), C_1x_0(\cdot) >_{L_2(0,\infty;Y_1)} -$$

$$2 < C_2x(\cdot), C_2x_0(\cdot) >_{L_2(0,\infty;Y_2)} +2 < Lx(\cdot), u_0(\cdot) >_{L_2(0,\infty;U)} +$$

$$2 < Lx_0(\cdot), u(\cdot) >_{L_2(0,\infty;U)} +2 < Ru(\cdot), u_0(\cdot) >_{L_2(0,\infty;U)} +c, \tag{3.94}$$

for some $c \in \mathbb{R}$. Next we infer that there exists a self-adjoint operator $T \in \mathcal{L}(H_0)$ and some element $y \in H_0$ such that (3.94) can be reformulated as

$$\int_0^\infty \mathcal{F}(x(t)+x_0(t), u(t)+u_0(t))dt =< Th, h >_{H_0} + < h, y >_{H_0} +c, \tag{3.95}$$

where $h := (x(\cdot), u(\cdot)) \in H_0$. Indeed, it follows from Lemma 3.2 that for any admissible $C \in \mathcal{L}(\mathcal{W}, Y)$ the map from H_0 to $L_2(0,\infty;Y)$ determined by $(x(\cdot), u(\cdot)) \mapsto Cx(\cdot)$ is linear and bounded. Since C_1, C_2 and L are admissible output operators and R is bounded, it is straightforward to conclude from (3.58) and Lemma 3.2 that we have a self-adjoint operator $T \in \mathcal{L}(H_0)$, such that for all $h := (x(\cdot), u(\cdot)) \in H_0$ there holds

$$\int_0^\infty \mathcal{F}(x(t), u(t))dt = < Th, h >_{H_0} . \tag{3.96}$$

Similarly, one can show that in (3.94),

$$2 < C_1 x(\cdot), C_1 x_0(\cdot) >_{L_2(0,\infty;Y_1)} -2 < C_2 x(\cdot), C_2 x_0(\cdot) >_{L_2(0,\infty;Y_2)} +$$

$$2 < Lx(\cdot), u_0(\cdot) >_{L_2(0,\infty;U)} +2 < Lx_0(\cdot), u(\cdot) >_{L_2(0,\infty;U)} +$$

$$2 < Ru(\cdot), u_0(\cdot) >_{L_2(0,\infty;U)}$$

is equal to $< h, y >_{H_0}$ for some $y \in H_0$ (use again Lemma 3.2 and Riesz's representation theorem for H_0). Hence we have (3.95).
It follows from (3.93) and the above that

$$\inf_{u(\cdot)\in U_{adm}} J(x_0, u(\cdot)) = \inf_{h\in H_0} < Th, h >_{H_0} + < h, y >_{H_0} +c. \tag{3.97}$$

In order to apply Lemma 3.6, we still have to show that T in (3.96) is coercive (of course using the frequency domain inequality of (iii)). Both components of $(x(\cdot), u(\cdot)) \in H_0$ are Fourier transformable and it follows from Lemma 3.5 that $\hat{x}(i\omega) \in D(A^{\mathcal{V}})$ for almost all $\omega \in \mathbb{R}$ and

$$\begin{aligned} i\omega\hat{x}(i\omega) &= A^{\mathcal{V}}\hat{x}(i\omega) + B\hat{u}(i\omega) \\ \widehat{Cx}(i\omega) &= C\hat{x}(i\omega), \end{aligned}$$

for almost all $\omega \in \mathbb{R}$, for any admissible $C \in \mathcal{L}(\mathcal{W}, Y)$. Hence, for all $h = (x(\cdot), u(\cdot)) \in H_0$ we have

$$< Th, h >_{H_0} = \int_0^\infty \mathcal{F}(x(t), u(t))dt$$

$$= \frac{1}{2\pi}\int_{-\infty}^\infty \mathcal{F}(\hat{x}(i\omega), \hat{u}(i\omega))d\omega \qquad \text{(Plancherel's Theorem,}$$

$$\text{see [78, section 4.8] and section 2.3, (2.16))}$$

$$\geq \epsilon\frac{1}{2\pi}\int_{-\infty}^\infty (\|\hat{x}(i\omega)\|_{\mathcal{V}}^2 + \|\hat{u}(i\omega)\|_U^2)d\omega \qquad \text{(using } (iii) \text{ and (3.86))}$$

$$= \epsilon\int_0^\infty (\|x(t)\|_{\mathcal{V}}^2 + \|u(t)\|_U^2)dt \qquad \text{(again using Plancherel's Theorem)}$$

$$= \epsilon\|(x(\cdot), u(\cdot))\|_{H_0}^2 = \epsilon\|h\|_{H_0}^2.$$

This proves that T is coercive and so Lemma 3.6 implies that there exists a unique pair $(\bar{x}(\cdot), \bar{u}(\cdot)) \in H$ that minimizes the cost function in (3.93). Using the definition of H in (3.91), this implies (i).

Finally, it follows from (3.81) and (3.82) that the optimal control $\bar{u}(\cdot)$ is given by the state-feedback $\bar{u}(\cdot) = -R^{-1}(B^*P + L)x(\cdot)$ and that (3.63) is satisfied.

The fact that there exists at most one $P \in \mathcal{L}(\mathcal{V})$ that satisfies item (ii) follows from the general uniqueness result of Lemma 2.33 and Remark 3.11.

■

Chapter 4

$\mathcal{H}_\infty$-control with state-feedback

In this chapter we formulate and solve the $\mathcal{H}_\infty$-control problem with state-feedback for the Pritchard-Salamon class. In particular, we show that under certain conditions there exists a state-feedback that solves the sub-optimal $\mathcal{H}_\infty$-control problem if and only if there exists a solution to a certain operator Riccati equation. The main result is given in Section 4.1 (see Theorem 4.4) and proved in Section 4.2.

The result is a generalization of the finite-dimensional results of Doyle et al. [26] and Tadmor [87] (see also the summary of results in Section 1.1) and the proof follows roughly the lines of the proof of Tadmor [87]. Pritchard-Salamon systems are *infinite-dimensional* systems with *unbounded* input and output operators and the consequences of these facts for the proof shall be stressed. As in the previous chapter on the LQ-problem, we shall deal with the unboundedness of input and output operators by considering initial conditions in the 'larger space' $\mathcal{V}$, so that in fact only the output operators are unbounded (with respect to $\mathcal{V}$).

In order to alleviate the notational complexity we shall first suppose that a certain simplifying assumption is satisfied. However, this assumption is not necesary, as will be shown in Section 4.3 (see Theorem 4.20).

The result in this chapter is the first step towards solving the $\mathcal{H}_\infty$-control problem with measurement-feedback in Chapter 5.

4.1 Problem formulation and main result

Let $\mathcal{W}, \mathcal{H}$ and $\mathcal{V}$ be real separable Hilbert spaces, satisfying $\mathcal{W} \hookrightarrow \mathcal{H} \hookrightarrow \mathcal{V}$, let $S(\cdot)$ be a C_0-semigroup on $\mathcal{V}$ which restricts to C_0-semigroups on $\mathcal{W}$ and $\mathcal{H}$ and suppose that $D(A^{\mathcal{V}}) \hookrightarrow \mathcal{W}$. Let $B_1 \in \mathcal{L}(W, \mathcal{V})$, $B_2 \in \mathcal{L}(U, \mathcal{V})$, $C_1 \in \mathcal{L}(\mathcal{W}, Z), D_{11} \in \mathcal{L}(W, Z)$ and $D_{12} \in \mathcal{L}(U, Z)$, where U, W and Z are also real separable Hilbert spaces. We identify the dual of $\mathcal{H}$ with itself and the duals of $\mathcal{W}$ and $\mathcal{V}$ are determined by the pivot space formulation of Section 2.5 (see formula (2.69)). Furthermore, the duals of U, W and Z are assumed to be identified with themselves.

Assuming that B_1 and B_2 are admissible input operators and that C_1 is an admissible output operator, we define the smooth Pritchard-Salamon system $\Sigma_G = \Sigma(S(\cdot), (B_1 \;\; B_2), C_1, (D_{11} \;\; D_{12}))$ of the form

$$\Sigma_G : \begin{cases} x(t) &= S(t)x_0 + \int_0^t S(t-s)(B_1 w(s) + B_2 u(s))ds \\ z(t) &= C_1 x(t) + D_{11} w(t) + D_{12} u(t), \end{cases} \tag{4.1}$$

where $x_0 \in \mathcal{V}, t \geq 0$ and $(w(\cdot), u(\cdot)) \in L_2^{loc}(0, \infty; W \times U)$. We call $x(t) \in \mathcal{V}$ the state of the system, $u(t) \in U$ is the control input, $w(t) \in W$ is the disturbance input and $z(t) \in Z$ is the to-be-controlled output.

The purpose of this chapter is to find a state-feedback that exponentially stabilizes the system, such that the corresponding closed-loop map from $w(\cdot)$ to $z(\cdot)$ satisfies a certain norm bound, or, to be more precise:
Suppose that $F \in \mathcal{L}(\mathcal{W}, U)$ is an admissible output operator for this system, such that the C_0-semigroup $S_{B_2F}(\cdot)$ is exponentially stable on $\mathcal{V}$. If we apply the admissible state-feedback $u(\cdot) = Fx(\cdot)$ to system (4.1) we obtain the Pritchard-Salamon system $\Sigma(S_{B_2F}(\cdot), B_1, (C_1 + D_{12}F), D_{11})$ (cf. Lemma 2.14):

$$\Sigma_{G_{B_2F}} : \begin{cases} x(t) &= S_{B_2F}(t)x_0 + \int_0^t S_{B_2F}(t-s)B_1 w(s)ds \\ z(t) &= (C_1 + D_{12}F)x(t) + D_{11}w(t). \end{cases} \tag{4.2}$$

It follows from Lemma 3.2 that the linear map G_{B_2F} (corresponding to $x_0 = 0$) defined by

$$(G_{B_2F}w)(t) = z(t) = (C_1 + D_{12}F) \int_0^t S_{B_2F}(t-s)B_1 w(s)ds + D_{11}w(t) \tag{4.3}$$

satisfies $G_{B_2F} \in \mathcal{L}(L_2(0,\infty; W), L_2(0,\infty; Z))$. It follows from Lemma 3.4 that $\|G_{B_2F}\| = \|\tilde{G}_{B_2F}(\cdot)\|_\infty$, where $\tilde{G}_{B_2F}$ is the closed-loop transfer function given by

$$\tilde{G}_{B_2F}(s) = (C_1 + D_{12}F)(sI - A_{B_2F}^{\mathcal{V}})^{-1}B_1 + D_{11} \in \mathcal{H}_\infty(\mathcal{L}(W, Z)).$$

Definition 4.1
A *γ-admissible state-feedback* for Σ_G given by (4.1) is an admissible output operator $F \in \mathcal{L}(\mathcal{W}, U)$ such that the C_0-semigroup $S_{B_2F}(\cdot)$ is exponentially stable on $\mathcal{V}$ and the corresponding linear map G_{B_2F} defined by (4.3) satisfies $\|G_{B_2F}\| < \gamma$.

Remark 4.2
We recall that if a C_0-semigroup is exponentially stable on $\mathcal{V}$, it need not be exponentially stable on $\mathcal{W}$ and vice versa (see [15]). For the moment we only ask for stability on $\mathcal{V}$, which is sufficient to render the closed-loop map given by (4.3) bounded. If, in addition, we know that there exist some admissible input and output operators F and G such that $S_{GF}(\cdot)$ is exponentially stable on both $\mathcal{W}$ and $\mathcal{V}$, then any admissible perturbation of $S(\cdot)$ is exponentially stable on $\mathcal{W}$ if and only if it is exponentially stable on $\mathcal{V}$ (see Theorem 2.20 item (iv)).

Finally, we note that the approach here is similar to the one that we had for the LQ-problem in Chapter 3. In particular, since we consider initial conditions in the 'larger space' $\mathcal{V}$ and stability on $\mathcal{V}$, we shall only have to worry about unboundedness of the output operators.

Below we shall relate the existence of a γ-admissible state-feedback to the solvability of an operator Riccati equation, but first we have to make two regularity assumptions and a simplifying assumption:

there is an $\epsilon > 0$ such that for all $(\omega, x, u) \in \mathbb{R} \times D(A^{\mathcal{V}}) \times U$ with

$$i\omega x = A^{\mathcal{V}}x + B_2 u, \text{ there holds } \|C_1 x + D_{12}u\|_Z^2 \geq \epsilon \|x\|_{\mathcal{V}}^2, \tag{4.4}$$

$$D_{12}'D_{12} \text{ is coercive,} \tag{4.5}$$

$$D_{11} = 0. \tag{4.6}$$

Remark 4.3
Assumptions (4.4) and (4.5) are the generalizations of the weak finite-dimensional regularity assumptions in [37]. In finite-dimensional terms, (4.4) means that the system $\Sigma(A, B_2, C_1, D_{12})$ has no invariant zeros on the imaginary axis (whence the term *invariant zeros condition*). The $\mathcal{H}_\infty$-control problem with assumptions (4.4) and (4.5) is usually called *regular*. We note that (4.4),(4.5) and (4.6) imply that the LQ-problem with stability is solvable, where the cost is given by $\|z(\cdot)\|^2_{L_2(0,\infty;Z)}$ (see Chapter 3).

It follows from Appendix C.1 that if $D_{12}'C_1 = 0$ is satisfied and $D_{12}'D_{12}$ is coercive, then (4.4) is *implied* by the detectability assumption that there exists an admissible output operator $G \in \mathcal{L}(Z, \mathcal{V})$ such that $S_{GC_1}(\cdot)$ is exponentially stable on $\mathcal{V}$.

The assumption that $D_{11} = 0$ is merely made to reduce the complexity of the formulas. In Section 4.3 we shall show that without this assumption we can still relate the existence of a γ-admissible state-feedback to the solvability of an operator Riccati equation. Hence, we have a perfect generalization of the weakest a priori regularity assumptions that have been used in the finite-dimensional case.

Finally, in Section 4.3 we show that the regularity assumptions (4.4)-(4.5) can be removed as well, however at the expense of introducing an extra, regularizing parameter into the problem.

Next, we present the main result of this chapter. It is a perfect generalization of the finite-dimensional state-feedback result in [37].

Theorem 4.4

Let Σ_G be the smooth Pritchard-Salamon system given by (4.1) and suppose that assumptions (4.4),(4.5) and (4.6) hold. The following are equivalent statements:

(*i*) *there exists a γ-admissible state-feedback for Σ_G,*

(*ii*) *there exists an admissible output operator $F \in \mathcal{L}(\mathcal{W}, U)$ such that $S_{B_2F}(\cdot)$ is exponentially stable on $\mathcal{V}$ and there exists a $\delta > 0$ such that for all $w(\cdot) \in L_2(0,\infty;W)$ there exists a $u(\cdot) \in L_2(0,\infty;U)$ such that $x(\cdot)$ defined by*

$$x(t) := \int_0^t S(t-s)(B_1w(s) + B_2u(s))ds \tag{4.7}$$

satisfies $x(\cdot) \in L_2(0,\infty;\mathcal{V})$ and $z_{w,u}(\cdot)$ defined by

$$z_{w,u}(\cdot) := C_1x(\cdot) + D_{12}u(\cdot) \tag{4.8}$$

satisfies

$$\|z_{w,u}(\cdot)\|^2_{L_2(0,\infty;Z)} \leq (\gamma^2 - \delta^2)\|w(\cdot)\|^2_{L_2(0,\infty;W)},$$

(*iii*) *there exists a $P \in \mathcal{L}(\mathcal{V}, \mathcal{V}')$ with $P = P' \geq 0$ satisfying*

$$\begin{aligned} &< Px, (A^{\mathcal{V}} - B_2(D_{12}'D_{12})^{-1}D_{12}'C_1)y >_{<\mathcal{V}',\mathcal{V}>} + \\ &< (A^{\mathcal{V}} - B_2(D_{12}'D_{12})^{-1}D_{12}'C_1)x, Py >_{<\mathcal{V},\mathcal{V}'>} + \\ &< P(\gamma^{-2}B_1B_1' - B_2(D_{12}'D_{12})^{-1}B_2')Px, y >_{<\mathcal{V}',\mathcal{V}>} + \\ &< (I - D_{12}(D_{12}'D_{12})^{-1}D_{12}')C_1x, \\ &(I - D_{12}(D_{12}'D_{12})^{-1}D_{12}')C_1y >_Z = 0 \end{aligned} \tag{4.9}$$

for all $x, y \in D(A^{\mathcal{V}})$, such that $A_P^{\mathcal{V}}$ given by $D(A_P^{\mathcal{V}}) = D(A^{\mathcal{V}})$ and

$$A_P^{\mathcal{V}} = A^{\mathcal{V}} - B_2(D_{12}'D_{12})^{-1}D_{12}'C_1 + (\gamma^{-2}B_1B_1' - B_2(D_{12}'D_{12})^{-1}B_2')P$$

is the infinitesimal generator of a C_0-semigroup $S_P(\cdot)$ which is exponentially stable on $\mathcal{V}$.

In this case, a γ-admissible state-feedback is given by $u(\cdot) = Fx(\cdot)$, where $F \in \mathcal{L}(\mathcal{W}, U)$ is the admissible output operator given by

$$F = -(D_{12}'D_{12})^{-1}(B_2'P + D_{12}'C_1). \tag{4.10}$$

Moreover, if there exists an operator P satisfying item (iii), then it is unique.

Remark 4.5

In the proof of this result, which is given in Section 4.2, we shall partly use the approach of [87] for the finite-dimensional case. There, Tadmor relates the $\mathcal{H}_\infty$-control problem with state-feedback to the solvability of a certain sup-inf-problem (see (4.15)), which can be solved using some arguments from game theory. We also refer to [70] and [86] where several parts of the proof in [87] have been worked out in detail. It is because of this approach that we are able to prove the equivalence with *(ii)* (this was also realized in [86]). The nice thing about this fact is that it can be used in the proof of the measurement-feedback result in Chapter 5: if there exists a dynamic measurement-feedback that solves the problem, then there also exists a static state-feedback that solves the problem and a solution to the state-feedback Riccati equation. Hence, it also shows that if there exists a γ-admissible *dynamic* state-feedback, then there also exists a γ-admissible *static* state-feedback. It follows from Section 5.4 that this assertion remains true without any a priori assumptions.

Remark 4.6

We could also have formulated the Riccati equation in terms of the operator $P_0 := c^{-1}P \in \mathcal{L}(\mathcal{V})$, where $c = i_{\mathcal{H}} \mid_{\mathcal{V}^d} (i_{\mathcal{V}})^{-1}$ is the isometry from $\mathcal{V}$ to $\mathcal{V}'$ (cf. (2.69)). It is easy to see that $B_1^* = B_1'c$ and $B_2^* = B_2'c$. Supposing that $D_{12}'[C_1 \;\; D_{12}] = [0 \;\; I]$ (this simplifies the formulas considerably), (4.9) can be reformulated as

$$< P_0x, A^{\mathcal{V}}y >_{\mathcal{V}} + < A^{\mathcal{V}}x, P_0y >_{\mathcal{V}} + < P_0(\gamma^{-2}B_1B_1^* - B_2B_2^*)P_0x, y >_{\mathcal{V}}$$

$$+ < C_1x, C_1y >_Z = 0 \text{ for all } x, y \in D(A). \tag{4.11}$$

We have chosen the formulation of (4.9) because it is more convenient in the derivation of the measurement-feedback result. In the proof of Theorem 4.4, we shall in fact derive the existence of a P_0 as above and then define P by $P := cP_0$.

4.2 Proof of the state-feedback result

It is easy to see that item (i) implies item (ii). Therefore, we shall only prove the implications $(ii) \Rightarrow (iii)$ and $(iii) \Rightarrow (i)$. We shall first prove these implications for the special case that

$$D_{12}'[C_1 \ \ D_{12}] = [0 \ \ I]. \tag{4.12}$$

Then we remove this extra assumption by introducing an admissible preliminary feedback, using the results of Lemma 2.14. It is easy to see that (4.12) implies that (4.5) holds and that the to-be-controlled output satisfies

$$< z, z >_Z = < C_1 x, C_1 x >_Z + < u, u >_U, \tag{4.13}$$

for all $x \in W$ and $u \in U$. So for the moment we have that (4.4) holds and that (4.12) is satisfied.

a) Proof of $(ii) \Rightarrow (iii)$

In this part we assume that (ii) is satisfied. Consider the system (4.1), where the initial condition is given by $x_0 = \xi$ for some $\xi \in \mathcal{V}$ and $w(\cdot)$ is some element in $L_2(0,\infty;W)$. Furthermore, let $u \in L_2(0,\infty;U)$ be any input such that the corresponding state from (4.1) satisfies $x(\cdot) \in L_2(0,\infty;\mathcal{V})$ (the existence of such $u(\cdot)$ is guaranteed by the assumption that (ii) is satisfied). Note that the to-be-controlled output $z(\cdot) \in L_2(0,\infty;Z)$ (see Lemma 3.2) and so the following cost functional is well-defined:

$$J(\xi, w(\cdot), u(\cdot)) := \|z(\cdot)\|^2_{L_2(0,\infty;Z)} - \gamma^2 \|w(\cdot)\|^2_{L_2(0,\infty;W)}. \tag{4.14}$$

Because of (4.12) and (4.13), $\|z(\cdot)\|^2_{L_2(0,\infty;Z)}$ is given by

$$\|z(\cdot)\|^2_{L_2(0,\infty;Z)} = \int_0^\infty (< C_1 x(\tau), C_1 x(\tau) >_Z + < u(\tau), u(\tau) >_U)\, d\tau.$$

Using the assumption that (ii) holds, we will solve the following *sup-inf-problem*:

$$\sup_{w(\cdot) \in L_2(0,\infty;W)} \ \inf_{u(\cdot) \in L_2(0,\infty;U)} J(\xi, w(\cdot), u(\cdot)), \tag{4.15}$$

where $u(\cdot) \in L_2(0,\infty;U)$ must be such that $x(\cdot) \in L_2(0,\infty;\mathcal{V})$. The solution of this problem will lead us to the solution of (4.9) with the required properties.

First we consider the following optimal control problem.

$$\mathcal{P}: \begin{cases} \text{Given } (\xi, w(\cdot)) \in \mathcal{V} \times L_2(0,\infty;W), \text{ find the infimum of} \\ \{\|z(\cdot)\|^2_{L_2(0,\infty;Z)} \mid (x(\cdot), z(\cdot)) \text{ satisfy (4.1) with } x_0 = \xi \\ \text{and } u \in L_2(0,\infty;U) \text{ is such that } x(\cdot) \in L_2(0,\infty;\mathcal{V})\}. \end{cases}$$

To solve this problem we need Lemma 3.10. We know that there exists an admissible output operator $F \in \mathcal{L}(\mathcal{W}, U)$ such that $S_{B_2F}(\cdot)$ is exponentially stable on $\mathcal{V}$. Remark 2.19 then implies that the pair $(A^{\mathcal{V}}, B_2)$ is exponentially stabilizable. Furthermore, we conclude from assumption (4.4) combined with assumption (4.12) that condition (*iii*) in Theorem 3.10 is satisfied with

$$\mathcal{F}(x,u) = \|C_1x + D_{12}u\|_Z^2 = \|C_1x\|_Z^2 + \|u\|_U^2.$$

Hence, using Remark 3.14, formula (3.66), this theorem implies the existence of some $L \in \mathcal{L}(\mathcal{V})$ with $L = L^* \leq 0$ such that

$$< A^{\mathcal{V}}x, Ly >_{\mathcal{V}} + < Lx, A^{\mathcal{V}}y >_{\mathcal{V}} +$$

$$< LB_2B_2^*Lx, y >_{\mathcal{V}} - < C_1x, C_1y >_Z = 0 \tag{4.16}$$

for all $x, y \in D(A^{\mathcal{V}})$ and $A_L^{\mathcal{V}} = A^{\mathcal{V}} + B_2B_2^*L$ is the generator of a C_0-semigroup $S_L(\cdot)$, which is exponentially stable on $\mathcal{V}$. It follows that $A_L^{\mathcal{V}*}$ is the infinitesimal generator of the exponentially stable semigroup $S_L^*(\cdot)$ on $\mathcal{V}$ and the following is well-defined for all $w(\cdot) \in L_2(0,\infty;W)$.

$$r(t) := \int_t^\infty S_L^*(\tau - t)LB_1w(\tau)d\tau. \tag{4.17}$$

We conclude from Appendix B.4 that $r(\cdot) \in L_2(0,\infty;\mathcal{V})$, $r(\cdot)$ is strongly continuous (w.r.t. the topology of $\mathcal{V}$) and $r(t) \xrightarrow{\mathcal{V}} 0$ as $t \to \infty$.
Next, given $\xi \in \mathcal{V}$ and $w(\cdot) \in L_2(0,\infty;W)$, we define

$$\bar{x}(t) := S_L(t)\xi + \int_0^t S_L(t-s)(B_2B_2^*r(s) + B_1w(s))ds, \tag{4.18}$$

(this is (4.1) with $x_0 = \xi$ and $u(\cdot) = B_2^*(Lx(\cdot) + r(\cdot))$). It follows from Appendix B.1 that $\bar{x}(\cdot) \in L_2(0,\infty;\mathcal{V})$ is strongly continuous and $\bar{x}(t) \xrightarrow{\mathcal{V}} 0$ as $t \to \infty$.
Now we define

$$\eta(\cdot) := L\bar{x}(\cdot) + r(\cdot), \tag{4.19}$$

(so $\eta(\cdot) \in L_2(0,\infty;\mathcal{V})$, $\eta(\cdot)$ is strongly continuous and $\eta(t) \xrightarrow{\mathcal{V}} 0$ as $t \to \infty$).

In the following lemma we prove an important property of $\eta(\cdot)$ for smooth disturbances and smooth initial conditions (η satisfies a 'Hamilton-Jacobi equation' corresponding to the optimal control problem $\mathcal{P}$, as explained in [87] for the finite-dimensional case).

Lemma 4.7

Suppose that $w(\cdot) \in L_2(-\infty,\ 0; W)$ is continuously differentiable such that $\dot{w}(\cdot) \in L_1(-\infty, 0; W)$ and let $\xi \in D(A^{\mathcal{V}})$. Then we have $r(t) \in D(A_L^{\mathcal{V}*})$, $\bar{x}(t) \in D(A_L^{\mathcal{V}})$, $r(\cdot)$ and $\bar{x}(\cdot)$ are both continuously differentiable (w.r.t. the topology of $\mathcal{V}$) and there holds

$$< \xi, \frac{d}{dt}\eta(t) >_{\mathcal{V}} = < C_1\xi, C_1\bar{x}(t) >_Z - < A^{\mathcal{V}}\xi, \eta(t) >_{\mathcal{V}} . \tag{4.20}$$

Proof

First of all, if $w(\cdot) \in L_2(-\infty\ , 0; W)$ is such that $\dot{w}(\cdot) \in L_1(-\infty, 0; W)$, then $r(t) \in D((A_L^{\mathcal{V}})^*)$ for all $t \geq 0$, $r(\cdot)$ is continuously differentiable w.r.t. the topology of $\mathcal{V}$ and

$$\dot{r} = -(A_L^{\mathcal{V}})^* r - LB_1 w. \tag{4.21}$$

The proof of these facts follows from Appendix B.5 with $X = \mathcal{V}$, $T(t) = S_L^*(t)$ and

$$s(t) := r(-t) = \int_{-t}^{\infty} S_L^*(\tau + t) LB_1 w(\tau) d\tau = \int_{-\infty}^{t} S_L^*(t - \tau) LB_1 w(-\tau) d\tau.$$

Now since $r(\cdot)$ and $w(\cdot)$ are both continuously differentiable we can infer from Appendix B.3 that $\bar{x}(\cdot)$ given by (4.18) satisfies $\bar{x}(t) \in D(A_L^{\mathcal{V}})$ for all $t \geq 0$, that $\bar{x}(\cdot)$ is continuously differentiable w.r.t. the topology of $\mathcal{V}$ and that

$$\dot{\bar{x}} = A_L^{\mathcal{V}} \bar{x} + B_2 B_2^* r + B_1 w. \tag{4.22}$$

Since $L \in \mathcal{L}(\mathcal{V})$, it follows that $\eta(\cdot)$ given by $\eta(\cdot) = L\bar{x}(\cdot) + r(\cdot)$ is continuously differentiable w.r.t. the topology of $\mathcal{V}$. Thus, we can calculate

$$< x, \frac{d}{dt}\eta(t) >_{\mathcal{V}} = < x, L\dot{\bar{x}}(t) >_{\mathcal{V}} + < x, \dot{r}(t) >_{\mathcal{V}} =$$

$$= < Lx, (A^{\mathcal{V}} + B_2 B_2^* L)\bar{x}(t) + B_2 B_2^* r(\cdot) + B_1 w(t) >_{\mathcal{V}}$$

$$- < (A^{\mathcal{V}} + B_2 B_2^* L)x, r(t) >_{\mathcal{V}} - < x, LB_1 w(t) >_{\mathcal{V}}$$

using (4.21) and (4.22)

$$= < C_1 x, C_1 \bar{x}(t) >_Z - < A^{\mathcal{V}} x, \eta(t) >_{\mathcal{V}},$$

where in the last step we have used the Riccati equation for L given by (4.16).

■

Using Lemma 4.7 we can prove the following result, which is a type of maximum principle solution of the optimal control problem $\mathcal{P}$ ($\eta(\cdot)$ in (4.20) plays the role of the 'adjoint variable').

Lemma 4.8

The optimal control for problem $\mathcal{P}$ is given by

$$\bar{u}(\cdot) := B_2^*\eta(\cdot) \tag{4.23}$$

and the infimum is in fact a minimum.

Proof

For arbitrary $\xi \in \mathcal{V}$ and $w(\cdot) \in L_2(0,\infty;W)$ we define $r(\cdot)$, $\bar{x}(\cdot)$, $\eta(\cdot)$ and $\bar{u}(\cdot)$ as in (4.17), (4.18), (4.19) and (4.23). Suppose that we have a $u(\cdot) \in L_2(0,\infty;U)$ such that $x(\cdot)$ in problem $\mathcal{P}$ satisfies $x(\cdot) \in L_2(0,\infty;\mathcal{V})$. We wish to prove that

$$< \eta(0), \xi >_{\mathcal{V}} = -\int_0^\infty (< C_1\bar{x}(\tau), C_1x(\tau) >_Z$$

$$+ < \eta(\tau), B_2u(\tau) >_{\mathcal{V}} + < \eta(\tau), B_1w(\tau) >_{\mathcal{V}})\, d\tau. \tag{4.24}$$

In the finite-dimensional case one differentiates $< \eta(t), x(t) >_{\mathcal{V}}$ with respect to t and then (after some manipulations using Lemma 4.7, formula (4.20)) integrating from 0 to ∞ gives the result. However, (4.20) only holds for smooth $w(\cdot)$ and $\xi \in D(A^{\mathcal{V}})$. Moreover, for $x(\cdot)$ to be continuously differentiable we would need that $u(\cdot)$ be smooth. We overcome this difficulty by introducing sequences $\xi_n \in D(A^{\mathcal{V}})$, $u_n(\cdot) \in CCD(0,\infty;U)$ and $w_n(\cdot) \in CCD(0,\infty;W)$ such that $\xi_n \xrightarrow{\mathcal{V}} x$, $\|u_n(\cdot) - u(\cdot)\|_{L_2(0,\infty;U)} \to 0$ and $\|w_n(\cdot) - w(\cdot)\|_{L_2(0,\infty;W)} \to 0$ as $n \to \infty$. Using these sequences we derive a formula which provides the result via a limiting argument (note that these sequences exist, since $D(A^{\mathcal{V}})$ is dense in $\mathcal{V}$ and $CCD(0,\infty;Y)$ is dense in $L_2(0,\infty;Y)$, where Y is any separable Hilbert space):
To this end, we define for $t \in [0,T]$, $T > 0$

$$x_n(t) := S(t)\xi_n + \int_0^t S(t-s)\,(B_1w_n(s) + B_2u_n(s))\, ds.$$

Since $\xi_n \in D(A^{\mathcal{V}})$, $u_n(\cdot) \in CCD(0,\infty;U)$ and $w_n(\cdot) \in CCD(0,\infty;W)$ it follows from Appendix B.3 that $x_n(\cdot)$ is continuously differentiable and

$$\dot{x}_n = A^{\mathcal{V}}x_n + B_1w_n + B_2u_n; \quad x_n(0) = \xi_n, x_n(t) \in D(A^{\mathcal{V}}).$$

Furthermore, Appendix B.6 tells us that $\|x_n(\cdot) - x(\cdot)\|_{L_2(0,T;\mathcal{V})} \to 0$ and $x_n(T) \xrightarrow{\mathcal{V}} x(T)$ as $n \to \infty$.
Similarly, the sequences $\xi_n(\cdot)$ and $w_n(\cdot)$ induce sequences $r_n(\cdot)$, $\bar{x}_n(\cdot)$ and $\eta_n(\cdot)$ via (4.17), (4.18) and (4.19) which are also continuously differentiable w.r.t. the topology of $\mathcal{V}$ and which converge to the corresponding $r(\cdot)$, $\bar{x}(\cdot)$ and $\eta(\cdot)$

in the L_2-norm and pointwise on $[0,T]$ (see Appendices B.1,B.3,B.6 and B.7). Furthermore, we have

$$< x, \frac{d}{dt}\eta_n(t) >_{\mathcal{V}} = < C_1 x, C_1 \bar{x}_n(t) >_Z - < A^{\mathcal{V}} x, \eta_n(t) >_{\mathcal{V}},$$

for all $x \in D(A^{\mathcal{V}})$ (cf. Lemma 4.7). Using these results, we obtain

$$\frac{d}{dt} < \eta_n(t), x_n(t) >_{\mathcal{V}} = < C_1 \bar{x}_n(t), C_1 x_n(t) >_Z -$$

$$< \eta_n(t), A^{\mathcal{V}} x_n(t) >_{\mathcal{V}} + < \eta_n(t), A^{\mathcal{V}} x_n(t) + B_1 w_n(t) + B_2 u_n(t) >_{\mathcal{V}} =$$

$$< C_1 \bar{x}_n(t), C_1 x_n(t) >_Z + < \eta_n(t), B_2 u_n(t) + B_1 w_n(t) >_{\mathcal{V}} .$$

Integrating this from 0 to T we obtain

$$< \eta_n(T), x_n(T) >_{\mathcal{V}} - < \eta_n(0), x_n(0) >_{\mathcal{V}} = \int_0^T (< C_1 \bar{x}_n(\tau), C_1 x_n(\tau) >_Z$$

$$+ < \eta_n(\tau), B_2 u_n(\tau) >_{\mathcal{V}} + < \eta_n(\tau), B_1 w_n(\tau) >_{\mathcal{V}}) d\tau. \tag{4.25}$$

We claim that the limit as $n \to \infty$ of this last expression can be obtained by just removing the subscript n. Here the only difficulty is the 'unbounded term' $\int_0^T (< C_1 \bar{x}_n(\tau), C_1 x_n(\tau) >_Z d\tau$, but using Appendix B.6 (and, in particular, formula (B.20)), we conclude that the limit of (4.25) as $n \to \infty$ is given by

$$< \eta(T), x(T) >_{\mathcal{V}} - < \eta(0), x(0) >_{\mathcal{V}} = \int_0^T (< C_1 \bar{x}(\tau), C_1 x(\tau) >_Z +$$

$$< \eta(\tau), B_2 u(\tau) >_{\mathcal{V}} + < \eta(\tau), B_1 w(\tau) >_{\mathcal{V}}) d\tau.$$

In order to obtain (4.24), we let T tend to infinity in the above expression, using the facts that $x(T) \xrightarrow{\mathcal{V}} 0$, $\eta(T) \xrightarrow{\mathcal{V}} 0$ as $T \to \infty$ (see Appendices B.1,B.4), $\eta(\cdot), B_2 u(\cdot), B_2 w(\cdot) \in L_2(0,\infty;\mathcal{V})$ and $C_1 \bar{x}(\cdot), C_1 x(\cdot) \in L_2(0,\infty;Z)$ (see Lemma 3.2).

Recall that we want to minimize

$$\|z(\cdot)\|^2_{L_2(0,\infty;Z)} = \int_0^\infty (< C_1 x(\tau), C_1 x(\tau) >_Z + < u(\tau), u(\tau) >_U)\, d\tau.$$

Now using (4.24), we find that

$$\|z(\cdot)\|^2_{L_2(0,\infty;Z)} + 2 < \eta(0), \xi > =$$

$$\int_0^\infty (< C_1 x(\tau) - C_1 \bar{x}(\tau), C_1 x(\tau) - C_1 \bar{x}(\tau) >_Z$$

$$+ < u(\tau) - \bar{u}(\tau), u(\tau) - \bar{u}(\tau) >_U - < C_1 \bar{x}(\tau), C_1 \bar{x}(\tau) >_Z -$$

$$< \bar{u}(\tau), \bar{u}(\tau) >_U -2 < \eta(\tau), B_1 w(\tau) >_{\mathcal{V}})d\tau$$

and this expression is minimized for $u(\cdot) = \bar{u}(\cdot) = B_2^* \eta(\cdot)$. ■

Before we continue with the proof of the necessity part, we introduce the operators $\mathcal{F} : \mathcal{V} \times L_2(0,\infty;W) \to L_2(0,\infty;\mathcal{V}) \times L_2(0,\infty;U) \times L_2(0,\infty;\mathcal{V})$ and $\mathcal{G} : \mathcal{V} \times L_2(0,\infty;W) \to L_2(0,\infty;Z)$ defined by

$$\mathcal{F}(\xi, w(\cdot)) := (\bar{x}(\cdot), \bar{u}(\cdot), \eta(\cdot)) \tag{4.26}$$

$$\mathcal{G}(\xi, w(\cdot)) := \bar{z}(\cdot) = C_1 \bar{x}(\cdot) + D_{12} \bar{u}(\cdot). \tag{4.27}$$

$\mathcal{F}$ and $\mathcal{G}$ are well-defined, bounded linear operators: indeed, since $S_L^*(\cdot)$ is exponentially stable on $\mathcal{V}$, the map from $w(\cdot)$ to $r(\cdot)$ in (4.17) is linear and bounded from $L_2(0,\infty;W)$ to $L_2(0,\infty;\mathcal{V})$ and therefore the map from $(\xi, w(\cdot))$ to $\bar{x}(\cdot)$ in (4.18) is linear and bounded from $\mathcal{V} \times L_2(0,\infty;W)$ to $L_2(0,\infty;\mathcal{V})$. Now since $L \in \mathcal{L}(\mathcal{V})$, $\eta(\cdot) = L\bar{x}(\cdot) + r(\cdot)$ and $\bar{u}(\cdot) = B_2^* \eta(\cdot)$, it follows that $\mathcal{F}$ is linear and bounded. The linearity of $\mathcal{G}$ is now obvious and the boundedness follows from the above with Lemma 3.2.

Next, we consider the problem of finding the so-called worst disturbance, i.e., we set $u(\cdot) = \bar{u}(\cdot)$ in (4.15) and we consider solving the optimization problem

$$\sup_{w(\cdot) \in L_2(0,\infty;W)} J(\xi, w(\cdot), \bar{u}(\cdot)) =$$

$$\sup_{w(\cdot) \in L_2(0,\infty;W)} \|\bar{z}(\cdot)\|^2_{L_2(0,\infty;Z)} - \gamma^2 \|w(\cdot)\|^2_{L_2(0,\infty;W)} =$$

$$\sup_{w(\cdot) \in L_2(0,\infty;W)} \|\mathcal{G}(\xi, w(\cdot))\|^2_{L_2(0,\infty;Z)} - \gamma^2 \|w(\cdot)\|^2_{L_2(0,\infty;W)}. \tag{4.28}$$

In the following lemma we need the assumption that (ii) is satisfied. Defining

$$\mathcal{C}(\xi, w(\cdot)) := \gamma^2 \|w(\cdot)\|^2_{L_2(0,\infty;W)} - \|\mathcal{G}(\xi, w(\cdot))\|^2_{L_2(0,\infty;Z)} \tag{4.29}$$

we obtain the following result:

Lemma 4.9

Let $\mathcal{C}$ be defined by (4.29). Then $\mathcal{C}(0, w(\cdot)) \geq 0$ and $\mathcal{C}(0, w(\cdot))^{\frac{1}{2}}$ defines a norm on $L_2(0,\infty;W)$, which is equivalent to the usual L_2-norm on $L_2(0,\infty;W)$.

Proof

According to our assumption that (ii) is satisfied, there exists a $\delta > 0$ such that for all $w(\cdot) \in L_2(0,\infty;W)$ there exists a $u(\cdot) \in L_2(0,\infty;U)$ such that $x(\cdot)$ given by

$$x(t) = \int_0^t S(t-s)(B_1 w(s) + B_2 u(s))ds$$

satisfies $x(\cdot) \in L_2(0,\infty;\mathcal{V})$ and $z_{w,u}(\cdot) = C_1x(\cdot) + D_{12}u(\cdot)$ satisfies

$$\|z_{w,u}(\cdot)\|^2_{L_2(0,\infty;Z)} \leq (\gamma^2 - \delta^2)\|w(\cdot)\|^2_{L_2(0,\infty;W)}. \tag{4.30}$$

Note that $\mathcal{G}(0, w(\cdot))$ denotes the output that corresponds to the *optimal* control in problem $\mathcal{P}$ with $\xi = 0$, and $z_{w,u}(\cdot)$ just denotes the output that corresponds to *some* stabilizing control in (4.1) with $\xi = 0$. Thus for all $w(\cdot) \in L_2(0,\infty;W)$ there holds

$$\|\mathcal{G}(0, w(\cdot))\|_{L_2(0,\infty;Z)} \leq \|z_{w,u}(\cdot)\|_{L_2(0,\infty;Z)}. \tag{4.31}$$

Hence we conclude from (4.30) and (4.31) that

$$\gamma^2\|w(\cdot)\|^2_{L_2(0,\infty;W)} - \|\mathcal{G}(0, w(\cdot))\|^2_{L_2(0,\infty;W)} \geq \delta^2\|w(\cdot)\|^2_{L_2(0,\infty;W)} \tag{4.32}$$

for all $w(\cdot) \in L_2(0,\infty;W)$. It follows from the definition of $\mathcal{C}$ in (4.29) and (4.32) that $\mathcal{C}(0, w(\cdot)) \geq 0$ and that

$$\gamma\|w(\cdot)\|_2 \geq \mathcal{C}(0, w(\cdot))^{\frac{1}{2}} \geq \delta\|w(\cdot)\|_2 \text{ for all } w(\cdot) \in L_2(0,\infty;W). \tag{4.33}$$

Finally, we note that $\mathcal{C}(0, w(\cdot))$ is a quadratic form in $w(\cdot)$ and so (4.33) implies that $\mathcal{C}(0, w(\cdot))^{\frac{1}{2}}$ defines a norm on $L_2(0,\infty;W)$ and that it is equivalent to the L_2-norm on $L_2(0,\infty;W)$. ■

Finding the worst disturbance is now equivalent to finding the infimum of $\mathcal{C}(\xi, w(\cdot))$ for $w(\cdot) \in L_2(0,\infty;W)$. In doing so, we deviate from the approach in [87] by using the following lemma from Yakubovich (see e.g. [101] or [61]):

Lemma 4.10
Consider a Hilbert space V with a quadratic form

$$J(v) = \langle Kv, v \rangle \quad \textit{with } K \in \mathcal{L}(V); \quad K = K^*. \tag{4.34}$$

Let M_0 be a closed subspace of V and M the translation of M_0 by some element m in V (i.e. $M = M_0 + m$).
Suppose that the following condition holds:

$$\inf_{v_0 \in M_0} \frac{\langle Kv_0, v_0 \rangle}{\langle v_0, v_0 \rangle} > 0. \tag{4.35}$$

Then there exists a unique element $u \in M$ such that

$$J(u) = \inf_{v \in M} J(v),$$

where u is of the form $u = m^* + m$ with $m^* \in M_0$ and $m^* = Tm$ for some $T \in \mathcal{L}(V)$.

The following result shows that the infimum of $\mathcal{C}(\xi, w(\cdot))$ is in fact a minimum.

Lemma 4.11
Given $\xi \in \mathcal{V}$ there exists a unique $w^*(\cdot) \in L_2(0,\infty;W)$ such that

$$\inf_{w(\cdot)\in L_2(0,\infty;W)} \mathcal{C}(\xi, w(\cdot)) = \mathcal{C}(\xi, w^*(\cdot)) \tag{4.36}$$

and there exists an $H \in \mathcal{L}(\mathcal{V}, L_2(0,\infty;W))$ such that

$$w^*(\cdot) = H\xi. \tag{4.37}$$

Proof
We apply Lemma 4.10 with $V := \mathcal{V} \times L_2(0,\infty;W)$, $M_0 := \{(0, w(\cdot)) \mid w(\cdot) \in L_2(0,\infty;W)\} \subset V$, $m := (\xi, 0)$, $M := \{(\xi, w(\cdot)) \mid w(\cdot) \in L_2(0,\infty;W)\} = (\xi,0) + M_0 \subset V$ and (using $v = (\xi, w(\cdot))$)

$$J(v) = J(\xi, w(\cdot)) := \mathcal{C}(\xi, w(\cdot)) =$$

$$\gamma^2 \|w(\cdot)\|^2_{L_2(0,\infty;W)} - \|\mathcal{G}(\xi, w(\cdot))\|^2_{L_2(0,\infty;Z)} =$$

$$\gamma^2 < w(\cdot), w(\cdot) >_{L_2(0,\infty;W)} - < \mathcal{G}(\xi, w(\cdot)), \mathcal{G}(\xi, w(\cdot)) >_{L_2(0,\infty;Z)}=$$

$$\gamma^2 < w(\cdot), w(\cdot) >_{L_2(0,\infty;W)} - < \mathcal{G}^*\mathcal{G}(\xi, w(\cdot)), (\xi, w(\cdot)) >_V=$$

$$< K(\xi, w(\cdot)), (\xi, w(\cdot)) >_V = < Kv, v >_V,$$

where we have defined $Kv = K(\xi, w(\cdot)) := (0, \gamma^2 w(\cdot)) - \mathcal{G}^*\mathcal{G}(\xi, w(\cdot))$ (so that $K = K^* \in \mathcal{L}(V)$).
Condition (4.35) is satisfied because there exists a $\delta > 0$ such that for $v_0 = (0, w(\cdot)) \in M_0$ there holds

$$< Kv_0, v_0 >_V = J(0, w(\cdot)) = \mathcal{C}(0, w(\cdot)) \geq \delta^2 \|w(\cdot)\|^2_{L_2(0,\infty;W)}$$

(using Lemma 4.9)

$$= \delta^2 < v_0, v_0 >_V .$$

Hence, Lemma 4.10 implies that there exists a unique element of M of the form $m^* + m = (0, w^*(\cdot)) + (\xi, 0) = (\xi, w^*(\cdot))$ such that

$$\inf_{w(\cdot)\in L_2(0,\infty;W)} \mathcal{C}(\xi, w(\cdot)) = \inf_{v\in M} J(v) = J(m^* + m) = \mathcal{C}(\xi, w^*(\cdot)).$$

Moreover, there exists a $T \in \mathcal{L}(V)$ such that

$$m^* = (0, w^*(\cdot)) = Tm = T(\xi, 0)$$

and this readily implies the lemma. ∎

As mentioned before, the minimizing $w^*(\cdot)$ from Lemma 4.11 is called the worst disturbance because it maximizes $J(\xi, w(\cdot), \bar{u}(\cdot))$ in (4.28). In the following lemma we find a characterization of it in terms of $\eta(\cdot)$. In the sequel we shall use the definitions

$$(x^*(\cdot), u^*(\cdot), \eta^*(\cdot)) := \mathcal{F}(\xi, w^*(\cdot)) \tag{4.38}$$

and

$$z^*(\cdot) := \mathcal{G}(\xi, w^*(\cdot)), \tag{4.39}$$

for a given $\xi \in \mathcal{V}$.

Lemma 4.12
Let $\xi \in \mathcal{V}$ be given and let $w^*(\cdot)$ denote the unique worst disturbance from Lemma 4.11. For arbitrary $w(\cdot) \in L_2(0,\infty;W)$ we define $\eta(\cdot)$ as in (4.19) (see also (4.26)). Considering $\eta(\cdot)$ as a function of $w(\cdot)$, it follows that $w^*(\cdot)$ is the unique element of $L_2(0,\infty;W)$ satisfying

$$w(\cdot) = -\gamma^{-2} B_1^* \eta(\cdot). \tag{4.40}$$

Proof
First of all we shall prove that $w^*(\cdot)$ satisfies (4.40). Note that according to (4.38) this is equivalent to proving that $w^*(\cdot) = -\gamma^{-2}B_1^*\eta^*(\cdot)$. Denoting $w^0(\cdot) := -\gamma^{-2}B_1^*\eta^*(\cdot)$, we shall therefore show that $w^0(\cdot) = w^*(\cdot)$.

Define $(x^0(\cdot), u^0(\cdot), \eta^0(\cdot)) := \mathcal{F}(\xi, w^0(\cdot))$ and $z^0(\cdot) := \mathcal{G}(\xi, w^0(\cdot))$. Consider the time interval $[0,T]$ for some $T > 0$. We claim that

$$\begin{aligned}&\gamma^2\|w^*(\cdot)\|^2_{L_2(0,T;W)} - \|z^*(\cdot)\|^2_{L_2(0,T;Z)} - \gamma^2\|w^0(\cdot)\|^2_{L_2(0,T;W)} + \\ &\|z^0(\cdot)\|^2_{L_2(0,T;Z)} + < \eta^*(T), x^*(T) >_{\mathcal{V}} - < \eta^*(T), x^0(T) >_{\mathcal{V}} = \\ &\gamma^2\|w^0(\cdot) - w^*(\cdot)\|^2_{L_2(0,T;W)} + \|z^0(\cdot) - z^*(\cdot)\|^2_{L_2(0,T;Z)}.\end{aligned} \tag{4.41}$$

In the finite-dimensional case, this result follows from a completion of the squares argument, using the differentiability of $x^0(\cdot), x^*(\cdot)$ and $\eta^*(\cdot)$. However, here these funcions are not differentiable in general (only for smooth input functions and smooth initial conditions). As before, this difficulty can be circumvented by introducing sequences $\xi_n \in D(A^{\mathcal{V}})$ and $w_n^*(\cdot) \in CCD(0,\infty;W)$, such that $\xi_n \xrightarrow{\mathcal{V}} \xi$ and $\|w_n^*(\cdot) - w^*(\cdot)\|_{L_2(0,\infty;W)} \to 0$ as $n \to \infty$. Furthermore, the property of $\eta(\cdot)$ in Lemma 4.7 is needed. A limiting argument (with n tending to ∞) gives formula (4.41). This method was also used in the proof Lemma 4.8 and so we can safely omit the details. In (4.41), we can now take the limit as $T \to \infty$ and obtain (using $\eta^*(T) \xrightarrow{\mathcal{V}} 0$, $x^*(T) \xrightarrow{\mathcal{V}} 0$ and $x^0(T) \xrightarrow{\mathcal{V}} 0$

as $T \to \infty$, see Appendices B.6,B.7 and the argument in the proof of Lemma 4.8)

$$\gamma^2\|w^*(\cdot)\|^2_{L_2(0,\infty;W)} - \|z^*(\cdot)\|^2_{L_2(0,\infty;Z)} - \gamma^2\|w^0(\cdot)\|^2_{L_2(0,\infty;W)} +$$

$$\|z^0(\cdot)\|^2_{L_2(0,\infty;Z)} = \gamma^2\|w^0(\cdot) - w^*(\cdot)\|^2_{L_2(0,\infty;W)} + \|z^0(\cdot) - z^*(\cdot)\|^2_{L_2(0,\infty;Z)}.$$

This can be reformulated as

$$\mathcal{C}(\xi, w^*(\cdot)) - \mathcal{C}(\xi, w^0(\cdot)) =$$

$$\gamma^2\|w^0(\cdot) - w^*(\cdot)\|^2_{L_2(0,\infty;W)} + \|z^0(\cdot) - z^*(\cdot)\|^2_{L_2(0,\infty;Z)},$$

from which we conclude that $w^*(\cdot) = w^0(\cdot) = -\gamma^{-2}B_1^*\eta^*(\cdot)$.

Next we shall give (a sketch of) the proof of uniqueness (we follow the finite-dimensional proof in [70, 86]). Suppose that there exists another element $\bar{w}(\cdot) \in L_2(0,\infty;W)$ that satisfies (4.40). Defining $(\bar{x}(\cdot), \bar{u}(\cdot), \bar{\eta}(\cdot)) := \mathcal{F}(\xi, \bar{w}(\cdot))$ and $\bar{z}(\cdot) := \mathcal{G}(\xi, \bar{w}(\cdot))$ we shall show that $\bar{w}(\cdot) = w^*(\cdot)$. Using some arguments similar to those in the proof of (4.41), it can be shown that for arbitrary $T > 0$

$$\gamma^2\|w^*(\cdot) - \bar{w}(\cdot)\|^2_{L_2(0,T;W)} - \|z^*(\cdot) - \bar{z}(\cdot)\|^2_{L_2(0,T;Z)} +$$

$$2 < \eta^*(T) - \bar{\eta}(T), x^*(T) - \bar{x}(T) >_{\mathcal{V}} =$$

$$-\gamma^2\|w^*(\cdot) - \bar{w}(\cdot)\|^2_{L_2(0,T;W)} + \|z^*(\cdot) - \bar{z}(\cdot)\|^2_{L_2(0,T;Z)}.$$

Using the fact that $\eta^*(T) \xrightarrow{\mathcal{V}} 0$, $x^*(T) \xrightarrow{\mathcal{V}} 0$, $\bar{\eta}(T) \xrightarrow{\mathcal{V}} 0$ and $\bar{x}(T) \xrightarrow{\mathcal{V}} 0$ as $T \to \infty$, it follows that

$$\gamma^2\|w^*(\cdot) - \bar{w}(\cdot)\|^2_{L_2(0,\infty;W)} - \|z^*(\cdot) - \bar{z}(\cdot)\|^2_{L_2(0,\infty;Z)} = 0. \tag{4.42}$$

Since

$$z^*(\cdot) - \bar{z}(\cdot) = \mathcal{G}(\xi, w^*(\cdot)) - \mathcal{G}(\xi, \bar{w}(\cdot)) = \mathcal{G}(0, w^*(\cdot) - \bar{w}(\cdot))$$

(using the linearity of $\mathcal{G}$), (4.42) can be reformulated as

$$\gamma^2\|w^*(\cdot) - \bar{w}(\cdot)\|^2_{L_2(0,\infty;W)} - \|\mathcal{G}(0, w^*(\cdot) - \bar{w}(\cdot))\|^2_{L_2(0,\infty;Z)} = 0.$$

Finally, Lemma 4.9 implies that $w^*(\cdot) = \bar{w}(\cdot)$ and so uniqueness is proved. ∎

Summary 4.13
Before continuing, we recapitulate the results obtained so far: we have solved the sup-inf optimization problem (4.15)

$$\sup_{w(\cdot)\in L_2(0,\infty;W)} \inf_{u(\cdot)\in L_2(0,\infty;U)} J(\xi, w(\cdot), u(\cdot)),$$

for fixed $\xi \in \mathcal{V}$ and J given by (4.14). The infimization part with respect to $u \in L_2(0,\infty;U)$ (for fixed $w(\cdot) \in L_2(0,\infty;W)$ and $\xi \in \mathcal{V}$) was treated in Lemma 4.8 and we characterized the optimal control $\bar{u}(\cdot) = B_2^*\eta(\cdot)$, where $\eta(\cdot)$ is given by (4.19) (and of course $\eta(\cdot)$ depends on $w(\cdot)$ and ξ). Furthermore, the infimum was in fact a minimum. Then in Lemmas 4.11 and 4.12 we treated the part of finding the worst disturbance in

$$\sup_{w(\cdot)\in L_2(0,\infty;W)} J(\xi, w(\cdot), \bar{u}(\cdot))$$

(compare with (4.28) and (4.29)). We have shown that in fact the maximum is attained by a unique $w^*(\cdot)$. Furthermore, $w^*(\cdot) = H\xi$ for some $H \in \mathcal{L}(\mathcal{V}, L_2(0,\infty;W))$ and $w^*(\cdot)$ is the unique solution to $w(\cdot) = -\gamma^{-2}B_1^*\eta(\cdot)$. Finally, we recall the definitions of (4.38) and (4.39): $(x^*(\cdot), u^*(\cdot), \eta^*(\cdot)) = \mathcal{F}(\xi, w^*(\cdot))$ and $z^*(\cdot) = \mathcal{G}(\xi, w^*(\cdot))$ (we shall use these definitions throughout the rest of this section).

The next step is the feedback synthesis part: we show that $\eta^*(\cdot)$ (the adjoint variable $\eta(\cdot)$ corresponding to the worst disturbance) satisfies $\eta^*(\cdot) = -P_0x^*(\cdot)$ for some $P_0 \in \mathcal{L}(\mathcal{V})$. This implies that the optimal control and the worst disturbance are in fact of a feedback form, the gain depending on P_0. Furthermore, we show that this P_0 is nonnegative definite and that it is a stabilizing solution to the Riccati equation (4.11). Let us define $P_0 : \mathcal{V} \to \mathcal{V}$ by

$$P_0\xi := -L\xi - \int_0^\infty S_L^*(\tau)LB_1(H\xi)(\tau)d\tau. \tag{4.43}$$

First of all, we show that P_0 is bounded and derive the relationship between $\eta^*(\cdot)$ and $x^*(\cdot)$:

Lemma 4.14
Let $\xi \in \mathcal{V}$ be arbitrary and let L, H, $w^*(\cdot), \eta^*(\cdot)$ and $x^*(\cdot)$ be defined as in Summary 4.13. The operator P_0 defined by (4.43) satisfies $P_0 \in \mathcal{L}(\mathcal{V})$ and there holds

$$\eta^*(\cdot) = -P_0x^*(\cdot).$$

Proof

Since $L \in \mathcal{L}(\mathcal{V})$, $H \in \mathcal{L}(\mathcal{V}, L_2(0,\infty;W))$ (Lemma 4.11) and $S_L^*(\cdot)$ is exponentially stable on $\mathcal{V}$, we conclude that $P_0 \in \mathcal{L}(\mathcal{V})$.

Next we note that $\eta^*(\cdot)$ and $x^*(\cdot)$ are determined by (4.17),(4.18) and (4.19) where $w(\cdot)$ is replaced by $w^*(\cdot) = H\xi$. This implies in particular that $\eta^*(\cdot)$ and $x^*(\cdot)$ are both continuous functions (with respect to the topology of $\mathcal{V}$) and there holds

$$\eta^*(0) = L\xi + \int_0^\infty S_L^*(\tau) L B_1 (H\xi)(\tau) d\tau$$

and so

$$\eta^*(0) = -P_0\xi = -P_0 x^*(0). \tag{4.44}$$

In order to prove $\eta^*(t) = -P_0 x^*(t)$ for $t > 0$ we use an argument from [86]. Having arrived at $x^*(t)$, at some time $t > 0$, we can consider a new sup-inf optimization problem starting at time t with the initial value $x^*(t) \in \mathcal{V}$, etc. For this problem we also find the optimal control and the worst disturbance as in Lemma 4.11 (use time-invariance). It follows that this new worst disturbance also satisfies equation (4.40) and so the uniqueness result of Lemma 4.12 implies that the new worst disturbance is given by $w^*(\tau)$ for all $\tau \in (t,\infty)$. Hence the adjoint variable and the state corresponding to the new worst disturbance are given by $\eta^*(\cdot)$ and $x^*(\cdot)$, respectively. Since the new sup-inf optimization problem starts at time t with initial condition $x^*(t)$ it follows from (4.44) that $\eta^*(t) = -P_0 x^*(t)$. Since t was arbitrary the result follows. ∎

Next we show that $u^*(\cdot)$ and $w^*(\cdot)$ are in feedback form and that the corresponding closed-loop map is exponentially stable on $\mathcal{V}$:

Lemma 4.15

Let $\xi \in \mathcal{V}$ be arbitrary and suppose that $w^*(\cdot), \eta^*(\cdot), x^*(\cdot)$ and $u^*(\cdot)$ are defined as in Summary 4.13. There holds $u^*(\cdot) = -B_2^* P_0 x^*(\cdot)$ and $w^*(\cdot) = \gamma^{-2} B_1^* P_0 x^*(\cdot)$. Furthermore, $\begin{pmatrix} \gamma^{-2} B_1^* P_0 \\ -B_2^* P_0 \end{pmatrix} \in \mathcal{L}(\mathcal{V}, W \times U)$ is an admissible output operator for the system (4.1) and the C_0-semigroup

$$S_P(\cdot) := S_{\begin{pmatrix} B_1 & B_2 \end{pmatrix} \begin{pmatrix} \gamma^{-2} B_1^* P_0 \\ -B_2^* P_0 \end{pmatrix}}(\cdot)$$

is exponentially stable on $\mathcal{V}$.

Proof

The statements for $u^*(\cdot)$ and $w^*(\cdot)$ follow immediately from Lemmas 4.8, 4.12

and 4.14. In fact, since $x^*(\cdot)$ and $\eta^*(\cdot)$ are both continuous functions with respect to the topology of $\mathcal{V}$ (as explained in the proof of Lemma 4.14), $u^*(\cdot)$ and $w^*(\cdot)$ are also continous (as elements in L_2 they can be chosen as continuous functions).

It is easy to see that $\begin{pmatrix} \gamma^{-2}B_1^*P_0 \\ -B_2^*P_0 \end{pmatrix}$ is an admissible output operator for (4.1) because it is *bounded* on $\mathcal{V}$, and so $S_P(\cdot)$ is a well defined C_0-semigroup on both $\mathcal{W}$ and $\mathcal{V}$ (see Lemma 2.13). Furthermore, we have

$$x^*(t) = S(t)\xi + \int_0^t S(t-s)\,(B_1w^*(s) + B_2u^*(s))\,ds =$$

$$S(t)\xi + \int_0^t S(t-s)\,(\gamma^{-2}B_1B_1^*P_0x^*(s) - B_2B_2^*P_0x^*(s))\,ds.$$

Therefore, $x^*(\cdot) = S_P(\cdot)\xi$ (see Lemma 2.14) and since $x^*(\cdot)$ is in $L_2(0,\infty;\mathcal{V})$ for arbitrary intitial conditions $\xi \in \mathcal{V}$ we conclude that $S_P(\cdot)$ is exponentially stable on $\mathcal{V}$ (use Datko's result given in Lemma A.1). ■

Finally, we show that P_0 is nonnegative definite and that it satisfies the Riccati equation (4.11).

Lemma 4.16
The operator $P_0 \in \mathcal{L}(\mathcal{V})$ defined by (4.43) satisfies

$$P_0 = P_0^* \geq 0$$

and

$$< P_0x, A^{\mathcal{V}}y >_{\mathcal{V}} + < A^{\mathcal{V}}x, P_0y >_{\mathcal{V}} + < P_0(\gamma^{-2}B_1B_1^* - B_2B_2^*)P_0x, y >_{\mathcal{V}}$$
$$+ < C_1x, C_1y >_{\mathcal{Z}} = 0 \text{ for all } x, y \in D(A^{\mathcal{V}}). \tag{4.11}$$

Proof
First we note that the infinitesimal generator of $S_P(\cdot)$ on $\mathcal{V}$ is given by $D(A_P^{\mathcal{V}}) = D(A^{\mathcal{V}})$ and $A_P^{\mathcal{V}}x = (A^{\mathcal{V}} + \gamma^{-2}B_1B_1^*P_0 - B_2B_2^*P_0)x$ for all $x \in D(A_P^{\mathcal{V}})$ (cf. Lemma 2.13 and Lemma 4.15).

Now let $\xi, y \in D(A^{\mathcal{V}}) = D(A_P^{\mathcal{V}})$. Then $x^*(t) = S_P(t)\xi \in D(A_P^{\mathcal{V}})$ for all $t \geq 0$, $x^*(\cdot)$ is strongly continuously differentiable (w.r.t. the topology on $\mathcal{V}$) and, from Lemma 4.14, $\eta^*(\cdot) = -P_0x^*(\cdot)$. Thus we have

$$< y, \frac{d}{dt}\eta^*(t) >_{\mathcal{V}} = < y, -P_0A_P^{\mathcal{V}}S_P(t)\xi >_{\mathcal{V}}. \tag{4.45}$$

Now $\eta^*(\cdot)$ also satisfies (4.19), where $\bar{x}(\cdot)$ is replaced by $x^*(\cdot)$ and $r(\cdot)$ is given by (4.17) with $w^*(\cdot)$. Since $w^*(\cdot) = \gamma^{-2}B_1^*P_0S_P(\cdot)\xi$ it is strongly continuously differentiable and $\dot{w}^*(\cdot) = \gamma^{-2} B_1^* P_0 S_P(\cdot)\, A_P^{\mathcal{V}} \xi \in L_1(0,\infty; W)$, so we can appeal to Lemma 4.7 to obtain

$$< y, \frac{d}{dt}\eta^*(t) >_{\mathcal{V}} = < C_1 y, C_1 x^*(t) >_Z - < A^{\mathcal{V}} y, \eta^*(t) >_{\mathcal{V}} . \tag{4.46}$$

Equating (4.45) and (4.46) for $t = 0$ gives

$$< y, P_0 A_P^{\mathcal{V}} \xi >_{\mathcal{V}} + < A^{\mathcal{V}} y, P_0 \xi >_{\mathcal{V}} + < C_1 y, C_1 \xi >_Z = 0 \tag{4.47}$$

and so

$$< y, P_0 A^{\mathcal{V}} \xi >_{\mathcal{V}} + < A^{\mathcal{V}} y, P_0 \xi >_{\mathcal{V}} +$$

$$< y, P_0(\gamma^{-2} B_1 B_1^* - B_2 B_2^*) P_0 \xi >_{\mathcal{V}} + < C_1 y, C_1 \xi >_Z = 0. \tag{4.48}$$

Using (4.47) we will now prove that P_0 is self-adjoint. For all $\xi, y \in D(A^{\mathcal{V}}) = D(A_P^{\mathcal{V}})$ we have

$$< y, (P_0 - P_0^*) A_P^{\mathcal{V}} \xi >_{\mathcal{V}} + < A_P^{\mathcal{V}} y, (P_0 - P_0^*) \xi >_{\mathcal{V}} = < y, P_0 A_P^{\mathcal{V}} \xi >_{\mathcal{V}} +$$

$$< A^{\mathcal{V}} y, P_0 \xi >_{\mathcal{V}} + < (\gamma^{-2} B_1 B_1^* - B_2 B_2^*) P_0 y, P_0 \xi >_{\mathcal{V}} - < P_0 y, A^{\mathcal{V}} \xi >_{\mathcal{V}}$$

$$- < P_0 A_P^{\mathcal{V}} y, \xi >_{\mathcal{V}} - < P_0 y, (\gamma^{-2} B_1 B_1^* - B_2 B_2^*) P_0 \xi >_{\mathcal{V}}$$

$$= < C_1 \xi, C_1 y >_Z - < C_1 y, C_1 \xi >_Z = 0.$$

This implies that $\frac{d}{dt} < S_P(t) y, (P_0 - P_0^*) S_P(t) \xi >_{\mathcal{V}} = 0$ and so we see that for all $\xi, y \in D(A^{\mathcal{V}})$, $< S_P(t) y, (P_0 - P_0^*) S_P(t) \xi >_{\mathcal{V}} = < y, (P_0 - P_0^*) \xi >_{\mathcal{V}}$. But since $S_P(\cdot)$ is exponentially stable on $\mathcal{V}$ and $D(A^{\mathcal{V}})$ is dense in $\mathcal{V}$, this implies that $P_0 = P_0^*$.
Now (4.11) follows immediately from (4.48).
Finally, we have to show that P_0 is nonnegative definite. Using $P_0 = P_0^*$ and (4.11) we compute for $\xi \in D(A^{\mathcal{V}}) = D(A_P^{\mathcal{V}})$

$$\frac{d}{dt} < x^*(t), P_0 x^*(t) >_{\mathcal{V}} =$$

$$< A_P^{\mathcal{V}} x^*(t), P_0 x^*(t) >_{\mathcal{V}} + < P_0 x^*(t), A_P^{\mathcal{V}} x^*(t) >_{\mathcal{V}} =$$

$$\gamma^{-2} < B_1^* P_0 x^*(t), B_1^* P_0 x^*(t) >_W - < C_1 x^*(t), C_1 x^*(t) >_Z -$$

$$< B_2^* P_0 x^*(t), B_2^* P_0 x^*(t) >_U \quad \text{(from (4.11))}$$

$$= \gamma^2 < w^*(t), w^*(t) >_W - < C_1 x^*(t), C_1 x^*(t) >_Z - < u^*(t), u^*(t) >_U$$

(from Lemma 4.15)

$$= \gamma^2 < w^*(t), w^*(t) >_W - < z^*(t), z^*(t) >_Z \tag{4.49}$$

(using (4.13)).

Since $S_P(\cdot)$ is exponentially stable on $\mathcal{V}$, we may integrate (4.49) over $(0, \infty)$ to obtain

$$< \xi, P_0\xi >_{\mathcal{V}} = \|z^*(\cdot)\|^2_{L_2(0,\infty;Z)} - \gamma^2\|w^*(\cdot)\|^2_{L_2(0,\infty;W)} = -C(\xi, w^*(\cdot)) \tag{4.50}$$

(in the last step we used the definition of $C(\xi, w(\cdot))$ in (4.29) and the fact that $z^*(\cdot) = \mathcal{G}(\xi, w^*(\cdot))$). Since $w^*(\cdot)$ is such that

$$\inf_{w(\cdot)\in L_2(0,\infty;W)} \mathcal{C}(\xi, w(\cdot)) = \mathcal{C}(\xi, w^*(\cdot))$$

(this is the result of Lemma 4.11), it follows that

$$\mathcal{C}(\xi, w^*(\cdot)) \leq \mathcal{C}(\xi, 0) = -\|\mathcal{G}(\xi, 0)\|^2_{L_2(0,\infty;Z)} \leq 0. \tag{4.51}$$

Finally, combination of (4.50) and (4.51) implies that $< \xi, P_0\xi >_{\mathcal{V}} \geq 0$ for all $\xi \in D(A^{\mathcal{V}})$ and since $D(A^{\mathcal{V}})$ is dense in $\mathcal{V}$ we infer that $P_0 \geq 0$. ∎

So far, we have proved that there exists a $P_0 \in \mathcal{L}(\mathcal{V})$ with $P_0 = P_0^* \geq 0$ satisfying (4.11), with the additional property that the C_0-semigroup $S_P(\cdot)$ is exponentially stable on $\mathcal{V}$. Defining $P := cP_0$, where $c = i_{\mathcal{H}} \mid_{\mathcal{V}'} (i_{\mathcal{V}})^{-1}$ is the isometry from $\mathcal{V}$ to $\mathcal{V}'$, it follows that $P \in \mathcal{L}(\mathcal{V}, \mathcal{V}')$ is nonnegative definite, that P satisfies (4.9) and that

$$S_P(\cdot) = S_{\begin{pmatrix} B_1 & B_2 \end{pmatrix}\begin{pmatrix} \gamma^{-2}B_1^*P_0 \\ -B_2^*P_0 \end{pmatrix}}(\cdot) = S_{\begin{pmatrix} B_1 & B_2 \end{pmatrix}\begin{pmatrix} \gamma^{-2}B_1'P \\ -B_2'P \end{pmatrix}}(\cdot),$$

is exponentially stable on $\mathcal{V}$ (note that assumption (4.12) simplifies the formulas in Theorem 4.4 considerably). This concludes part a) (i.e. the proof of the implication $(ii) \Rightarrow (iii)$), under assumption (4.12).

b) Proof of $(iii) \Rightarrow (i)$

We suppose that (iii) holds and define $P_0 := c^{-1}P$ (recall that $c = i_{\mathcal{H}} \mid_{\mathcal{V}'} (i_{\mathcal{V}})^{-1}$ is the isometry from $\mathcal{V}$ to $\mathcal{V}'$). It follows that $P_0 \in \mathcal{L}(\mathcal{V})$ is nonnegative definite and that it satisfies the Riccati equation (4.11) such that $A^{\mathcal{V}} + \gamma^{-2}B_1B_1^*P_0 - B_2B_2^*P_0$ is the generator of the C_0-semigroup $S_P(\cdot)$ which is exponentially stable on $\mathcal{V}$ (again, we note that (4.12) simplifies the formulas considerably).

We want to find an admissible output operator $F \in \mathcal{L}(\mathcal{W}, U)$ such that the C_0-semigroup $S_{B_2F}(\cdot)$ is exponentially stable on $\mathcal{V}$ and the closed-loop map G_{B_2F} defined by (4.3) satisfies $\|G_{B_2F}\| < \gamma$. In the following lemma we show that the feedback $F = -B_2'P = -B_2^*P_0$ satisfies these conditions. We note that because of the definition of P_0 and the simplifying assumption (4.12), there holds $F = -B_2'P = -(D_{12}'D_{12})^{-1}(B_2'P + D_{12}'C_1)$, just as in (4.10).

Lemma 4.17

The state-feedback $u(\cdot) = Fx(\cdot) = -B_2'Px(\cdot) = -B_2^*P_0x(\cdot)$, with P and P_0 as above, is such that $S_{B_2F}(\cdot)$ is exponentially stable on $\mathcal{V}$ and guarantees the closed-loop inequality $\|G_{B_2F}\| < \gamma$.

Proof

Note that $F = -B_2^*P_0 \in \mathcal{L}(\mathcal{V}, U)$ is an admissible output operator because it is bounded on $\mathcal{V}$. First we show that the semigroup $S_{B_2F}(\cdot)$ generated by $A^{\mathcal{V}} - B_2B_2^*P_0$ is exponentially stable on $\mathcal{V}$. From the Riccati equation (4.11) we infer that for all $x, y \in D(A^{\mathcal{V}})$

$$< (A^{\mathcal{V}} - B_2B_2^*P_0)x, P_0y >_{\mathcal{V}} + < P_0x, (A^{\mathcal{V}} - B_2B_2^*P_0)y >_{\mathcal{V}}=$$

$$- < C_1x, C_1y >_Z - < P_0(B_2B_2^* + \gamma^{-2}B_1B_1^*)P_0x, y >_{\mathcal{V}}\leq$$

$$- < \gamma^{-2}P_0B_1B_1^*P_0x, y >_{\mathcal{V}} .$$

Since $A^{\mathcal{V}} - B_2B_2^*P_0 = A_P^{\mathcal{V}} - \gamma^{-2}B_1B_1^*P_0$, and it is given that the semigroup generated by $A_P^{\mathcal{V}}$ is exponentially stable on $\mathcal{V}$, we see that the pair $(A^{\mathcal{V}} - B_2B_2^*P_0, B_1^*P_0)$ is exponentially detectable on $\mathcal{V}$. Hence, the fact that $S_{B_2F}(\cdot)$ is exponentially stable on $\mathcal{V}$ follows from Zabczyk's Lyapunov type result in Lemma A.3. Next we show that G_{B_2F} given by

$$\begin{cases} x(t) = \int_0^t S_{B_2F}(t-s)B_1w(s)ds \\ (G_{B_2F}w)(t) = z(t) = (C_1 - D_{12}B_2^*P_0)x(t) \quad t \geq 0, \end{cases}$$

where $w(\cdot) \in L_2(0,\infty; W)$, satisfies $\|G_{B_2F}\| < \gamma$.

First of all, we note that since $S_{B_2F}(\cdot)$ is exponentially stable on $\mathcal{V}$ it follows that $x(\cdot) \in L_2(0,\infty;\mathcal{V})$, $x(t) \xrightarrow{\mathcal{V}} 0$ as $t \to \infty$ (see Appendix B.1) and we have $G_{B_2F} \in \mathcal{L}(L_2(0,\infty; W), L_2(0,\infty; Z))$ (see Lemma 3.2).

Suppose that, in addition, $w(\cdot) \in L_2(0,\infty; W)$ is continuously differentiable and denote $w_0(\cdot) := w(\cdot) - \gamma^{-2}B_1^*P_0x(\cdot) \in L_2(0,\infty; W)$. Using Appendix B.3, we calculate

$$\frac{d}{dt} < x(t), P_0x(t) >_{\mathcal{V}}=$$

$$< (A^{\mathcal{V}} - B_2B_2^*P_0)x(t), P_0x(t) >_{\mathcal{V}} + < B_1w(t), P_0x(t) >_{\mathcal{V}} +$$

$$< P_0x(t), (A^{\mathcal{V}} - B_2B_2^*P_0)x(t) >_{\mathcal{V}} + < P_0x(t), B_1w(t) >_{\mathcal{V}}=$$

$$- < C_1x(t), C_1x(t) >_Z - < B_2^*P_0x(t), B_2^*P_0x(t) >_U -$$

$$\gamma^2 < w(t) - \gamma^{-2}B_1^*P_0x(t), w(t) - \gamma^{-2}B_1^*P_0x(t) >_W +$$

$$\gamma^2 < w(t), w(t) >_W \text{ (using (4.11))} = - < z(t), z(t) >_Z$$

$$-\gamma^2 < w_0(t), w_0(t) >_W + \gamma^2 < w(t), w(t) >_W,$$

using (4.13) and the definition of $w_0(\cdot)$ above. Since $x(t) \xrightarrow{\mathcal{V}} 0$ as $t \to \infty$ (see Appendix B.2) and $x(0) = 0$, we may integrate this over $(0, \infty)$ to obtain

$$\|z(\cdot)\|^2_{L_2(0,\infty;Z)} - \gamma^2 \|w(\cdot)\|^2_{L_2(0,\infty;W)} = -\gamma^2 \|w_0(\cdot)\|^2_{L_2(0,\infty;W)}. \tag{4.52}$$

Furthermore, using Lemma 2.14, $x(t)$ can also be written as

$$x(t) = \int_0^t S_P(t-s) B_1 w_0(s) ds,$$

where $S_P(\cdot)$ is the semigroup generated by $A_P^{\mathcal{V}} = A^{\mathcal{V}} - B_2 B_2^* P_0 + \gamma^{-2} B_1 B_1^* P_0$, which is exponentially stable on $\mathcal{V}$. The exponential stability of $S_P(\cdot)$ on $\mathcal{V}$ implies the boundedness of the mapping $w_0(\cdot) \in L_2(0, \infty; W) \mapsto x(\cdot) \in L_2(0, \infty; \mathcal{V})$ and therefore also of the mapping $w_0(\cdot) \in L_2(0, \infty; W) \mapsto w(\cdot) = w_0(\cdot) + \gamma^{-2} B_1^* P_0 x(\cdot) \in L_2(0, \infty; W)$. Hence, there exists a positive constant $\delta > 0$ independent of $w(\cdot)$ or $w_0(\cdot)$, such that

$$\|w(\cdot)\|_{L_2(0,\infty;W)} \leq \gamma/\delta \|w_0(\cdot)\|_{L_2(0,\infty;W)}.$$

Combining this with (4.52) yields

$$\|z(\cdot)\|^2_{L_2(0,\infty;Z)} = \gamma^2 \|w(\cdot)\|^2_{L_2(0,\infty;W)} - \gamma^2 \|w_0(\cdot)\|^2_{L_2(0,\infty;W)}$$

$$\leq (\gamma^2 - \delta^2) \|w(\cdot)\|^2_{L_2(0,\infty;W)}. \tag{4.53}$$

As before, (4.53) can be extended to the general case of $w(\cdot) \in L_2(0, \infty; W)$ by the introduction of an approximating sequence of smooth $w_n(\cdot)$ which converges to $w(\cdot)$. Using Appendix B.6 (4.53) then holds for $w_n(\cdot)$ and we can take the limit for $n \to \infty$ to obtain

$$\|(G_{B_2F} w)(\cdot)\|^2_{L_2(0,\infty;Z)} = \|z(\cdot)\|^2_{L_2(0,\infty;Z)} \leq (\gamma^2 - \delta^2) \|w(\cdot)\|^2_{L_2(0,\infty;W)},$$

for all $w(\cdot) \in L_2(0, \infty; W)$. Hence, $\|G_{B_2F}\| < \gamma$. ■

This concludes part *b*) (the proof of (*iii*) $\Rightarrow$ (*i*)), under assumption (4.12).

To complete the proof of Theorem 4.4, we show that the implications (*ii*) $\Rightarrow$ (*iii*) and (*iii*) $\Rightarrow$ (*i*) remain true without the extra assumption (4.12).
The idea is to apply a preliminary feedback

$$u(\cdot) = \tilde{F} x(\cdot) + (D_{12}' D_{12})^{-\frac{1}{2}} v(\cdot), \tag{4.54}$$

where $v(\cdot)$ is a new control input and $\tilde{F}$ is given by

$$\tilde{F} := -(D_{12}'D_{12})^{-1}D_{12}'C_1. \tag{4.55}$$

Note that $\tilde{F} \in \mathcal{L}(\mathcal{W}, U)$ is an admissible output operator because $C_1 \in \mathcal{L}(\mathcal{W}, Z)$ is admissible and $(D_{12}'D_{12})^{-1}D_{12}' \in \mathcal{L}(Z, U)$, so that we can use Lemma 2.14. With the input $u(\cdot) = \tilde{F}x(\cdot) + (D_{12}'D_{12})^{-\frac{1}{2}}v(\cdot)$, the system (4.1) transforms into the smooth Pritchard-Salamon system $\Sigma_{\bar{G}} = \Sigma(\bar{S}(\cdot), (\bar{B}_1 \;\; \bar{B}_2), \bar{C}_1, (0 \;\; \bar{D}_{12}))$ of the form

$$\begin{cases} x(t) &= \bar{S}(t)x_0 + \int_0^t \bar{S}(t-s)(\bar{B}_1 w(s) + \bar{B}_2 v(s))ds \\ z(t) &= \bar{C}_1 x(t) + \bar{D}_{12}v(t), \end{cases} \tag{4.56}$$

where

$$\begin{aligned} \bar{S}(\cdot) &= S_{B_2\tilde{F}}(\cdot) \\ \bar{B}_2 &= B_2(D_{12}'D_{12})^{-\frac{1}{2}} \\ \bar{B}_1 &= B_1 \\ \bar{C}_1 &= (I - D_{12}(D_{12}'D_{12})^{-1}D_{12}')C_1 = C_1 + D_{12}\tilde{F} \\ \bar{D}_{12} &= D_{12}(D_{12}'D_{12})^{-\frac{1}{2}}. \end{aligned}$$

Note that we need the assumption that $D_{12}'D_{12}$ is coercive to guarantee the existence of a bounded inverse and that $\bar{C}_1$ is an admissible output operator for both $S(\cdot)$ and $\bar{S}(\cdot)$. Lemma 2.14 implies that with the input

$$v(\cdot) := (D_{12}'D_{12})^{\frac{1}{2}}(u(\cdot) - \tilde{F}x(\cdot)) \tag{4.57}$$

in the transformed system (4.56), we obtain the original system (4.1). Furthermore, we see that with the state-feedbacks $u(\cdot) = F_1 x(\cdot)$ in system (4.1) and $v(\cdot) = F_2 x(\cdot)$ in system (4.56), where $F_1 \in \mathcal{L}(\mathcal{W}, U)$ and $F_2 \in \mathcal{L}(\mathcal{W}, U)$ are admissible output operators related by

$$F_1 = (D_{12}'D_{12})^{-\frac{1}{2}}F_2 + \tilde{F} \text{ or } F_2 = (D_{12}'D_{12})^{\frac{1}{2}}(F_1 - \tilde{F}),$$

we have

$$\bar{S}_{\bar{B}_2F_2}(\cdot) = S_{B_2F_1}(\cdot) \text{ and } \bar{G}_{B_2F_2} = G_{B_2F_1}.$$

It is easy to see that the system $\Sigma_{\bar{G}}$ given by (4.56) satisfies the assumptions (4.4) and (4.12) (assumption (4.4) is invariant under state-feedback, as is shown in Appendix C).

Now suppose that *(ii)* is satisfied for Σ_G. It follows from the above that there exists an admissible output operator $F \in \mathcal{L}(\mathcal{W}, U)$ such that $\bar{S}_{\bar{B}_2F}(\cdot)$ is exponentially stable on $\mathcal{V}$. Furthermore, using the relation $v(\cdot) = (D_{12}'D_{12})^{\frac{1}{2}}(u(\cdot) - \tilde{F}x(\cdot))$ it is easy to see that *(ii)* is also satisfied for $\Sigma_{\bar{G}}$.

Hence, using the fact that we have proved Theorem 4.4 under the extra assumption that (4.12) holds, we conclude that there exists a nonnegative definite $P \in \mathcal{L}(\mathcal{V}, \mathcal{V}')$ such that for all $x, y \in D(A^{\mathcal{V}})$

$$< Px, \bar{A}^{\mathcal{V}} y >_{<\mathcal{V}',\mathcal{V}>} + < \bar{A}^{\mathcal{V}} x, Py >_{<\mathcal{V},\mathcal{V}'>} +$$

$$< P(\gamma^{-2}\bar{B}_1\bar{B}_1' - \bar{B}_2\bar{B}_2')Px, y >_{<\mathcal{V}',\mathcal{V}>} + < \bar{C}_1 x, \bar{C}_1 y >_Z = 0$$

and $\bar{A}^{\mathcal{V}} + (\gamma^{-2}\bar{B}_2\bar{B}_2' - \bar{B}_1\bar{B}_1')P$ is the generator of an exponentially stable semigroup exponentially stable on $\mathcal{V}$ (cf. Theorem 4.4). It is now clear from the definition of Σ_G in (4.56) and the perturbation results of Lemma 2.13 that item *(iii)* of Theorem 4.4 is satisfied.

Now suppose that *(iii)* is satisfied for Σ_G. Defining $\Sigma_{\bar{G}}$ as in 4.56 it follows immediately from the perturbation results in Lemma 2.13 that $\Sigma_{\bar{G}}$ also satisfies item *(iii)* of Theorem 4.4. Hence, the feedback $v(\cdot) = F_2 x(\cdot)$, with $F_2 := -\bar{B}_2' P$, applied to (4.56) is such that $\bar{S}_{\bar{B}_2 F_2}(\cdot)$ is exponentially stable on $\mathcal{V}$ and $\|\bar{G}_{\bar{B}_2 F_2}\| < \gamma$. It follows from the above that the feedback $u(\cdot) = F_1 x(\cdot)$ applied to system (4.1), with $F_1 = -(D_{12}'D_{12})^{-1}(B_2'P + D_{12}'C_1)x(\cdot)$, is such that $S_{B_2F_1}(\cdot)$ is exponentially stable on $\mathcal{V}$ and $\|G_{B_2F_1}\| < \gamma$, so that item *(i)* is satisfied for Σ_G.

Finally, the uniqueness of P follows immediately from Lemma 2.33 and this concludes the proof of Theorem 4.4.

4.3 Relaxation of the a priori assumptions

In this section we show how we can remove the assumptions (4.4), (4.5) and (4.6). In the first part we show how an $\mathcal{H}_\infty$-control problem with state-feedback with feedthrough from disturbance to the to-be-controlled output (i.e. $D_{11} \neq 0$) can be reduced to an $\mathcal{H}_\infty$-control problem without this feedthrough. In the second part we show that assumptions (4.4) and (4.5) can be removed by introducing a 'regularizing parameter'. In order to reduce the notational burden we assume in this section that $\gamma = 1$ (as usual, the general case can be obtained by scaling).

4.3.1 Feedthrough from disturbance to output

Let

$$\Sigma_G = \Sigma(S(\cdot), (B_1 \;\; B_2), C_1, (\; D_{11} \;\; D_{12} \;))$$

given by

$$\Sigma_G : \begin{cases} x(t) &= S(t)x_0 + \int_0^t S(t-s)(B_1 w(s) + B_2 u(s))ds \\ z(t) &= C_1 x(t) + D_{11} w(t) + D_{12} u(t), \end{cases} \tag{4.58}$$

be a smooth Pritchard-Salamon system as in (4.1).

In principle, we could try to derive a result for Σ_G as we had for the system Σ_G with $D_{11} = 0$ in Theorem 4.4, but there is a problem with condition *(ii)* of that theorem. Consider the finite-dimensional system

$$\begin{cases} \dot{x} &= -x \\ z &= w + u. \end{cases}$$

It is easy to see that for all $w(\cdot) \in L_2(0,\infty)$ there exists a $u(\cdot) \in L_2(0,\infty)$ such that $x(\cdot) \in L_2(0,\infty)$ and $z(\cdot) = 0$ (take $u(\cdot) = -w(\cdot)$), but there exists no stabilizing *state-feedback* controller that makes the closed-loop norm equal to 0 (for $x_0 = 0$ we have $u(\cdot) = Fx(\cdot) = 0$ so that $z(\cdot) = w(\cdot)$).

However, for Σ_G we *can* derive the equivalence of the existence of an admissible state-feedback controller with the solvability of a certain Riccati equation. First of all, we recall that for an admissible state-feedback controller there holds $\|G_{B_2F}\| = \|\tilde{G}_{B_2F}\|_\infty < 1$ and because of (2.23) and (4.3) this implies that $\|D_{11}\| < 1$. Apparently, $\|D_{11}\| < 1$ is a necessary condition for the existence of an admissible state-feedback. Now suppose that $\|D_{11}\| < 1$ and define the transformed Pritchard-Salamon system without feedthrough

$$\Sigma_{\hat{G}} = \Sigma(\hat{S}(\cdot), (\hat{B}_1 \;\; \hat{B}_2), \hat{C}_1, (\;0 \;\; \hat{D}_{12}\;))$$

given by

$$\Sigma_{\hat{G}} : \begin{cases} \hat{x}(t) &= \hat{S}(t)\hat{x}_0 + \int_0^t \hat{S}(t-s)(\hat{B}_1 \hat{w}(s) + \hat{B}_2 \hat{u}(s))ds \\ \hat{z}(t) &= \hat{C}_1 \hat{x}(t) + \hat{D}_{12} \hat{u}(t), \end{cases}$$

where

$$\left.\begin{aligned} \hat{S}(\cdot) &:= S_{B_1(I-D_{11}'D_{11})^{-1}D_{11}'C_1}(\cdot) \\ \hat{B}_1 &:= B_1(I - D_{11}'D_{11})^{-\frac{1}{2}} \\ \hat{B}_2 &:= B_2 + B_1(I - D_{11}'D_{11})^{-1}D_{11}'D_{12} \\ \hat{C}_1 &:= (I - D_{11}D_{11}')^{-\frac{1}{2}}C_1 \\ \hat{D}_{12} &:= (I - D_{11}D_{11}')^{-\frac{1}{2}}D_{12}. \end{aligned}\right\} \tag{4.59}$$

We note that because of Lemma 2.13, $\Sigma_{\hat{G}}$ is also a smooth Pritchard-Salamon system. The following holds.

Lemma 4.18
Let $F \in \mathcal{L}(\mathcal{W}, U)$ be an admissible output operator. The state-feedback $u(\cdot) = Fx(\cdot)$ is admissible for Σ_G if and only if $\|D_{11}\| < 1$ and $\hat{u}(\cdot) = F\hat{x}(\cdot)$ is admissible for $\Sigma_{\hat{G}}$.

The proof of this lemma will follow from (the proof of) Lemma 5.17, where we shall transform an $\mathcal{H}_\infty$-control problem with measurement-feedback with feedthrough from the disturbance to the to-be-controlled output to an equivalent problem without this feedthrough term. We note that similar transformations have been used in [36, 79] for the finite-dimensional case.

In order to derive a result like Theorem 4.4 for Σ_G, we show that if the regularity assumptions (4.4) and (4.5) are satisfied for Σ_G, then the corresponding assumptions for $\Sigma_{\hat{G}}$ are also satisfied:

Lemma 4.19
Suppose that $\|D_{11}\| < 1$ and that the pair $(A^{\mathcal{V}}, B_2)$ is exponentially stabilizable. If Σ_G satisfies the regularity assumptions (4.4) and (4.5), then

(i) $\hat{D}_{12}'\hat{D}_{12}$ is coercive,

(ii) there exists an $\epsilon > 0$ such that for all $(\omega, x, u) \in \mathbb{R} \times D(\hat{A}^{\mathcal{V}}) \times U$ satisfying $i\omega x = \hat{A}^{\mathcal{V}}x + \hat{B}_2 u$, there holds

$$\|\hat{C}_1 x + \hat{D}_{12} u\|_Z^2 \geq \epsilon \|x\|_{\mathcal{V}}^2.$$

Proof
It follows immediately from the definition of $\hat{D}_{12}$ that coerciveness of $D_{12}'D_{12}$ implies that $\hat{D}_{12}'\hat{D}_{12}$ is coercive. Furthermore, it is straightforward to show that for all $\omega \in \mathbb{R}$

$$\begin{pmatrix} I & B_1 D_{11}'(I - D_{11}D_{11}')^{-\frac{1}{2}} \\ 0 & (I - D_{11}D_{11}')^{\frac{1}{2}} \end{pmatrix} \begin{pmatrix} i\omega I - \hat{A}^{\mathcal{V}} & -\hat{B}_2 \\ \hat{C}_1 & \hat{D}_{12} \end{pmatrix} =$$

$$\begin{pmatrix} i\omega I - A^{\mathcal{V}} & -B_2 \\ C_1 & D_{12} \end{pmatrix}$$

(use Lemma 2.13). Since

$$\begin{pmatrix} I & B_1 D_{11}'(I - D_{11}D_{11}')^{-\frac{1}{2}} \\ 0 & (I - D_{11}D_{11}')^{\frac{1}{2}} \end{pmatrix} \in \mathcal{L}(\mathcal{V} \times Z) \text{ and}$$

$$\begin{pmatrix} I & B_1 D_{11}'(I - D_{11}D_{11}')^{-\frac{1}{2}} \\ 0 & (I - D_{11}D_{11}')^{\frac{1}{2}} \end{pmatrix}^{-1} \in \mathcal{L}(\mathcal{V} \times Z),$$

the rest of the proof follows from Appendix C.1. ■

The next theorem now follows from Theorem 4.4 and Lemmas 4.18 and 4.19 and requires no further proof.

Theorem 4.20
Suppose that the smooth Pritchard-Salamon system Σ_G given by (4.58) satisfies the regularity assumptions (4.4) and (4.5). The following are equivalent statements:

(*i*) *there exists an admissible state-feedback controller $u(\cdot) = Fx(\cdot)$ for Σ_G,*

(*ii*) *$\|D_{11}\| < 1$ and with the definitions of (4.59) there exists a $P \in \mathcal{L}(\mathcal{V}, \mathcal{V}')$ with $P = P' \geq 0$ satisfying*

$$< Px, (\hat{A}^{\mathcal{V}} - \hat{B}_2(\hat{D}_{12}'\hat{D}_{12})^{-1}\hat{D}_{12}'\hat{C}_1)y >_{<\mathcal{V}',\mathcal{V}>} +$$

$$< (\hat{A}^{\mathcal{V}} - \hat{B}_2(\hat{D}_{12}'\hat{D}_{12})^{-1}\hat{D}_{12}'\hat{C}_1)x, Py >_{<\mathcal{V},\mathcal{V}'>} +$$

$$< P(\gamma^{-2}\hat{B}_1\hat{B}_1' - \hat{B}_2(\hat{D}_{12}'\hat{D}_{12})^{-1}\hat{B}_2')Px, y >_{<\mathcal{V}',\mathcal{V}>} +$$

$$< (I - \hat{D}_{12}(\hat{D}_{12}'\hat{D}_{12})^{-1}\hat{D}_{12}')\hat{C}_1x, (I - \hat{D}_{12}(\hat{D}_{12}'\hat{D}_{12})^{-1}\hat{D}_{12}')\hat{C}_1y >_Z = 0$$

for all $x, y \in D(\hat{A}^{\mathcal{V}}) = D(A^{\mathcal{V}})$, such that $\hat{A}_P^{\mathcal{V}}$ given by $D(\hat{A}_P^{\mathcal{V}}) = D(A^{\mathcal{V}})$ and

$$\hat{A}_P^{\mathcal{V}} = \hat{A}^{\mathcal{V}} - \hat{B}_2(\hat{D}_{12}'\hat{D}_{12})^{-1}\hat{D}_{12}'\hat{C}_1 + (\gamma^{-2}\hat{B}_1\hat{B}_1' - \hat{B}_2(\hat{D}_{12}'\hat{D}_{12})^{-1}\hat{B}_2')P$$

is the infinitesimal generator of a C_0-semigroup $\hat{S}_P(\cdot)$ which is exponentially stable on $\mathcal{V}$.

Furthermore, in this case the state-feedback $u(\cdot) = Fx(\cdot)$, where $F \in \mathcal{L}(\mathcal{W}, U)$ is the admissible output operator given by

$$F = -(\hat{D}_{12}'\hat{D}_{12})^{-1}(\hat{B}_2'P + \hat{D}_{12}'\hat{C}_1),$$

is an admissible state-feedback.

4.3.2 How to 'remove' the regularity assumptions

Here we show that the regularity assumptions (4.4) and (4.5) can be removed. Actually we do not really remove the regularity assumptions, because we have to allow for an extra parameter in the problem. Similar tricks have been used in [52, 104] for the finite-dimensional case and in [74, 88] for infinite-dimensional $\mathcal{H}_\infty$-control problems.

Consider the smooth Pritchard-Salamon system Σ_G of the form (4.58) and define the transformed Pritchard-Salamon system

$$\Sigma_{G_\epsilon} = \Sigma(S(\cdot), (B_1 \quad B_2), C_{1\epsilon}, (\ D_{11\epsilon} \quad D_{12\epsilon}\ ));$$

$$\Sigma_{G_\epsilon} : \begin{cases} x(t) &= S(t)x_0 + \int_0^t S(t-s)(B_1 w(s) + B_2 u(s))ds \\ z_\epsilon(t) &= C_{1\epsilon}x(t) + D_{11\epsilon}w(t) + D_{12\epsilon}u(t), \end{cases} \tag{4.60}$$

where

$$C_{1\epsilon} = \begin{pmatrix} C_1 \\ \epsilon I_{\mathcal{V}} \\ 0 \end{pmatrix}, \ D_{11\epsilon} = \begin{pmatrix} D_{11} \\ 0 \\ 0 \end{pmatrix} \text{ and } D_{12\epsilon} = \begin{pmatrix} D_{12} \\ 0 \\ \epsilon I_U \end{pmatrix},$$

for some $\epsilon > 0$. It is easy to see that Σ_{G_ϵ} is again a smooth Pritchard-Salamon system. Furthermore, it satisfies the regularity assumptions (4.4) and (4.5) because $D'_{12\epsilon}D_{12\epsilon} \geq \epsilon^2 I_U$ and for all $x \in \mathcal{W}$ and $u \in U$ we have $\|C_{1\epsilon}x + D_{12\epsilon}u\|^2_{Z\times\mathcal{V}\times U} \geq \epsilon^2\|x\|^2_{\mathcal{V}}$. Furthermore, comparison of the original system (4.58) and the transformed system (4.60) shows that if $u(\cdot) \in L_2(0,\infty;U)$ and $w(\cdot) \in L_2(0,\infty;W)$ are such that $x(\cdot) \in L_2(0,\infty;\mathcal{V})$, then the to-be-controlled output z_ϵ satisfies

$$\|z_\epsilon(\cdot)\|^2_{L_2(0,\infty;Z\times\mathcal{V}\times U)} =$$

$$\|z(\cdot)\|^2_{L_2(0,\infty;Z)} + \epsilon^2\|x(\cdot)\|^2_{L_2(0,\infty;\mathcal{V})} + \epsilon^2\|u(\cdot)\|^2_{L_2(0,\infty;U)}. \tag{4.61}$$

We have the following result.

Theorem 4.21
Let $F \in \mathcal{L}(\mathcal{W}, U)$ be an admissible output operator. If the state-feedback $u(\cdot) = Fx(\cdot)$ is admissible for Σ_G, then there exists an $\epsilon_1 > 0$ such that for all $0 \leq \epsilon \leq \epsilon_1$ the same state-feedback is admissible for Σ_{G_ϵ}. Conversely, if for some $\epsilon \geq 0$, $u(\cdot) = Fx(\cdot)$ is admissible for Σ_{G_ϵ} then the same state-feedback is admissible for Σ_G.

Proof

The stability of the closed-loop systems does not depend on the operators corresponding to the to-be-controlled output, so we only have to worry about the norms of the closed-loop maps, G_{B_2F} and $(G_\epsilon)_{B_2F}$ (cf. (4.3)).

Suppose that $u(\cdot) = Fx(\cdot)$ is an admissible state-feedback for Σ_G. Then there exists some $\delta > 0$ such that $\|G_{B_2F}\|^2 < 1-\delta$, so that for zero initial conditions and $u(\cdot) = Fx(\cdot)$ we have $\|z(\cdot)\|^2_{L_2(0,\infty;Z)} \leq (1-\delta)\|w(\cdot)\|^2_{L_2(0,\infty;W)}$. Furthermore, it follows from Lemma 3.2 that there exists an $\epsilon_1 > 0$ such that for zero initial conditions $\epsilon_1^2\|x(\cdot)\|^2_{L_2(0,\infty;\mathcal{V})} + \epsilon_1^2\|u(\cdot)\|^2_{L_2(0,\infty;U)} \leq \frac{1}{2}\delta\|w(\cdot)\|^2_{L_2(0,\infty;W)}$. Now (4.61) implies that for all $0 \leq \epsilon \leq \epsilon_1$ and all $w(\cdot) \in L_2(0,\infty;W)$

$$\|z_\epsilon(\cdot)\|^2_{L_2(0,\infty;Z\times\mathcal{V}\times U)} \leq \|z_{\epsilon_1}(\cdot)\|^2_{L_2(0,\infty;Z\times\mathcal{V}\times U)} \leq (1-\frac{1}{2}\delta)\|w(\cdot)\|^2_{L_2(0,\infty;W)}$$

and therefore $\|(G_\epsilon)_{B_2F}\| < 1$ for all $0 \leq \epsilon \leq \epsilon_1$.

The second part of the theorem follows from the observation that (4.61) implies that $\|G_{B_2F}\| \leq \|(G_\epsilon)_{B_2F}\|$. ∎

Since Σ_{G_ϵ} satisfies the regularity assumptions, Theorems 4.20 and 4.21 can be used to derive a result of the following form: there exists an admissible state-feedback $u(\cdot) = Fx(\cdot)$ for Σ_G if and only if there exists a stabilizing solution of a Riccati equation parametrized by a sufficiently small ϵ. This means that in general the Riccati equation is parametrized by *two* parameters (γ and ϵ), which is less attractive from an applications point of view. For the finite-dimensional case, other approaches to remove the regularity conditions can be found in Stoorvogel [86] and Scherer [83]. The extension of these ideas to the infinite-dimensional case remains an open problem.

Chapter 5

$\mathcal{H}_\infty$-control with measurement-feedback

In this chapter we present the main result of the book: the state-space solution to the sub-optimal $\mathcal{H}_\infty$-control problem with measurement-feedback for Pritchard-Salamon systems. Under some regularity assumptions and some simplifying assumptions (the *a priori assumptions*), we show in Section 5.1 that the existence of a dynamic measurement-feedback that solves the $\mathcal{H}_\infty$-control problem is equivalent to the solvability of two coupled Riccati equations. In the case that the problem is solvable, we give a parametrization of all (sub-optimal) controllers that solve the problem.

The result is a generalization of the finite-dimensional result given by Doyle et al. [26], Glover and Doyle [37] and Tadmor [87] and in the proof we shall use several ideas that have been used for the finite-dimensional case in [26, 37, 36, 86] (see also the summary of results in Section 1.1). In particular, the structure of the proof emerges from a combination of ideas in Doyle et al. [26], Tadmor [87] and Stoorvogel [86]. The generalization of these ideas to the Pritchard-Salamon class is not straightforward, due to the fact that we have to deal with *infinite-dimensional* systems with *unbounded* input and output operators (this is of course the raison d'être of the book). Most of the difficulties concerning infinite-dimensionality and unboundedness of input and output operators can be taken care of by using the previous chapters. In fact, in this chapter we shall see how the preliminary results of Chapter 2 can be used (in particular, duality and stability theory). Furthermore, we shall exploit the state-feedback result from Chapter 4. Another crucial element in the solution to the $\mathcal{H}_\infty$-control problem with measurement-feedback is 'Redheffer's Lemma', which is presented separately in Section 5.2. The actual proof of the main result is then given in Section 5.3. Finally, in Section 5.4 we show how the a priori assumptions can be removed.

5.1 Problem formulation and main result

Let $\mathcal{W}, \mathcal{H}$ and $\mathcal{V}$ be real separable Hilbert spaces, satisfying

$$\mathcal{W} \hookrightarrow \mathcal{H} \hookrightarrow \mathcal{V}. \tag{5.1}$$

We identify $\mathcal{H}$ with its dual and we use the pivot space formulation of Section 2.5 to represent the duals of $\mathcal{W}$ and $\mathcal{V}$ so that

$$\begin{array}{ccccc} \mathcal{V}^d & \hookrightarrow & \mathcal{H}^d & \hookrightarrow & \mathcal{W}^d \\ \downarrow i_{\mathcal{H}} \, |_{\mathcal{V}^d} & & \downarrow i_{\mathcal{H}} & & \downarrow \overline{i_{\mathcal{H}}} \\ \mathcal{V}' & \hookrightarrow & \mathcal{H}' = \mathcal{H} & \hookrightarrow & \mathcal{W}' \end{array}$$

(see also (2.69)). Furthermore, we suppose that U, W, Y and Z are also real separable Hilbert spaces and let $B_1 \in \mathcal{L}(W, \mathcal{V}), B_2 \in \mathcal{L}(U, \mathcal{V}), C_1 \in \mathcal{L}(\mathcal{W}, Z), C_2 \in \mathcal{L}(\mathcal{W}, Y)$, $D_{11} \in \mathcal{L}(W, Z), D_{12} \in \mathcal{L}(U, Z)$, $D_{21} \in \mathcal{L}(W, Y)$ and $D_{22} \in \mathcal{L}(U, Y)$. We assume that B_1 and B_2 are both admissible input operators and that C_1 and C_2 are admissible output operators, all with respect to $(\mathcal{W}, \mathcal{V})$ (cf. Definition 2.1). We shall identify the duals of U, Y, W and Z with themselves, so that $B_1' \in \mathcal{L}(\mathcal{V}', W), B_2' \in \mathcal{L}(\mathcal{V}', U), C_1' \in \mathcal{L}(Z, \mathcal{W}'), C_2' \in \mathcal{L}(Y, \mathcal{W}'), D_{11}' \in \mathcal{L}(Z, W), D_{12}' \in \mathcal{L}(Z, U)$, $D_{21}' \in \mathcal{L}(Y, W)$ and $D_{22}' \in \mathcal{L}(Y, U)$. We consider the smooth Pritchard-Salamon system (see also Figure 1.1)

$$\Sigma_G = \Sigma(S(\cdot), (B_1 \;\; B_2), \begin{pmatrix} C_1 \\ C_2 \end{pmatrix}, \begin{pmatrix} D_{11} & D_{12} \\ D_{21} & D_{22} \end{pmatrix})$$

given by

$$\Sigma_G \begin{cases} x(t) &= S(t)x_0 + \int_0^t S(t-s)(B_1 w(s) + B_2 u(s))ds \\ z(t) &= C_1 x(t) + D_{11} w(t) + D_{12} u(t) \\ y(t) &= C_2 x(t) + D_{21} w(t) + D_{22} u(t), \end{cases} \tag{5.2}$$

where $x_0 \in \mathcal{V}, t \geq 0, w(\cdot) \in L_2^{loc}(0, \infty; W)$ and $u(\cdot) \in L_2^{loc}(0, \infty; U)$ (we recall that smoothness implies that $D(A^{\mathcal{V}}) \hookrightarrow \mathcal{W}$). We call $x(t)$ the state of the system, $u(t) \in U$ the control input, $w(t) \in W$ the disturbance input and $z(t) \in Z$ the to-be-controlled output.

Now suppose that we have a controller $\Sigma_K = \Sigma(T(\cdot), N, L, R)$ of the form

$$\Sigma_K \begin{cases} p(t) &= T(t)p_0 + \int_0^t T(t-s) N u_2(s) ds \\ y_2(t) &= Lp(t) + Ru_2(t) \end{cases} \tag{5.3}$$

for $p_0 \in \mathcal{V}_K$ and $u_2(\cdot) \in L_2^{loc}(0,\infty;Y)$. Here $\mathcal{W}_K$ and $\mathcal{V}_K$ are also real separable Hilbert spaces, satisfying $\mathcal{W}_K \hookrightarrow \mathcal{V}_K$, $T(\cdot)$ is a C_0-semigroup on $\mathcal{V}_K$ which restricts to a C_0-semigroup on $\mathcal{W}_K$, the generators of $T(\cdot)$ on $\mathcal{W}_K$ and $\mathcal{V}_K$ are denoted by $M^{\mathcal{W}_K}$ and $M^{\mathcal{V}_K}$ and $N \in \mathcal{L}(Y,\mathcal{V}_K)$, $L \in \mathcal{L}(\mathcal{W}_K, U)$ and $R \in \mathcal{L}(Y,U)$ (see also Section 2.7). Furthermore, we suppose that the representations of $\mathcal{W}_K^d$ and $\mathcal{V}_K^d$ are given by $\{\mathcal{W}_K', j\}$ and $\{\mathcal{V}_K', k\}$ such that $k = j\mid_{\mathcal{V}_K'}$ and $\mathcal{V}_K' \hookrightarrow \mathcal{W}_K'$ (cf. (2.66) and (2.67)).

If $x_0 = 0$, we can express (5.2) as:

$$\begin{pmatrix} z(\cdot) \\ y(\cdot) \end{pmatrix} = \begin{pmatrix} G_{11} & G_{12} \\ G_{21} & G_{22} \end{pmatrix} \begin{pmatrix} w(\cdot) \\ u(\cdot) \end{pmatrix}, \tag{5.4}$$

where G_{ij} represent the corresponding linear maps defined in (2.111).

The linear map K from $L_2^{loc}(0,\infty;Y)$ to $L_2^{loc}(0,\infty;U)$ is given by

$$(Ku_2)(t) := L\int_0^t T(t-s)Nu_2(s)ds + Ru_2(t). \tag{5.5}$$

In Section 2.7 we have explained how to make sense of the feedback interconnection of (5.2) and (5.3) with $u_2(\cdot) = y(\cdot)$ and $u(\cdot) = y_2(\cdot)$ as in Figure 5.1. We define the real Hilbert spaces $\tilde{\mathcal{V}} = \mathcal{V} \times \mathcal{V}_K$ and $\tilde{\mathcal{W}} = \mathcal{W} \times \mathcal{W}_K$ and we

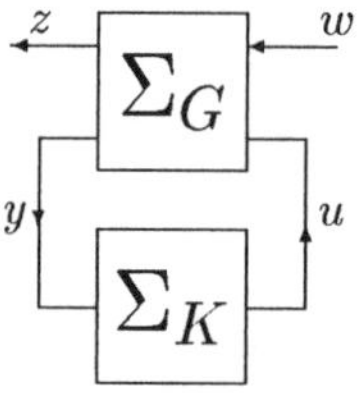

Figure 5.1: $\Sigma_{G_{zw}} = \Sigma_{\mathcal{F}(G,K)}$.

note that $\tilde{\mathcal{W}} \hookrightarrow \tilde{\mathcal{V}}$. We define $\tilde{S}(\cdot), \tilde{B}, \tilde{C}$ and $\tilde{D}$ as in (2.113) and we suppose that $I - D_{22}R$ (or, equivalently, $I - RD_{22}$) is boundedly invertible, so that the interconnection is well-posed (see also Section 2.7). We can consider the combination of (5.2) and (5.3) as a Pritchard-Salamon system of the form (2.114). We define $\mathcal{J}$ and $\bar{D}$ as in (2.88) and (2.115) and so the interconnection of (5.2) and (5.3) is given by the Pritchard-Salamon system $\Sigma_{zw} = \Sigma(\mathcal{S}(\cdot), \mathcal{B}, \mathcal{C}, \mathcal{D})$:

$$\Sigma_{G_{zw}} \begin{cases} \begin{pmatrix} x(t) \\ p(t) \end{pmatrix} = \mathcal{S}(\cdot)\begin{pmatrix} x_0 \\ p_0 \end{pmatrix} + \int_0^t \mathcal{S}(t-s)\mathcal{B}w(s)ds \\ \\ z(t) = \mathcal{C}\begin{pmatrix} x(t) \\ p(t) \end{pmatrix} + \mathcal{D}w(t), \end{cases} \tag{5.6}$$

where

$$\left.\begin{aligned}
\mathcal{S}(\cdot) &= \tilde{S}_{\tilde{B}\bar{D}\mathcal{J}\tilde{C}}(\cdot)\\
\mathcal{B} &= \begin{pmatrix} B_1 \\ 0 \end{pmatrix} + \tilde{B}\bar{D}\mathcal{J}\begin{pmatrix} D_{21} \\ 0 \end{pmatrix}\\
\mathcal{C} &= (C_1 \;\; 0) + (D_{12} \;\; 0)\bar{D}\mathcal{J}\tilde{C}\\
\mathcal{D} &= (D_{12} \;\; 0)\bar{D}\mathcal{J}\begin{pmatrix} D_{21} \\ 0 \end{pmatrix} + D_{11}.
\end{aligned}\right\} \tag{5.7}$$

Supposing that $x_0 = 0$ and $p_0 = 0$, we can formulate (5.6) differently, using (5.4) with $u(\cdot) = Ky(\cdot)$:

$$z(\cdot) = (G_{zw}w)(\cdot) = \mathcal{F}(G, K)w(\cdot) = (G_{11} + G_{12}K(I - G_{22}K)^{-1}G_{21})w(\cdot) \tag{5.8}$$

(see also Section 2.7). If $\mathcal{S}(\cdot)$ is exponentially stable on $\tilde{\mathcal{V}}$ and $\tilde{\mathcal{W}}$, then $\mathcal{F}(G, K) \in \mathcal{L}(L_2(0,\infty; W), L_2(0,\infty; Z))$ (see Lemma 2.23) and there holds

$$\hat{z}(s) = \tilde{G}_{zw}(s)\hat{w}(s) \text{ for all } s \in \mathbb{C}^+,$$

where $\tilde{G}_{zw}(\cdot)$ is the closed-loop transfer function given by

$$\tilde{G}_{zw}(s) = \mathcal{C}(sI - \mathcal{A}^{\tilde{\mathcal{V}}})^{-1}\mathcal{B} + \mathcal{D}.$$

Furthermore, $\tilde{G}_{zw}(\cdot) \in \mathcal{H}_\infty(\mathcal{L}(W, Z))$ and we have

$$\|\tilde{G}_{zw}(\cdot)\|_\infty = \|\mathcal{F}(G, K)\|$$

(see (2.77)).

We can define the *optimal* $\mathcal{H}_\infty$*-control problem* as the problem of finding a controller Σ_K of the form (5.3) such that $\mathcal{S}(\cdot)$ in (5.6) is exponentially stable on $\tilde{\mathcal{W}}$ and $\tilde{\mathcal{V}}$ and $\|\tilde{G}_{zw}(\cdot)\|_\infty = \|\mathcal{F}(G, K)\|$ is minimized. As in the finite-dimensional state-space approach to the problem, we consider the *sub-optimal* $\mathcal{H}_\infty$*-control problem.* In order to explain what the sub-optimal problem is, we first give a definition.

Definition 5.1
Let Σ_G be the Pritchard-Salamon system given by (5.2) and suppose that the controller Σ_K of the form (5.3) is such that $I - D_{22}R$ (or, equivalently, $I - RD_{22}$) is boundedly invertible. Furthermore, suppose that $\mathcal{S}(\cdot)$ in (5.6) is exponentially stable on $\tilde{\mathcal{W}}$ and $\tilde{\mathcal{V}}$ and that $\|\tilde{G}_{zw}(\cdot)\|_\infty = \|\mathcal{F}(G, K)\| < \gamma$, for some $\gamma > 0$. In this case, we call Σ_K a γ*-admissible controller* for Σ_G.

The sub-optimal $\mathcal{H}_\infty$-problem is to find, for a given $\gamma > 0$, a γ-admissible controller for Σ_G. In Theorem 5.4 we show that the existence of a γ-admissible controller is equivalent to the solvability of two coupled Riccati equations (in this case, we give a parametrization of *all* γ-admissible controllers). The solution to the sub-optimal problem is a first step toward solving the optimal $\mathcal{H}_\infty$-control problem, in the sense that the optimal case can be approximated via a procedure called γ-iteration (in this procedure, γ is decreased until solutions to the Riccati equations with the required properties fail to exist).

Remark 5.2
As far as the exponential stability of the closed-loop C_0-semigroup is concerned, we note an important difference with the state-feedback case in Chapter 4, Definition 4.1. There we required the closed-loop C_0-semigroup $S_{B_2F}(\cdot)$ to be exponentially stable on $\mathcal{V}$, whereas here we want that the closed-loop C_0-semigroup $\mathcal{S}(\cdot)$ be exponentially stable on both $\tilde{\mathcal{W}} = \mathcal{W} \times \mathcal{W}_K$ and $\tilde{\mathcal{V}} = \mathcal{V} \times \mathcal{V}_K$. The reason for this is that at a certain point we need to use a duality argument (see Lemma 5.11), so that we can use the state-feedback result for the transpose of Σ_G. The point is that if $\mathcal{S}(\cdot)$ is only exponentially stable on $\tilde{\mathcal{V}}$, it follows that $\mathcal{S}'(\cdot)$ is only exponentially stable on $\tilde{\mathcal{V}}'$, while we actually need stability on the 'larger space', in this case $\tilde{\mathcal{W}}'$.

In the statement of the main result, we shall use several assumptions. First of all, we shall use the simplifying assumption

$$D_{11} = 0 \quad \text{and} \quad D_{22} = 0 \tag{5.9}$$

(this means that there is neither feedthrough from the disturbance w to the to-be-controlled output z, nor feedthrough from the input u to the measured output y). This assumption is introduced merely to simplify the formulas; it is not essential to the argument. In Section 5.4 we discuss how to remove this assumption. We note that $D_{22} = 0$ implies that the closed-loop system is automatically well-posed (see also Section 2.7). Furthermore, we shall use the following regularity assumptions:

$$\text{there exists an } \epsilon > 0 \text{ such that for all } (\omega, x, u) \in \mathbb{R} \times D(A^{\mathcal{V}}) \times U \text{ with}$$

$$i\omega x = A^{\mathcal{V}} x + B_2 u, \;\; \text{there holds } \|C_1 x + D_{12} u\|_Z^2 \geq \epsilon \|x\|_{\mathcal{V}}^2, \tag{5.10}$$

$$D_{12}' D_{12} \text{ is coercive,} \tag{5.11}$$

$$\text{there exists an } \epsilon > 0 \text{ such that for all } (\omega, x', y) \in \mathbb{R} \times D((A^{\mathcal{W}})') \times Y \text{ with}$$

$$i\omega x' = (A^{\mathcal{W}})' x' + C_2' y, \;\; \text{there holds } \|B_1' x' + D_{21}' y\|_W^2 \geq \epsilon \|x'\|_{\mathcal{W}'}^2 \tag{5.12}$$

$$D_{21} D_{21}' \text{ is coercive.} \tag{5.13}$$

Remark 5.3
Note that assumptions (5.10) and (5.11) were also present in the state-feedback result Theorem 4.4. The $\mathcal{H}_\infty$-control problem with these assumptions is usually called *regular*. Assumptions (5.12) and (5.13) are dual to the first two assumptions, in the sense that they correspond to the state-feedback assumptions for the transpose of Σ_G given by

$$\Sigma_{G^\natural}\begin{cases} x'(t) &= S'(t)x_0' + \int_0^t S'(t-s)(C_1'z(s) + C_2'y(s))ds \\ w(t) &= B_1'x'(t) + D_{11}'z(t) + D_{21}'y(t) \\ u(t) &= B_2'x'(t) + D_{12}'z(t) + D_{22}'y(t), \end{cases}$$

where $x_0' \in \mathcal{W}', t \geq 0, z(\cdot) \in L_2^{loc}(0,\infty; Z)$ and $y(\cdot) \in L_2^{loc}(0,\infty; Y)$ (the notion of transpose was defined at the end of Section 2.6). Assumptions (5.10)-(5.13) are the infinite-dimensional analogues of the weakest assumptions under which the regular version of the finite-dimensional $\mathcal{H}_\infty$-control problem has been solved (see e.g. [37]).

The next theorem is a generalization of the solution to the regular finite-dimensional $\mathcal{H}_\infty$-control problem in [26] and [37] and it is essentially the main result of the book. It shows that the sub-optimal $\mathcal{H}_\infty$-control problem is solvable if and only two coupled Riccati equations are solvable. Furthermore, it gives a parametrization of all sub-optimal controllers.

Theorem 5.4
Consider the smooth Pritchard-Salamon system Σ_G given by (5.2), let $\gamma > 0$ and suppose that $D_{11} = 0, D_{22} = 0$ and that (5.10)-(5.13) hold. Then Σ_G has a γ-admissible controller Σ_K of the form (5.3) if and only if the following conditions hold:

(*i*) *there exists a $P \in \mathcal{L}(\mathcal{V}, \mathcal{V}')$ with $P = P' \geq 0$ satisfying*

$$< Px, (A^{\mathcal{V}} - B_2(D_{12}'D_{12})^{-1}D_{12}'C_1)y >_{<\mathcal{V}',\mathcal{V}>} +$$

$$< (A^{\mathcal{V}} - B_2(D_{12}'D_{12})^{-1}D_{12}'C_1)x, Py >_{<\mathcal{V},\mathcal{V}'>} +$$

$$< P(\gamma^{-2}B_1B_1' - B_2(D_{12}'D_{12})^{-1}B_2')Px, y >_{<\mathcal{V}',\mathcal{V}>} +$$

$$< (I - D_{12}(D_{12}'D_{12})^{-1}D_{12}')C_1x,$$

$$(I - D_{12}(D_{12}'D_{12})^{-1}D_{12}')C_1y >_Z = 0 \tag{5.14}$$

for all $x, y \in D(A^{\mathcal{V}})$*, such that the* C_0*-semigroup* $S_P(\cdot)$ *defined by*

$$S_P(\cdot) = S_{\begin{pmatrix} B_1 & B_2 \end{pmatrix}\begin{pmatrix} \gamma^{-2}B_1'P \\ -(D_{12}'D_{12})^{-1}(B_2'P + D_{12}'C_1) \end{pmatrix}}(\cdot), \quad (5.15)$$

is exponentially stable on $\mathcal{W}$ *and* $\mathcal{V}$*,*

(*ii*) *there exists a* $Q \in \mathcal{L}(\mathcal{W}', \mathcal{W})$ *with* $Q = Q' \geq 0$ *satisfying*

$$< Qx, ((A^{\mathcal{W}})' - C_2'(D_{21}D_{21}')^{-1}D_{21}B_1')y >_{<\mathcal{W},\mathcal{W}'>} +$$

$$< ((A^{\mathcal{W}})' - C_2'(D_{21}D_{21}')^{-1}D_{21}B_1')x, Qy >_{<\mathcal{W}',\mathcal{W}>} +$$

$$< Q(\gamma^{-2}C_1'C_1 - C_2'(D_{21}D_{21}')^{-1}C_2)Qx, y >_{<\mathcal{W},\mathcal{W}'>} +$$

$$< (I - D_{21}'(D_{21}D_{21}')^{-1}D_{21})B_1'x,$$

$$(I - D_{21}'(D_{21}D_{21}')^{-1}D_{21})B_1'y >_{\mathcal{W}} = 0 \quad (5.16)$$

for all $x, y \in D((A^{\mathcal{W}})')$*, such that the* C_0*-semigroup* $S_Q(\cdot)$ *defined via its adjoint as*

$$S_Q'(\cdot) = S'_{\begin{pmatrix} C_1' & C_2' \end{pmatrix}\begin{pmatrix} \gamma^{-2}C_1Q \\ -(D_{21}D_{21}')^{-1}(C_2Q + D_{21}B_1') \end{pmatrix}}(\cdot), \quad (5.17)$$

is exponentially stable on $\mathcal{W}$ *and* $\mathcal{V}$*,*

(*iii*) $$r_\sigma^{\mathcal{W}'}(PQ) = r_\sigma^{\mathcal{V}}(QP) < \gamma^2. \quad (5.18)$$

Moreover, in this case there exist infinitely many γ*-admissible controller-s which can be parametrized as follows. A controller* Σ_K *of the form* (5.3) *with* $(T(\cdot), N)$ *admissibly stabilizable and* $(L, T(\cdot))$ *admissibly detectable is* γ*-admissible if and only if the corresponding linear map denoted by*

$$(Ku)(t) = L\int_0^t T(t-s)Nu(s)ds + Ru(t)$$

satisfies $K = \mathcal{F}(\tilde{K}, \Lambda)$*. Here* $\tilde{K}$ *is the linear map corresponding to the Pritchard-Salamon system*

$$\Sigma_{\tilde{K}} = \Sigma(\bar{S}_{-\bar{B}_1(\bar{D}_{21})^{-1}\bar{C}_2 - \bar{B}_2(\bar{D}_{12})^{-1}\bar{C}_1}(\cdot), \begin{pmatrix} \bar{B}_1(\bar{D}_{21})^{-1} & \bar{B}_2(\bar{D}_{12})^{-1} \end{pmatrix},$$

$$\begin{pmatrix} -(\bar{D}_{12})^{-1}\bar{C}_1 \\ -(\bar{D}_{21})^{-1}\bar{C}_2 \end{pmatrix}, \begin{pmatrix} 0 & (\bar{D}_{12})^{-1} \\ (\bar{D}_{21})^{-1} & 0 \end{pmatrix});$$

$$\Sigma_{\tilde{K}} \begin{cases} p(t) &= \bar{S}_{-\bar{B}_1(\bar{D}_{21})^{-1}\bar{C}_2 - \bar{B}_2(\bar{D}_{12})^{-1}\bar{C}_1}(t)p_0 + \\ & \quad \int_0^t \bar{S}_{-\bar{B}_1(\bar{D}_{21})^{-1}\bar{C}_2 - \bar{B}_2(\bar{D}_{12})^{-1}\bar{C}_1}(t-s) \\ & \quad (\bar{B}_1(\bar{D}_{21})^{-1}y(s) + \bar{B}_2(\bar{D}_{12})^{-1}v(s))ds \\ u(t) &= -(\bar{D}_{12})^{-1}\bar{C}_1 p(t) + (\bar{D}_{12})^{-1}v(t) \\ r(t) &= -(\bar{D}_{21})^{-1}\bar{C}_2 p(t) + (\bar{D}_{21})^{-1}y(t), \end{cases} \tag{5.19}$$

where

$$\begin{array}{rcl} \bar{C}_1 &:=& (D_{12}'D_{12})^{-1/2}(B_2'P + D_{12}'C_1) \\ \bar{C}_2 &:=& C_2 + \gamma^{-2}D_{21}B_1'P \\ \bar{B}_1 &:=& (\bar{P}(C_2' + \gamma^{-2}PB_1D_{21}') + B_1D_{21}')(D_{21}D_{21}')^{-1/2} \\ \bar{B}_2 &:=& B_2 + \gamma^{-2}\bar{P}(PB_2 + C_1'D_{12}) \\ \bar{D}_{12} &:=& (D_{12}'D_{12})^{1/2} \\ \bar{D}_{21} &:=& (D_{21}D_{21}')^{1/2} \\ \bar{S}(\cdot) &:=& S_{\gamma^{-2}B_1B_1'P + \gamma^{-2}\bar{P}(PB_2 + C_1'D_{12})(D_{12}'D_{12})^{-1}(B_2'P + D_{12}'C_1)}(\cdot) \\ \bar{P} &:=& Q(I - \gamma^{-2}PQ)^{-1}. \end{array} \tag{5.20}$$

Furthermore, the parameter Λ is the linear map corresponding to a Pritchard-Salamon system $\Sigma_\Lambda = \Sigma(S_\Lambda(\cdot), B_\Lambda, C_\Lambda, D_\Lambda)$ with state-space $\mathcal{W}_\Lambda \hookrightarrow \mathcal{V}_\Lambda$:

$$\Sigma_\Lambda \begin{cases} \lambda(t) &= \int_0^t S_\Lambda(t-s)B_\Lambda r(s)ds \\ v(t) &= C_\Lambda \lambda(t) + D_\Lambda r(t) \end{cases} \tag{5.21}$$

such that $S_\Lambda(\cdot)$ is a C_0-semigroup which is exponentially stable on $\mathcal{V}_\Lambda$ and $\mathcal{W}_\Lambda$ and $\|\Lambda\| < \gamma$. In this case, the controller Σ_{K_Λ} defined by the interconnection of (5.19) and (5.21) is also γ-admissible.
Finally, if the solutions to the Riccati equations (5.14) and (5.16) exist, then they are unique.

Remark 5.5
If $(S(\cdot), B_2)$ is admissibly stabilizable and $(C_2, S(\cdot))$ is admissibly detectable, we can easily find a stabilizing measurement-feedback controller Σ_K (see Theorem 2.30). Hence, in this case there is always some γ such that $\|\mathcal{F}(G, K)\| < \gamma$. The theorem now implies that for γ large enough there always exist nonnegative definite operators P and Q that satisfy items $(i), (ii)$ and (iii) of the theorem.

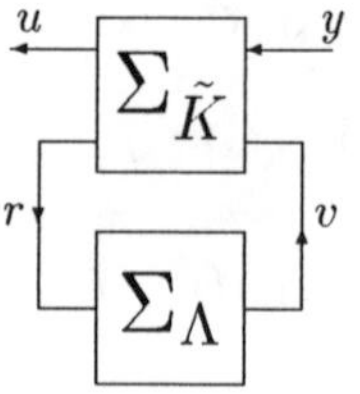

Figure 5.2: $K = \mathcal{F}(\tilde{K}, \Lambda)$.

Remark 5.6
The controller parametrization can be depicted as in Figure 5.2.

Remark 5.7
Theorem 5.4 is a perfect generalization of the finite-dimensional results for the $\mathcal{H}_\infty$-control problem in [26, 37] to the Pritchard-Salamon class. The reason for the possiblity of such a generalization is of course the fact that Pritchard-Salamon systems have the nice system theoretic properties that were treated in Chapter 2. The proof given in Section 5.3 is based on the state-feedback result in Theorem 4.4, some duality arguments using Section 2.5 and an important auxiliary result ('Redheffer's Lemma'), that will be given in the next section. In the proof we shall show that the state-feedback result from Chapter 4 can be used to derive the first Riccati equation (5.14) from the assumption that there exists a γ-admissible controller Σ_K for Σ_G. Furthermore, we shall show that if Σ_K is admissible for Σ_G, then $\Sigma_{K^\natural}$ is admissible for $\Sigma_{G^\natural}$ (see (2.79) and Remark 5.3) and so the state-feedback result can be used once again to derive the second 'dual' Riccati equation (5.16). We recall that for the state-feedback Riccati equation the difficulties as far as unboundedness of the input and output operators are concerned were taken care of by considering the problem on the 'larger space' $\mathcal{V}$, so that we only had to deal with the unboundedness of the output operator C_1. A similar approach is taken for the dual Riccati equation: the 'larger space' is $\mathcal{W}'$ (because $\mathcal{V}' \hookrightarrow \mathcal{W}'$) so that B_1' is the corresponding unbounded output operator for the transposed system $\Sigma_{G^\natural}$.

As mentioned above, the assumption that $D_{11} = 0$ and $D_{22} = 0$ is introduced merely to simplify the formulas. In Section 5.4 we show how this assumption can be removed. In Section 5.4 we shall also show that the regularity assumptions (5.10)-(5.13) can be removed, however at the expense of introducing an extra parameter.

5.2 Redheffer's Lemma

In this section we shall give a Pritchard-Salamon version of a lemma that has been crucial in the state-space approach to $\mathcal{H}_\infty$-control problems for finite-dimensional systems (see [26, Lemma 15]). In some sense it is also just a preliminary result, but its importance justifies a separate section.

The statement and proof of Redheffer's Lemma is different from the corresponding finite-dimensional result [26, Lemma 15]. In order to allow for systems in the Pritchard-Salamon class we shall treat the lemma in the time domain, rather than in the frequency domain. Redheffer's Lemma shall be used in the proof of the coupling condition (5.18) and the controller parametrization.

We start with a lemma that is, in some sense, a converse of the Small Gain Theorem. It is related to the finite-dimensional result that if $G(\cdot) \in \mathbb{R}L_\infty, \|G(\cdot)\|_\infty < 1$ and $(I - G(\cdot))^{-1} \in \mathbb{R}\mathcal{H}_\infty$ it follows that $G(\cdot) \in \mathbb{R}\mathcal{H}_\infty$ (see e.g. [86, Lemma 2.11])

Lemma 5.8
Let $\Sigma_G = \Sigma(S(\cdot), B, C, D)$ be a Pritchard-Salamon system of the form (2.7) *so that G denotes the linear map given by:*

$$(Gu)(\cdot) = C \int_0^{\cdot} S(\cdot - s)Bu(s)ds + Du(\cdot).$$

Now suppose that $U = Y$ and that D is such that $(I - D)^{-1} \in \mathcal{L}(U)$, so that $(I-G)^{-1}$ exists (see Lemma 2.26). If, in addition, $(I-G)^{-1} \in \mathcal{L}(L_2(0,\infty;U))$ and $\|Gu\|_2 \le \|u\|_2$ for all $u \in D(G)$, then $G \in \mathcal{L}(L_2(0,\infty;U))$.

In the proof below we shall use the notion of causality (see e.g. [22, Chapter 3]): a map $G : L_2^{loc}(0,\infty;U) \to L_2^{loc}(0,\infty;Y)$ is *causal* (or *nonanticipative*) if for all $u_1(\cdot), u_2(\cdot) \in L_2^{loc}(0,\infty;U)$ and for all $T > 0$

$$u_1(t) = u_2(t) \text{ for almost all } t \in [0,T] \quad \Rightarrow$$

$$(Gu_1)(t) = (Gu_2)(t) \text{ for almost all } t \in [0,T].$$

The input output map G corresponding to a Pritchard-Salamon system $\Sigma_G = \Sigma(S(\cdot),\ B, C, D)$ is of course causal.

Proof
Note first that for arbitrary $u_1 \in L_2(0,\infty;U)$, we have $Gu_1 \in L_2^{loc}(0,\infty;U)$. We only have to prove that $Gu_1 \in L_2(0,\infty;U)$, because that implies that $D(G) = L_2(0,\infty;U)$ and since G is closed as an operator from $D(G) \subset L_2(0,\infty;U)$ to $L_2(0,\infty;U)$, it then follows that $G \in \mathcal{L}(L_2(0,\infty;U))$ (see the remark after Definition 2.22).

Let $T > 0$ be arbitrary and define $y_{2T} \in L_2(0,\infty;U)$ as follows:

$$y_{2T}(t) := \begin{cases} ((I-G)u_1)(t) & \text{for almost all } t \in [0,T] \\ 0 & \text{for } t > T. \end{cases} \tag{5.22}$$

Since $(I-G)^{-1}$ is bounded we can define $u_{2T} \in L_2(0,\infty;U)$ by

$$u_{2T} := (I-G)^{-1} y_{2T}. \tag{5.23}$$

It follows that

$$((I-G)u_{2T})(t) = ((I-G)u_1)(t) \text{ for almost all } t \in [0,T]. \tag{5.24}$$

Since $(I-G)^{-1}$ is a causal system (it is the input output map corresponding to a Pritchard-Salamon system, see Lemma 2.26) this implies that

$$u_{2T}(t) = u_1(t) \text{ for almost all } t \in [0,T]. \tag{5.25}$$

Furthermore, since $u_{2T}, y_{2T} \in L_2(0,\infty;U)$ and $(I-G)^{-1}$ is bounded, (5.23) implies that

$$Gu_{2T} = u_{2T} - y_{2T} \in L_2(0,\infty;U) \tag{5.26}$$

and so $u_{2T} \in D(G)$.
Hence, using the fact that $\|Gu\|_2 \leq \|u\|_2$ for all $u \in D(G)$ we have

$$\|Gu_{2T}\|_2^2 = \|u_{2T} - y_{2T}\|_2^2 \leq \|u_{2T}\|_2^2. \tag{5.27}$$

Since $y_{2T}(t) = 0$ for $t > T$ (see (5.22)), we can express (5.27) as

$$\int_0^T \|u_{2T}(t) - y_{2T}(t)\|_U^2\,dt + \int_T^\infty \|u_{2T}(t)\|_U^2\,dt \leq \int_0^\infty \|u_{2T}(t)\|_U^2\,dt. \tag{5.28}$$

It follows from (5.24) and (5.25) that $(Gu_{2T})(t) = (Gu_1)(t)$ for almost all $t \in [0,T]$ and using this with (5.26) in (5.28) gives

$$\int_0^T \|Gu_1(t)\|_U^2\,dt \leq \int_0^\infty \|u_{2T}(t)\|_U^2\,dt - \int_T^\infty \|u_{2T}(t)\|_U^2\,dt =$$

$$\int_0^T \|u_{2T}(t)\|_U^2\,dt = \int_0^T \|u_1(t)\|_U^2\,dt, \tag{5.29}$$

where the last equality follows from (5.25).
T was arbitrary so (5.29) implies that

$$\int_0^T \|Gu_1(t)\|_U^2\,dt \leq \int_0^T \|u_1(t)\|_U^2\,dt \leq \|u_1\|_2^2 \text{ for all } T > 0$$

and this completes the proof. ∎

It follows from the proof that we could have stated Lemma 5.8 in more general terms: the essential ingredients are that G and $(I-G)^{-1}$ are both causal maps. Furthermore, instead of the $\|\cdot\|_2$-norm we could have used the $\|\cdot\|_p$-norm for $p<\infty$. We prefer the present formulation, because it is sufficient in the $\mathcal{H}_\infty$-context of the book.

Now consider the Pritchard-Salamon system (5.2) of the previous section:

$$\Sigma_G = \Sigma(S(\cdot), \begin{pmatrix} B_1 & B_2 \end{pmatrix}, \begin{pmatrix} C_1 \\ C_2 \end{pmatrix}, \begin{pmatrix} D_{11} & D_{12} \\ D_{21} & D_{22} \end{pmatrix});$$

$$\Sigma_G \begin{cases} x(t) &= S(t)x_0 + \int_0^t S(t-s)(B_1 w(s) + B_2 u(s))ds \\ z(t) &= C_1 x(t) + D_{11} w(t) + D_{12} u(t) \\ y(t) &= C_2 x(t) + D_{21} w(t) + D_{22} u(t) \end{cases} \tag{5.30}$$

and let G_{ij} represent the corresponding linear maps for zero initial conditions defined in (2.111). Furthermore, suppose that we have a second Pritchard-Salamon system $\Sigma_K = \Sigma(T(\cdot), N, L, R)$ of the form (5.3):

$$\Sigma_K \begin{cases} p(t) &= T(t)p_0 + \int_0^t T(t-s)Nu_2(s)ds \\ y_2(t) &= Lp(t) + Ru_2(t). \end{cases} \tag{5.31}$$

If $I - D_{22}R$ (or, equivalently, $I - RD_{22}$) has a bounded inverse, then the closed-loop system determined by $u_2(\cdot) = y(\cdot)$ and $u(\cdot) = y_2(\cdot)$ as in Figure 5.1 is given by $\Sigma_{G_{zw}} = \Sigma_{\mathcal{F}(G,K)}$ (see (5.6)-(5.7)):

$$\Sigma_{\mathcal{F}(G,K)} \begin{cases} \begin{pmatrix} x(t) \\ p(t) \end{pmatrix} &= \mathcal{S}(\cdot)\begin{pmatrix} x_0 \\ p_0 \end{pmatrix} + \int_0^t \mathcal{S}(t-s)\mathcal{B}w(s)ds \\ z(t) &= \mathcal{C}\begin{pmatrix} x(t) \\ p(t) \end{pmatrix} + \mathcal{D}w(t). \end{cases} \tag{5.32}$$

The following holds:

Lemma 5.9
Suppose that we have a Pritchard-Salamon system Σ_G of the form (5.30) satisfying the following assumptions:

the C_0-semigroup $S(\cdot)$ is exponentially stable on $\mathcal{W}$ and $\mathcal{V}$, (5.33)

the system Σ_G is inner, i.e. if $x_0 = 0$, then for all $w \in L_2(0,\infty;W), u$

$$\in L_2(0,\infty;U) \text{ there holds } \|z(\cdot)\|_2^2 + \|y(\cdot)\|_2^2 = \|w(\cdot)\|_2^2 + \|u(\cdot)\|_2^2, \quad (5.34)$$

$$\text{we have } Y = W, D_{21}^{-1} \in \mathcal{L}(Y) \text{ and } G_{21}^{-1} \in \mathcal{L}(L_2(0,\infty;Y)). \quad (5.35)$$

Furthermore, let Σ_K be a Pritchard-Salamon system of the form (5.31) such that $I - RD_{22}$ (or, equivalently, $I - D_{22}R$) has a bounded inverse. Then the C_0-semigroup $\mathcal{S}(\cdot)$ corresponding to the closed-loop system (5.32) is exponentially stable on $\tilde{\mathcal{W}} = \mathcal{W} \times \mathcal{W}_K$ and $\tilde{\mathcal{V}} = \mathcal{V} \times \mathcal{V}_K$ and $\|G_{zw}\| = \|\mathcal{F}(G,K)\| < 1$ if and only if the C_0-semigroup $T(\cdot)$ corresponding to Σ_K is exponentially stable on $\mathcal{W}_K$ and $\mathcal{V}_K$ and $\|K\| < 1$.

Remark 5.10

The lemma is a time-domain infinite-dimensional version of the frequency domain result [26, Lemma 15]. There, the proof is based on some typical finite-dimensional frequency domain techniques. In order to be able to cope with Pritchard-Salamon systems in their most general form, we have chosen a time-domain formulation.

Apart from the exponential stability, the sufficiency part of Lemma 5.9 follows from Redheffer's results in [76]. The necessity part might be derived by applying some results in [76], but this would give a proof which is much longer than the one we give here. The clue would be [76, formula (17)] and the relation between matrix and $*$-product inverses for isometric operators. Since Redheffer considers only bounded operators and K is not a priori bounded, the result should first be obtained for $L_2(0,T)$ and then somehow extended to $L_2(0,\infty)$.

We note that in [75] a finite-dimensional time-varying version of Lemma 5.9 is given. The proof of that result is based on the 'Small-Gain Instability Theorem' in [89, Theorem 1], where the inner product structure of L_2 is used to derive a result that is similar to Lemma 5.8. It is easy to see that our approach also allows for time-varying systems. Furthermore, it may be possible to give a nonlinear version of the result, due to the general nature of the proof of Lemma 5.8.

Proof

(sufficiency)

Suppose that $T(\cdot)$ is exponentially stable on $\mathcal{W}_K$ and $\mathcal{V}_K$ and that $\|K\| < 1$.
Since $S(\cdot)$ is exponentially stable on $\mathcal{W}$ and $\mathcal{V}$, it follows from Lemma 2.23 that

$$G = \begin{pmatrix} G_{11} & G_{12} \\ G_{21} & G_{22} \end{pmatrix}$$

is bounded and so assumption (5.34) implies that $\|G_{22}\| \leq 1$. Hence, it follows from the 'small gain result' in Theorem 2.28 that $\mathcal{S}(\cdot)$ is exponentially stable on $\tilde{\mathcal{W}}$ and $\tilde{\mathcal{V}}$.
Now let $w \in L_2(0,\infty;W)$ be an input for the closed-loop system (5.32) and let $u \in L_2(0,\infty;U)$, $y \in L_2(0,\infty;Y)$ and $z \in L_2(0,\infty;Z)$ have the corresponding values so that

$$\begin{pmatrix} z(\cdot) \\ y(\cdot) \end{pmatrix} = \begin{pmatrix} G_{11} & G_{12} \\ G_{21} & G_{22} \end{pmatrix} \begin{pmatrix} w(\cdot) \\ u(\cdot) \end{pmatrix} \tag{5.36}$$

and $u(\cdot) = Ky(\cdot)$. Now since $\|K\| < 1$, there exists some $\epsilon > 0$ such that $\|u\|_2^2 \leq (1-\epsilon)\|y\|_2^2$. We know that G is inner so that

$$\|z\|_2^2 = \|w\|_2^2 + \|u\|_2^2 - \|y\|_2^2 \leq \|w\|_2^2 - \epsilon\|y\|_2^2. \tag{5.37}$$

Since (5.36) holds with $u(\cdot) = Ky(\cdot)$, we have

$$y = (I - G_{22}K)^{-1}G_{21}w.$$

Since G_{21}^{-1} exists and is bounded and K and G_{22} are also bounded, it follows that

$$\|w\|_2^2 = \left\|G_{21}^{-1}(I - G_{22}K)y\right\|_2^2 \leq const\,\|y\|_2^2. \tag{5.38}$$

Combining (5.37) and (5.38) shows that there exists a $\delta > 0$ such that $\|z\|_2^2 = \|G_{zw}w\|_2^2 \leq (1-\delta)\|w\|_2^2$ for all $w \in L_2(0,\infty;W)$, so $\|\mathcal{F}(G,K)\| = \|G_{zw}\| < 1$.

(necessity)
Suppose that $\mathcal{S}(\cdot)$ is exponentially stable on $\tilde{\mathcal{W}}$ and $\tilde{\mathcal{V}}$ and that $\|\mathcal{F}(G,K)\| = \|G_{zw}\| < 1$.
We proceed in 7 steps:

1.*Note that $K(I - G_{22}K)^{-1}$ and $(I - G_{22}K)^{-1}$ are bounded:*
This follows from the fact that $\mathcal{S}(\cdot)$ is exponentially stable on $\tilde{\mathcal{W}}$ and $\tilde{\mathcal{V}}$, formula (2.120) and Theorem 2.27 (cf. Remark 2.31).

2.*Prove that $\|Ky\|_2 \leq \|y\|_2$ for all $y \in D(K)$:*
Let $y \in D(K)$ and define $u := Ky \in L_2(0,\infty;U)$. Define $w := G_{21}^{-1}(I - G_{22}K)y = G_{21}^{-1}y - G_{21}^{-1}G_{22}u$, so that $w \in L_2(0,\infty;W)$ since G_{21}^{-1} is bounded.
Define $z := G_{zw}w \in L_2(0,\infty;Z)$. Now it is easy to see that

$$\begin{pmatrix} z(\cdot) \\ y(\cdot) \end{pmatrix} = \begin{pmatrix} G_{11} & G_{12} \\ G_{21} & G_{22} \end{pmatrix} \begin{pmatrix} w(\cdot) \\ u(\cdot) \end{pmatrix}$$

is satisfied and so, using the assumption that G is inner, we have $\|z\|_2^2 + \|y\|_2^2 = \|w\|_2^2 + \|u\|_2^2$. $\|G_{zw}\| < 1$ implies that $\|z\|_2^2 \leq (1-\epsilon)\|w\|_2^2$ for some $\epsilon > 0$ and so $\|Ky\|_2^2 = \|u\|_2^2 = \|z\|_2^2 + \|y\|_2^2 - \|w\|_2^2 \leq \|y\|_2^2 - \epsilon\|w\|_2^2 \leq \|y\|_2^2$.

3. *Note that $D(G_{22}K) = D(K)$:*
Since G_{22} is bounded, we have $D(G_{22}K) \supseteq D(K)$.
Now suppose that $y \in D(G_{22}K)$. It follows that $(I - G_{22}K)y \in L_2(0,\infty;Y)$. Since $K(I - G_{22}K)^{-1}$ is bounded it follows that $Ky = K(I - G_{22}K)^{-1}(I - G_{22}K)y \in L_2(0,\infty;U)$, and so $y \in D(K)$.

4. *Use Lemma 5.8 to prove that $G_{22}K$ is bounded:*
Using 2, 3 and the fact that $\|G_{22}\| \leq 1$ we have $\|G_{22}Ky\|_2 \leq \|Ky\|_2 \leq \|y\|_2$ for all $y \in D(G_{22}K)$. From 1 we know that $(I - G_{22}K)^{-1}$ is bounded. Lemma 5.8 implies that $G_{22}K$ is bounded.

5. *Prove that K is bounded:*
This follows from 3 and 4.

6. *Prove that $T(\cdot)$ is exponentially stable on $\mathcal{W}_K$ and $\mathcal{V}_K$:*
Since $\mathcal{S}(\cdot)$ is exponentially stable, it follows from Theorem 2.27 that $(T(\cdot), N)$ is admissibly stabilizable and $(L, T(\cdot)$ is admissibly detectable. Since K is bounded, it follows from Lemma 2.23 that $T(\cdot)$ is exponentially stable on $\mathcal{W}_K$ and $\mathcal{V}_K$.

7. *Prove that $\|K\| < 1$:*
We conclude from 2 and 5 that $\|K\| \leq 1$. Now let $y \in L_2(0,\infty;Y)$, define $u := Ky$ and $w := G_{21}^{-1}(I - G_{22}K)y$. As in 2 we have $\|Ky\|_2^2 \leq \|y\|_2^2 - \epsilon \|w\|_2^2$. Since $y = (I - G_{22}K)^{-1}G_{21}w$, we see that $\|y\|_2^2 \leq const \|w\|_2^2$ so that there exists some $\delta > 0$ such that $\|Ky\|_2^2 \leq (1-\delta)\|y\|_2^2$, and so $\|K\| < 1$. ■

5.3 Proof of the measurement-feedback result

First of all, we note that without loss of generality we can take $\gamma = 1$. The general result can then be obtained by the scaling $\bar{B}_1 := \gamma^{-\frac{1}{2}}B_1, \bar{C}_1 := \gamma^{-\frac{1}{2}}C_1, \bar{B}_2 := \gamma^{\frac{1}{2}}B_2, \bar{C}_2 := \gamma^{\frac{1}{2}}C_2, \bar{P} := \gamma^{-1}P, \bar{Q} := \gamma^{-1}Q$ and $\bar{K} := \gamma^{-1}K$, just as in [26]. So from now on we take $\gamma = 1$. Furthermore, we recall the assumptions of Theorem 5.4: $D_{11} = 0$, $D_{22} = 0$ and that (5.10)-(5.13) are satisfied.

The proof is divided into four parts:

In part **a)** we derive the existence of P and Q satisfying items (i) and (ii) of Theorem 5.4, under the assumption that that there exists a $\gamma(=1)$-admissible controller of the form (5.3). In part **b)**, using the existence of P satisfying item (i), we construct a certain inner system, that is crucial for the derivation of the coupling condition (5.18) and the controller parametrization. Then we actually derive the coupling condition so that the necessity part of Theorem

5.4 is proved. In part **c)** we derive the controller parametrization and part **d)** is devoted to the sufficiency part of Theorem 5.4. The uniqueness of P and Q follows immediately from Lemma 2.33.

Part a): The derivation of P and Q.
In this part we assume that there exists a $\gamma(=1)$-admissible controller Σ_K of the form (5.3) for (5.2).

First of all, we note that Σ_G satisfies the assumptions (4.4)-(4.6) of the state-feedback result in Theorem 4.4. The fact that there exists an admissible controller for (5.2) implies that the pairs $(S(\cdot), B_2)$ and $(C_2, S(\cdot))$ are admissibly stabilizable and detectable, respectively (see Theorem 2.27). Furthermore, it is easy to see that there exists a $\delta > 0$ such that for all $w(\cdot) \in L_2(0,\infty;W)$ there exists a $u(\cdot) \in L_2(0,\infty;U)$ such that $x(\cdot)$ given by

$$x(t) = \int_0^t S(t-s)(B_1 w(s) + B_2 u(s))ds$$

satisfies $x(\cdot) \in L_2(0,\infty;\mathcal{V})$ and $z(\cdot) = C_1 x(\cdot) + D_{12} u(\cdot)$ satisfies

$$\|z(\cdot)\|^2_{L_2(0,\infty;Z)} \leq (\gamma^2 - \delta^2)\|w(\cdot)\|^2_{L_2(0,\infty;W)}.$$

Hence, item (ii) of Theorem 4.4 is satisfied and so the existence of P in item (i) follows immediately from that theorem (note that stability on $\mathcal{V}$ of $S_P(\cdot)$ implies stability on $\mathcal{W}$, using Theorem 2.20 item (iv)).

In order to prove the existence of Q, we use a duality argument given in the next lemma. In the lemma we use the notion of the transpose of a system, which was defined at the end of Section 2.6.

Lemma 5.11
Suppose that we have a system Σ_G of the form (5.2) satisfying $D_{11} = 0$ and $D_{22} = 0$ and let Σ_K be a controller of the form (5.3). Consider the closed-loop system $\Sigma_{\mathcal{F}(G,K)}$ as in (5.6) and its transpose denoted by $\Sigma_{(\mathcal{F}(G,K))^\natural}$. Furthermore, consider the transposed versions of Σ_G and Σ_K given by

$$\Sigma_{G^\natural} = \Sigma(S'(\cdot), \begin{pmatrix} C_1' & C_2' \end{pmatrix}, \begin{pmatrix} B_1' \\ B_2' \end{pmatrix}, \begin{pmatrix} 0 & D_{21}' \\ D_{12}' & 0 \end{pmatrix})$$

and

$$\Sigma_{K^\natural} = \Sigma(T'(\cdot), L', N', R').$$

Considering $\Sigma_{K^\natural}$ as a controller for $\Sigma_{G^\natural}$, there holds

$$(\mathcal{F}(G,K))^\natural = \mathcal{F}(G^\natural, K^\natural) \tag{5.39}$$

and Σ_K is admissible for Σ_G if and only if $\Sigma_{K^\natural}$ is admissible for $\Sigma_{G^\natural}$.

Proof

Recall the definition of $\mathcal{S}(\cdot), \mathcal{B}, \mathcal{C}$ and $\mathcal{D}$ in (5.7). We note that the closed-loop system is automatically well-posed and that the formulas can be somewhat simplified because of the assumption that $D_{11} = 0$ and $D_{22} = 0$ (see also formula (2.118)). It follows that

$$\Sigma_{(\mathcal{F}(G,K))^\natural} = \Sigma(\mathcal{S}'(\cdot), \mathcal{B}', \mathcal{C}', \mathcal{D}').$$

Using the procedure that we used to derive (5.6) we can give a description of the feedback interconnection of $\Sigma_{G^\natural}$ and $\Sigma_{K^\natural}$, which we shall denote by $\Sigma_{\mathcal{F}(G^\natural, K^\natural)}$.

As in (2.113) we define

$$\begin{aligned} \tilde{S}_1(\cdot) &:= \begin{pmatrix} S'(\cdot) & 0 \\ 0 & T'(\cdot) \end{pmatrix} \quad ; \quad \tilde{B}_1 := \begin{pmatrix} C_2' & 0 \\ 0 & L' \end{pmatrix} \\ \tilde{C}_1 &:= \begin{pmatrix} B_2' & 0 \\ 0 & N' \end{pmatrix} \quad ; \quad \tilde{D}_1 := \begin{pmatrix} 0 & 0 \\ 0 & R' \end{pmatrix}. \end{aligned} \tag{5.40}$$

Furthermore, we define

$$\bar{D}_1 := \begin{pmatrix} I & R' \\ 0 & I \end{pmatrix}.$$

The feedback interconnection of $\Sigma_{G^\natural}$ and $\Sigma_{K^\natural}$ is given by

$$\Sigma_{\mathcal{F}(G^\natural, K^\natural)} = \Sigma(\mathcal{S}_1(\cdot), \mathcal{B}_1, \mathcal{C}_1, \mathcal{D}_1);$$

$$\Sigma_{\mathcal{F}(G^\natural, K^\natural)} \begin{cases} \begin{pmatrix} x'(t) \\ p'(t) \end{pmatrix} = \mathcal{S}_1(t) \begin{pmatrix} x_0' \\ p_0' \end{pmatrix} + \int_0^t \mathcal{S}(t-s)\mathcal{B}_1 w(s)ds \\ z(t) = \mathcal{C}_1 \begin{pmatrix} x'(t) \\ p'(t) \end{pmatrix} + \mathcal{D}_1 w(t), \end{cases}$$

where

$$\mathcal{S}_1(\cdot) = (\tilde{S}_1)_{\tilde{B}_1 \bar{D}_1 \mathcal{J} \tilde{C}_1}(\cdot) \qquad \mathcal{B}_1 = \begin{pmatrix} C_2' R' D_{12}' + C_1' \\ L' D_{12}' \end{pmatrix}$$

$$\mathcal{C}_1 = (B_1' + D_{21}' R' B_2' \;\; D_{21}' N') \quad \mathcal{D}_1 = D_{21}' R' D_{12}'.$$

Using Theorem 2.17, it is not difficult to show that $\mathcal{S}_1(\cdot) = \mathcal{S}'(\cdot)$. Furthermore, we have $\mathcal{B}_1 = \mathcal{C}', \mathcal{C}_1 = \mathcal{B}'$ and $\mathcal{D}_1 = \mathcal{D}'$. Hence

$$\Sigma_{(\mathcal{F}(G,K))^\natural} = \Sigma_{\mathcal{F}(G^\natural, K^\natural)}$$

and the result easily follows from Lemma 2.25. ∎

Since we assume that there exists an admissible controller Σ_K for Σ_G, Lemma 5.11 implies that $\Sigma_{K\natural}$ is admissible for $\Sigma_{G\natural}$. Furthermore, because of the assumptions (5.12) and (5.13), $\Sigma_{G\natural}$ satisfies the assumptions of Theorem 4.4 (see also Remark 5.3). Hence, that theorem implies the existence of Q satisfying item (*ii*) of Theorem 5.4.

Part b): The derivation of the coupling condition.

First of all, we use the Riccati equation for P and completion of the squares to construct a certain inner system that shall be crucial in the sequel.

Lemma 5.12
Let Σ_G be given by (5.2) and suppose that $D_{11} = 0$ and $D_{22} = 0$. Furthermore, suppose that P satisfies item (i) of Theorem 5.4. Then the C_0-semigroup

$$S_I(\cdot) := S(\cdot)_{-B_2(D_{12}'D_{12})^{-1}(B_2'P+D_{12}'C_1)}$$

is exponentially stable on $\mathcal{W}$ and $\mathcal{V}$ and the system

$$\Sigma_{G_I} := \Sigma(S_I(\cdot), \begin{pmatrix} B_1 & B_2(D_{12}'D_{12})^{-1/2} \end{pmatrix},$$

$$\begin{pmatrix} C_1 - D_{12}(D_{12}'D_{12})^{-1}(B_2'P + D_{12}'C_1) \\ -B_1'P \end{pmatrix}, \begin{pmatrix} 0 & D_{12}(D_{12}'D_{12})^{-1/2} \\ I & 0 \end{pmatrix})$$

given by

$$\Sigma_{G_I} \begin{cases} x_I(t) &= S_I(t)x_{I0} + \int_0^t S_I(t-s)(B_1 w(s) + \\ & \quad B_2(D_{12}'D_{12})^{-1/2}u_0(s))ds \\ z(t) &= (C_1 - D_{12}(D_{12}'D_{12})^{-1}(B_2'P + D_{12}'C_1))x_I(t) \\ & \quad +D_{12}(D_{12}'D_{12})^{-1/2}u_0(t) \\ w_0(t) &= -B_1'Px_I(t) + w(t), \end{cases} \tag{5.41}$$

satisfies

$$\|z(\cdot)\|^2_{L_2(0,\infty;Z)} + \|w_0(\cdot)\|^2_{L_2(0,\infty;W)} = \|w(\cdot)\|^2_{L_2(0,\infty;W)} + \|u_0(\cdot)\|^2_{L_2(0,\infty;U)}, \tag{5.42}$$

for all $w(\cdot) \in L_2(0,\infty;W)$ and $u_0(\cdot) \in L_2(0,\infty;U)$, provided that $x_{I0} = 0$.

Proof
The fact that $S_I(\cdot)$ is exponentially stable on $\mathcal{W}$ and $\mathcal{V}$ follows from the state-feedback result Theorem 4.4, because $S_I(\cdot) = S_{B_2F}(\cdot)$, where F is the admissible output operator given by

$$F = -(D_{12}'D_{12})^{-1}(B_2'P + D_{12}'C_1)$$

(see also formula (4.10)). Furthermore, using the perturbation results of Lemma 2.13, we know that the infinitesimal generator of $S_I(\cdot)$ on $\mathcal{V}$ is given by

$$A_I^{\mathcal{V}} = A^{\mathcal{V}} - B_2(D_{12}'D_{12})^{-1}D_{12}'C_1 - B_2(D_{12}'D_{12})^{-1}B_2'P. \tag{5.43}$$

Formula (5.43) and the Riccati equation for P shall be used to derive the completion of the squares formula (5.42). Consider the system Σ_{G_I} in (5.41) with $x_{I0} = 0$ and suppose that $w \in L_2(0,\infty;W)$ and $u_0 \in L_2(0,\infty;U)$. We claim that for any $T > 0$ there holds

$$\|z(\cdot)\|^2_{L_2(0,T;Z)} + \|w_0(\cdot)\|^2_{L_2(0,T;W)} - \|w(\cdot)\|^2_{L_2(0,T;U)} - \|u_0(\cdot)\|^2_{L_2(0,T;W)} =$$

$$- < x_I(T), Px_I(T) >_{<\mathcal{V},\mathcal{V}'>} . \tag{5.44}$$

In order to prove this claim we first assume that in addition $w(\cdot)$ and $u_0(\cdot)$ are continuously differentiable. Then, using Appendix B.3, we can differentiate:

$$\frac{d}{dt} < x_I(t), Px_I(t) >_{<\mathcal{V},\mathcal{V}'>}=$$

$$< A_I^{\mathcal{V}}x_I(t) + B_1w(t) + B_2(D_{12}'D_{12})^{-1/2}u_0(t), Px_I(t) >_{<\mathcal{V},\mathcal{V}'>} +$$

$$< x_I(t), P(A_I^{\mathcal{V}}x_I(t) + B_1w(t) + B_2(D_{12}'D_{12})^{-1/2}u_0(t)) >_{<\mathcal{V},\mathcal{V}'>}=$$

$$< (A^{\mathcal{V}} - B_2(D_{12}'D_{12})^{-1}D_{12}'C_1)x_I(t), Px_I(t) >_{<\mathcal{V},\mathcal{V}'>} +$$

$$< Px_I(t), (A^{\mathcal{V}} - B_2(D_{12}'D_{12})^{-1}D_{12}'C_1)x_I(t) >_{<\mathcal{V}',\mathcal{V}>} +$$

$$2 < B_1w(t), Px_I(t) >_{<\mathcal{V},\mathcal{V}'>} +2 < B_2(D_{12}'D_{12})^{-1/2}u_0(t), Px_I(t) >_{<\mathcal{V},\mathcal{V}'>}$$

$$-2 < B_2(D_{12}'D_{12})^{-1}B_2'Px_I(t), Px_I(t) >_{<\mathcal{V},\mathcal{V}'>} \text{ (using (5.43))} =$$

$$2 < B_1w(t), Px_I(t) >_{<\mathcal{V},\mathcal{V}'>} +2 < B_2(D_{12}'D_{12})^{-1/2}u_0(t), Px_I(t) >_{<\mathcal{V},\mathcal{V}'>}$$

$$-2 < B_2(D_{12}'D_{12})^{-1}B_2'Px_I(t), Px_I(t) >_{<\mathcal{V},\mathcal{V}'>} -$$

$$< P(B_1B_1' - B_2(D_{12}'D_{12})^{-1}B_2')Px_I(t), x_I(t) >_{<\mathcal{V}',\mathcal{V}>} -$$

$$< (I - D_{12}(D_{12}'D_{12})^{-1}D_{12}')C_1x_I(t), (I - D_{12}(D_{12}'D_{12})^{-1}D_{12}')C_1x_I(t) >_Z$$

(using the Riccati equation (5.14) for P with $\gamma = 1$)

$$= 2 < B_1w(t), Px_I(t) >_{<\mathcal{V},\mathcal{V}'>} - < PB_1B_1'Px_I(t), x_I(t) >_{<\mathcal{V}',\mathcal{V}>} +$$

$$\|u_0(t)\|_U^2 - \|z(t)\|_Z^2 \text{ (using the formula for } z(t) \text{ in (5.41))}$$

$$= -\|z(t)\|_Z^2 - \|w_0(t)\|_W^2 + \|u_0(t)\|_U^2 + \|w(t)\|_W^2, \tag{5.45}$$

(using the formula for $w_0(t)$ in (5.41)). We note the relation with the derivation of (4.52), where a similar completion of the squares formula is proved. Now integrating (5.45) from 0 to T gives (5.44). Furthermore, we can extend (5.44) to general $w \in L_2(0,\infty;W)$ and $u_0 \in L_2(0,\infty;U)$ by introducing smooth sequences $w_n \in L_2(0,T;W)$ and $u_{0n} \in L_2(0,T;U)$ that converge to $w(\cdot)$ and $u_0(\cdot)$ on $[0,T]$, just as we did in the proof of (4.24). Then a limiting argument using Appendix B.6 gives the result. The next step is to let T tend to ∞ in (5.44): since $S_I(\cdot)$ is exponentially stable on $\mathcal{W}$ and $\mathcal{V}$, it follows from Appendix B.1 that $x(T) \overset{\mathcal{V}}{\to} 0$ as $T \to \infty$ and so we infer (5.42). ■

Next, we define a transformed system Σ_{G_P} and we show that this system has the same admissible controllers as Σ_G, using the completion of the squares result from Lemma 5.12. We define

$$S_0(\cdot) := S_{B_1 B_1' P}(\cdot) \tag{5.46}$$

and

$$\Sigma_{G_P} := \Sigma(S_0(\cdot), \begin{pmatrix} B_1 & B_2 \end{pmatrix},$$

$$\begin{pmatrix} (D_{12}'D_{12})^{-1/2}(B_2'P + D_{12}'C_1) \\ C_2 + D_{21}B_1'P \end{pmatrix}, \begin{pmatrix} 0 & (D_{12}'D_{12})^{1/2} \\ D_{21} & 0 \end{pmatrix})$$

or, in state-space description,

$$\Sigma_{G_P} \begin{cases} x_P(t) &= S_0(t)x_{P0} + \int_0^t S_0(t-s)(B_1 w_0(s) + B_2 u(s))ds \\ u_0(t) &= (D_{12}'D_{12})^{-1/2}(B_2'P + D_{12}'C_1)x_P(t) + (D_{12}'D_{12})^{1/2}u(t) \\ y(t) &= (C_2 + D_{21}B_1'P)x_P(t) + D_{21}w_0(t). \end{cases} \tag{5.47}$$

The corresponding linear map G_P can be partitioned as

$$G_P : \begin{pmatrix} u_0(\cdot) \\ y(\cdot) \end{pmatrix} = \begin{pmatrix} {G_P}^{11} & {G_P}^{12} \\ {G_P}^{21} & {G_P}^{22} \end{pmatrix} \begin{pmatrix} w_0(\cdot) \\ u(\cdot) \end{pmatrix}. \tag{5.48}$$

Lemma 5.13

Let Σ_G be given by 5.2 and suppose that $D_{11} = 0$ and $D_{22} = 0$. Furthermore, suppose that P satisfies item (i) of Theorem 5.4 and let Σ_{G_P} be given by (5.47). Finally, let Σ_K be a controller of the form (5.3). We have the following equivalence:

The controller Σ_K is admissible for Σ_G if and only if it is admissible for Σ_{G_P}.

Proof

Recall the definition of Σ_{G_I} in (5.41). The corresponding linear map for zero initial conditions G_I is given by

$$G_I : \begin{pmatrix} z(\cdot) \\ w_0(\cdot) \end{pmatrix} = \begin{pmatrix} G_I^{11} & G_I^{12} \\ G_I^{21} & G_I^{22} \end{pmatrix} \begin{pmatrix} w(\cdot) \\ u_0(\cdot) \end{pmatrix}. \tag{5.49}$$

Suppose that we have a controller Σ_K of the form (5.3). We consider the feedback interconnections in Figure 5.3 and we define $G_{zw} := \mathcal{F}(G, K)$. Further-

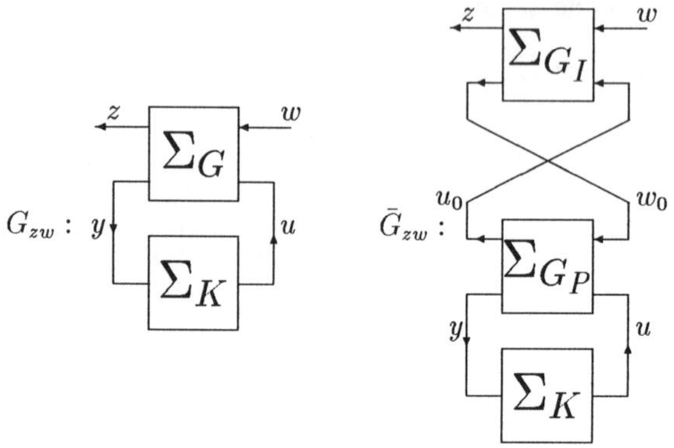

Figure 5.3: $G_{zw} = \bar{G}_{zw}$.

more, we define $\tilde{\mathcal{W}} := \mathcal{W} \times \mathcal{W}_K$, $\tilde{\mathcal{V}} := \mathcal{V} \times \mathcal{V}_K$, $\bar{\mathcal{W}} := \mathcal{W} \times \mathcal{W} \times \mathcal{W}_K$ and $\bar{\mathcal{V}} := \mathcal{V} \times \mathcal{V} \times \mathcal{V}_K$. As before, we can interpret $\Sigma_{G_{zw}}$ as a Pritchard-Salamon system on the state-spaces $\tilde{\mathcal{W}} \hookrightarrow \tilde{\mathcal{V}}$ (see also Figure 5.1 and (5.6)). The system on the right in Figure 5.3 shall be denoted by $\Sigma_{\bar{G}_{zw}}$ and it can be interpreted as a Pritchard-Salamon system on the state-spaces $\bar{\mathcal{W}} \hookrightarrow \bar{\mathcal{V}}$ in the obvious way. G_{zw} and $\bar{G}_{zw}$ denote the corresponding linear maps for zero initial conditions and we have $G_{zw} = \mathcal{F}(G, K)$ and $\bar{G}_{zw} = \mathcal{F}(G_I, \mathcal{F}(G_P, K))$. We have seen before that the feedback interconnection of Σ_G and Σ_K is well-posed because G_{22} has no feedthrough term. Similarly, the feedback interconnection of Σ_{G_P} and Σ_K, denoted by $\Sigma_{\mathcal{F}(G_P,K)}$, is well-posed because G_P^{22} has no feedthrough term, and the feedback interconnection of Σ_{G_I} and $\Sigma_{\mathcal{F}(G_P,K)}$ is well-posed because G_P^{22} has no feedthrough term.

We claim that the system on the left in Figure 5.3 is exponentially stable on $\tilde{\mathcal{W}}$ and $\tilde{\mathcal{V}}$ if and only the system on the right in Figure 5.3 is exponentially stable on $\bar{\mathcal{W}}$ and $\bar{\mathcal{V}}$ and that both closed-loop maps from w to z are the same, i.e. $G_{zw} = \bar{G}_{zw}$.

First of all, we reformulate the state equations of the system on the right with $w = 0$ (see the formulas for $\Sigma_{G_I}, \Sigma_{G_P}$ and Σ_K in (5.41), (5.47) and (5.3), respectively). Recall that $S_P(\cdot)$ is given by

$$S_P(\cdot) = S_{\left(B_1 \;\; B_2 \right)\left(\begin{array}{c} \gamma^{-2} B_1' P \\ -(D_{12}' D_{12})^{-1}(B_2' P + D_{12}' C_1) \end{array} \right)}(\cdot)$$

(see formula (5.15)). We have

$$x_I(t) = S_I(t) x_{I0} +$$

$$\int_0^t S_I(t-s)\left[B_2(D_{12}'D_{12})^{-1}(B_2'P + D_{12}'C_1)x_P(s) + B_2 u(s)\right] ds$$

(substituting $w(\cdot) = 0$ and the formula for $u_0(\cdot)$ in (5.41))

$$= S_P(t)x_{I0} + \int_0^t S_P(t-s)\left[-B_1 B_1' P x_I(s) + \right.$$

$$\left. B_2(D_{12}'D_{12})^{-1}(B_2'P + D_{12}'C_1)x_P(s) + B_2 u(s)\right] ds$$

(using the result on preliminary feedbacks in Lemma 2.14). Similarly, $x_P(t)$ in (5.47) can be expressed as

$$x_P(t) = S_P(t)x_{P0} + \int_0^t S_P(t-s)\left[-B_1 B_1' P x_I(s) + \right.$$

$$\left. B_2(D_{12}'D_{12})^{-1}(B_2'P + D_{12}'C_1)x_P(s) + B_2 u(s)\right] ds.$$

Denoting $e(t) := x_I(t) - x_P(t)$ it follows from the last two equations that

$$e(t) = S_P(t)(x_{I0} - x_{P0}) = S_P(t)e(0). \tag{5.50}$$

Furthermore, using Lemma 2.14, we can express $x_P(t)$ also as

$$x_P(t) =$$

$$S(t)x_{P0} + \int_0^t S(t-s)\left[B_1 B_1' P x_P(s) - B_1 B_1' P x_I(s) + B_2 u(s)\right] ds =$$

$$= S(t)x_{P0} + \int_0^t S(t-s)\left[B_2 u(s) - B_1 B_1' P e(s)\right] ds$$

and $y(t)$ in (5.47) can be expressed as

$$y(t) = C_2 x_P(t) - D_{21} B_1' P e(s).$$

Now since the equations for the controller Σ_K are given by

$$\left\{ \begin{array}{lcl} p(t) & = & T(t)p_0 + \int_0^t T(t-s) N u_2(s) ds \\ y_2(t) & = & Lp(t) + R u_2(t) \end{array} \right.$$

with $u_2(t) = y(t)$ and $u(t) = y_2(t)$, it follows from the beginning of Section 2.7 that

$$\begin{pmatrix} x_P(t) \\ p(t) \end{pmatrix} = \mathcal{S}(t)\begin{pmatrix} x_{P0} \\ p_0 \end{pmatrix} + \int_0^t \mathcal{S}(t-s)\begin{pmatrix} -(B_2RD_{21}+B_1)B_1'P \\ -ND_{21}B_1'P \end{pmatrix} e(s)ds, \tag{5.51}$$

where the C_0-semigroup $\mathcal{S}(\cdot)$ is precisely the C_0-semigroup corresponding to the closed-loop system on the left of Figure 5.3 (see also (5.6) and (5.7) with $D_{22} = 0$). Hence, the equations of the system on the right of Figure 5.3 with $w(t) = 0$ satisfy (5.50) and (5.51). Denoting the C_0-semigroup of the closed-loop system on the right of Figure 5.3 by $\bar{\mathcal{S}}(\cdot)$, we can express $x_I(\cdot), x_P(\cdot)$ and $p(\cdot)$ also as

$$\begin{pmatrix} x_I(\cdot) \\ x_P(\cdot) \\ p(\cdot) \end{pmatrix} = \bar{\mathcal{S}}(\cdot)\begin{pmatrix} x_{I0} \\ x_{P0} \\ p_0 \end{pmatrix}. \tag{5.52}$$

Since $S_P(\cdot)$ is exponentially stable on $\mathcal{V}$ and $\mathcal{W}$ we can use (5.50), (5.51) and (5.52) to show that $\mathcal{S}(\cdot)$ is exponentially stable on $\tilde{\mathcal{V}}$ and $\tilde{\mathcal{W}}$ if and only if $\bar{\mathcal{S}}(\cdot)$ is exponentially stable on $\bar{\mathcal{V}}$ and $\bar{\mathcal{W}}$. First we shall prove the sufficiency part of this statement. If $\bar{\mathcal{S}}(\cdot)$ is exponentially stable on $\bar{\mathcal{V}}$ and $\bar{\mathcal{W}}$ it follows from (5.52) that for arbitrary initial conditions $x_{I0} = x_{P0} \in \mathcal{V}$ and $p_0 \in \mathcal{V}_K$ (or $x_{I0} = x_{P0} \in \mathcal{W}$ and $p_0 \in \mathcal{W}_K$, respectively) there holds $x_I(\cdot), x_P(\cdot) \in L_2(0,\infty;\mathcal{V})$ (or $x_I(\cdot), x_P(\cdot) \in L_2(0,\infty;\mathcal{W})$, respectively) and so it follows from (5.50), (5.51) that $e(\cdot) = 0$ and

$$\mathcal{S}(\cdot)\begin{pmatrix} x_{P0} \\ p_0 \end{pmatrix} \in L_2(0,\infty;\tilde{\mathcal{V}})$$

$$(\text{or} \quad \mathcal{S}(\cdot)\begin{pmatrix} x_{P0} \\ p_0 \end{pmatrix} \in L_2(0,\infty;\tilde{\mathcal{W}}), \ \text{respectively}).$$

Hence Datko's result Lemma A.1 implies that $\mathcal{S}(\cdot)$ is exponentially stable on $\tilde{\mathcal{V}}$ and $\tilde{\mathcal{W}}$. For the necessity part we need Lemma (2.24). Let $(x_{I0}, x_{P0}, p_0) \in \bar{\mathcal{V}}$ (or $(x_{I0}, x_{P0}, p_0) \in \bar{\mathcal{W}}$, respectively) be arbitrary. If $\mathcal{S}(\cdot)$ is exponentially stable on $\tilde{\mathcal{V}}$ and $\tilde{\mathcal{W}}$ it follows from Lemma (2.24), formulas (5.50), (5.51) and the fact that $S_P(\cdot)$ is exponentially stable on $\mathcal{V}$ and $\mathcal{W}$ that $(e(\cdot), x_P(\cdot), p(\cdot)) \in L_2(0,\infty;\bar{\mathcal{V}})$ (or $(e(\cdot), x_P(\cdot), p(\cdot)) \in L_2(0,\infty;\bar{\mathcal{W}})$, respectively). Hence Datko's result Lemma A.1 implies that $\bar{\mathcal{S}}(\cdot)$ is exponentially stable on $\bar{\mathcal{V}}$ and $\bar{\mathcal{W}}$.

Next, we shall show that $\bar{G}_{zw} = G_{zw}$. Considering once more the system on the right in Figure 5.3 for nonzero $w(\cdot)$ and zero initial conditions, it is straightforward to show that the linear map $\bar{G}_{zw} = \mathcal{F}(G_I, \mathcal{F}(G_P, K))$ can be expressed as

$$\begin{pmatrix} x_P(t) \\ p(t) \\ z(t) \\ u(t) \end{pmatrix} = \begin{pmatrix} \int_0^t S(t-s)(B_1 w(s) + B_2 u(s) - B_1 B_1' P e(s)) ds \\ \int_0^t T(t-s) N y(s) ds \\ C_1 x_I(t) + D_{12} u(t) - D_{12}(D_{12}' D_{12})^{-1}(B_2' P + D_{12}' C_1) e(t) \\ Lp(t) + Ry(t) \end{pmatrix}$$

and that in fact $e(t) = x_I(t) - x_P(t) = 0$. Thus it follows that $\bar{G}_{zw} = G_{zw}$ and we have proved our claim. Therefore, Σ_K is admissible for Σ_G if and only if the system on the right in Figure 5.3 is exponentially stable on $\bar{\mathcal{W}}$ and $\bar{\mathcal{V}}$ and $\|\bar{G}_{zw}\| < 1$.

Finally, we want to apply Lemma 5.9 with Σ_G replaced by Σ_{G_I} (u becomes u_0 and y becomes w_0) and Σ_K replaced by $\Sigma_{\mathcal{F}(G_P,K)}$. We define $\mathcal{S}_P(\cdot)$ as the C_0-semigroup on $\tilde{\mathcal{W}} \hookrightarrow \tilde{\mathcal{V}}$ corresponding to the closed-loop system $\Sigma_{\mathcal{F}(G_P,K)}$. In Lemma 5.12 we have seen that Σ_{G_I} satisfies conditions (5.33) and (5.34). Furthermore, the linear map ${G_I}^{21}$ is given by

$${G_I}^{21} \begin{cases} x(t) &= \int_0^t S_I(t-s) B_1 \bar{w}(s) ds \\ \\ ({G_I}^{21}\bar{w})(t) &= -B_1' P x(t) + \bar{w}(t), \end{cases}$$

and it is easy to see that ${G_I}^{21}$ is invertible and that $({G_I}^{21})^{-1}$ is bounded (use Lemma 2.26 and the fact that $S_P(\cdot)$ is exponentially stable on $\mathcal{V}$). Hence, (5.35) is satisfied. Thus we have shown that G_I satisfies the assumptions of Lemma 5.9. Now Lemma 5.9 implies that the system on the right in Figure 5.3 is exponentially stable on $\bar{\mathcal{W}}$ and $\bar{\mathcal{V}}$ and $\|\mathcal{F}(G_I, \mathcal{F}(G_P, K))\| = \|\bar{G}_{zw}\| < 1$ if and only if the C_0-semigroup $\mathcal{S}_P(\cdot)$ corresponding to $\Sigma_{\mathcal{F}(G_P,K)}$ is exponentially stable on both $\tilde{\mathcal{W}}$ and $\tilde{\mathcal{V}}$ and $\|\mathcal{F}(G_P, K)\| < 1$, i.e. Σ_K is admissible for Σ_{G_P}. This completes the proof of Lemma 5.13. ∎

Now we are ready to derive the coupling condition (5.18) in Theorem 5.4. In the rest of this part **b)** we shall assume that Σ_G satisfies the assumptions of Theorem 5.4 and we suppose that there exists an admissible controller Σ_K for Σ_G. It follows from part **a)** that there exists an operator P that satisfies item (i) of Theorem 5.4. Defining Σ_{G_P} as in (5.47), it follows from Lemma 5.13 that Σ_K is also an admissible controller for Σ_{G_P}. Next, we use the duality argument from Lemma 5.11. The lemma implies that $\Sigma_{K^\natural}$ is admissible for $\Sigma_{(G_P)^\natural}$. Here

$$\Sigma_{(G_P)^\natural} = \Sigma(S_0'(\cdot), \begin{pmatrix} (PB_2 + C_1' D_{12})(D_{12}' D_{12})^{-1/2} & C_2' + P B_1 D_{21}' \end{pmatrix},$$

$$\left(\begin{pmatrix} B_1' \\ B_2' \end{pmatrix}, \begin{pmatrix} 0 & D_{21}' \\ (D_{12}'D_{12})^{1/2} & 0 \end{pmatrix} \right)$$

and

$$\Sigma_{(G_P)^\natural} \begin{cases} x'(t) &= S_0'(t)x_0' + \int_0^t S_0'(t-s)((PB_2 + C_1'D_{12}) \\ & \quad (D_{12}'D_{12})^{-1/2}w(s) + (C_2' + PB_1D_{21}')u(s))ds \\ z(t) &= B_1'x'(t) + D_{21}'u(t) \\ y(t) &= B_2'x'(t) + (D_{12}'D_{12})^{1/2}w(t). \end{cases} \tag{5.53}$$

Now we would like to apply Theorem 4.4 to $\Sigma_{(G_P)^\natural}$ and so we have to show that the assumptions of the theorem are satisfied. We recall that the assumption that Σ_G satisfies (2.8) implies that $D((A^{\mathcal{W}})') \hookrightarrow \mathcal{V}'$ (see Theorem 2.17). We shall show that

$$D((A_0^{\mathcal{W}})') \hookrightarrow \mathcal{V}' \tag{5.54}$$

and using assumptions (5.12)-(5.13) we shall show that $\Sigma_{(G_P)^\natural}$ satisfies assumptions (5.10)-(5.11), where $\Sigma_{(G_P)^\natural}$ should be considered to have the form (5.2) with

$$\begin{aligned} B_1 &= (PB_2 + C_1'D_{12})(D_{12}'D_{12})^{-1/2} \\ B_2 &= C_2' + PB_1D_{21}' \\ C_1 &= B_1' \\ C_2 &= B_2' \\ D_{11} &= 0 \\ D_{22} &= 0 \\ D_{12} &= D_{21}' \\ D_{21} &= (D_{12}'D_{12})^{1/2} \\ S(\cdot) &= S_0'(\cdot). \end{aligned} \tag{5.55}$$

Let us denote the infinitesimal generator of $S_0(\cdot)$ on $\mathcal{W}$ by $A_0^{\mathcal{W}}$. Then $(A_0^{\mathcal{W}})'$ is the infinitesimal generator of $S_0'(\cdot)$ on $\mathcal{W}'$ (see Lemma 2.15). Furthermore, it follows from the definition of $S_0(\cdot)$ (see (5.46)) and Theorem 2.17 that $S_0'(\cdot) = S'_{(B_1'P)'B_1'}(\cdot) = S'_{PB_1B_1'}(\cdot)$. Since $D((A^{\mathcal{W}})') \hookrightarrow \mathcal{V}'$ and $P \in \mathcal{L}(\mathcal{V}, \mathcal{V}')$ it follows from Lemma 2.13 that $D((A_0^{\mathcal{W}})') = D((A^{\mathcal{W}})')$ with equivalent graph norms and there holds $(A_0^{\mathcal{W}})'x' = ((A^{\mathcal{W}})' + PB_1B_1')x'$ for all $x' \in D((A^{\mathcal{W}})')$. Hence we have shown (5.54) and to prove the invariant zeros condition for $\Sigma_{(G_P)^\natural}$ we have to show that there exists an $\epsilon > 0$ such that for all $(\omega, x', y) \in \mathbb{R} \times D((A^{\mathcal{W}})') \times Y$ satisfying $i\omega x' = ((A^{\mathcal{W}})' + PB_1B_1')x' + (C_2' + PB_1D_{12}')y$, there holds

$$\|B_1'x' + D_{21}'y\|_W^2 \geq \epsilon \|x'\|_{\mathcal{W}'}^2. \tag{5.56}$$

To prove (5.56) we shall of course use assumption (5.12). Since there exists an admissible controller for Σ_G it follows that the pair $(C_2, S(\cdot))$ is admissibly detectable (see Theorem 2.27). Furthermore, Theorem 2.20 implies that the pair $(S'(\cdot), C_2')$ is admissibly stabilizable. Hence, it follows from Appendix C.1 item (v) that our assumption (5.12) is equivalent to the existence of some $\epsilon > 0$ such that for all $(\omega, x', y) \in \mathbb{R} \times D((A^{\mathcal{W}})') \times Y$

$$\left\| \begin{matrix} i\omega x' - (A^{\mathcal{W}})'x' - C_2'y \\ B_1'x' + D_{21}'y \end{matrix} \right\|^2_{\mathcal{W}'\times W} \geq \epsilon(\|x'\|^2_{\mathcal{W}'} + \|y\|^2_Y).$$

The relation between this last equation and (5.56) is given by

$$\begin{pmatrix} i\omega I - (A_0^{\mathcal{W}})' & -(C_2' + PB_1D_{12}') \\ B_1' & D_{21}' \end{pmatrix} =$$

$$\begin{pmatrix} I & -PB_1 \\ 0 & I \end{pmatrix} \begin{pmatrix} i\omega I - (A^{\mathcal{W}})' & -C_2' \\ B_1' & D_{21}' \end{pmatrix}$$

which is satisfied for all $\omega \in \mathbb{R}$. Since

$$\begin{pmatrix} I & -PB_1 \\ 0 & I \end{pmatrix}^{-1} = \begin{pmatrix} I & PB_1 \\ 0 & I \end{pmatrix} \in \mathcal{L}(\mathcal{W}' \times W),$$

there holds

$$\left\| \begin{pmatrix} I & -PB_1 \\ 0 & I \end{pmatrix} x \right\|_{\mathcal{W}'\times W} \geq const\|x\|_{\mathcal{W}'\times W} \quad \text{for all } x \in \mathcal{W}' \times W$$

and so (5.56) easily follows. Furthermore, because of (5.13) we may conclude that $\Sigma_{(G_P)^\natural}$ satisfies assumptions (5.10)-(5.11) (again, $\Sigma_{(G_P)^\natural}$ should be considered to have the form (5.2) with (5.55)).
Since $\Sigma_{(G_P)^\natural}$ satisfies all the assumptions of Theorem 4.4 and since $\Sigma_K^\natural$ is admissible for $\Sigma_{(G_P)^\natural}$ the theorem implies the existence of a nonnegative definite $\bar{P} \in \mathcal{L}(\mathcal{W}', \mathcal{W})$ such that

$$\begin{aligned} &< \bar{P}x, ((A_0^{\mathcal{W}})' - (C_2' + PB_1D_{21}')(D_{21}D_{21}')^{-1}D_{21}B_1')y >_{<\mathcal{W},\mathcal{W}'>} + \\ &< ((A_0^{\mathcal{W}})' - (C_2' + PB_1D_{21}')(D_{21}D_{21}')^{-1}D_{21}B_1')x, \bar{P}y >_{<\mathcal{W}',\mathcal{W}>} + \\ &< \bar{P}((PB_2 + C_1'D_{12})(D_{12}'D_{12})^{-1}(B_2'P + D_{12}'C_1) \\ &-(C_2' + PB_1D_{21}')(D_{21}D_{21}')^{-1}(C_2 + D_{21}B_1'P))\bar{P}x, y >_{<\mathcal{W},\mathcal{W}'>} + \\ &< (I - D_{21}'(D_{21}D_{21}')^{-1}D_{21})B_1'x, \\ &(I - D_{21}'(D_{21}D_{21}')^{-1}D_{21})B_1'y >_W = 0 \end{aligned} \tag{5.57}$$

for all $x, y \in D((A^{\mathcal{W}})')$, such that the C_0-semigroup $S_{\bar{P}}(\cdot)$ defined via its adjoint as

$$S'_{\bar{P}}(\cdot) = (S'_0)\begin{pmatrix} (PB_2 + C'_1 D_{12})(D'_{12}D_{12})^{-1/2} & (C'_2 + PB_1 D'_{21}) \end{pmatrix}$$

$$\begin{pmatrix} (D'_{12}D_{12})^{-1/2}(B'_2 P + D'_{12}C_1)\bar{P} \\ -(D_{21}D'_{21})^{-1}((C_2 + D_{21}B'_1 P)\bar{P} + D_{21}B'_1) \end{pmatrix}(\cdot), \qquad (5.58)$$

is exponentially stable on $\mathcal{W}$. In fact, Theorem 4.4 implies that $S'_{\bar{P}}(\cdot)$ is exponentially stable on the 'larger' space $\mathcal{W}'$, which is equivalent to exponential stability on $\mathcal{W}$ of $S_{\bar{P}}(\cdot)$ (see Lemma 2.15). Since $S_P(\cdot)$ is exponentially stable on $\mathcal{W}$ and $\mathcal{V}$ it follows from Theorem 2.20 item (iv) that $S_{\bar{P}}(\cdot)$ is also exponentially stable on $\mathcal{V}$.

So using the assumption that there exists an admissible controller Σ_K for Σ_G, we have derived the existence of $P, \bar{P}$ and Q satisfying the three Riccati equations (5.14), (5.16) and (5.57), combined with the corresponding stability properties. The point is that there exists an easy relationship between $P, \bar{P}$ and Q. It is shown in Appendix D, Lemma D.3 that $(I - PQ)^{-1} \in \mathcal{L}(\mathcal{W}')$, that $\bar{P} = Q(I - PQ)^{-1}$ and that in fact

$$r_\sigma^{\mathcal{W}'}(PQ) = r_\sigma^{\mathcal{V}}(QP) < 1.$$

The proof of the above result is given in the appendix because it is rather lenghty and contains of a lot of tedious calculations. This concludes part **b)** of the proof of Theorem 5.4.

Part c): The controller parametrization.

In this part we shall frequently use the results of Section 2.7; in particular, we shall exploit the fact that taking inverses, cascade interconnections and linear fractional transformations leaves the class of Pritchard-Salamon systems invariant.

Consider the smooth Pritchard-Salamon system Σ_G given by (5.2), satisfying $D_{11} = 0$ and $D_{22} = 0$. Furthermore, let us suppose that P satisfies item (i) of Theorem 5.4 and that we have a nonnegative definite $\bar{P} \in \mathcal{L}(\mathcal{W}', \mathcal{W})$ such that (5.57) is satisfied and $S_{\bar{P}}(\cdot)$ defined by (5.58) is exponentially stable on $\mathcal{W}$ and $\mathcal{V}$. We define Σ_{G_P} as in (5.47) and $\Sigma_{(G_P)^\natural}$ as in (5.53). Now we can define a system $\Sigma_{((G_P)^\natural)_{\bar{P}}}$ just as in (5.47), where

$$\Sigma_G = \Sigma(S(\cdot), (B_1 \;\; B_2), \begin{pmatrix} C_1 \\ C_2 \end{pmatrix}, \begin{pmatrix} D_{11} & D_{12} \\ D_{21} & D_{22} \end{pmatrix})$$

is replaced by

$$\Sigma_{(G_P)^\natural} = \Sigma(S'_0(\cdot), \begin{pmatrix} (PB_2 + C'_1 D_{12})(D'_{12}D_{12})^{-1/2} & C'_2 + PB_1 D'_{21} \end{pmatrix},$$

$$\begin{pmatrix} B_1' \\ B_2' \end{pmatrix}, \begin{pmatrix} 0 & D_{21}' \\ (D_{12}'D_{12})^{1/2} & 0 \end{pmatrix})$$

and P is replaced by $\bar{P}$. The following equivalences are an immediate consequence of Lemmas 5.11 and 5.13:

Σ_K is admissible for Σ_G
if and only if
$\Sigma_{K^\natural}$ is admissible for $\Sigma_{G^\natural}$
if and only if
$\Sigma_{K^\natural}$ is admissible for $\Sigma_{((G_P)^\natural)_{\bar{P}}}$
if and only if
Σ_K is admissible for $\Sigma_{(((G_P)^\natural)_{\bar{P}})^\natural}$.

The reason for all these transformations is that $\Sigma_{(((G_P)^\natural)_{\bar{P}})^\natural}$ has a very nice structure, as was also realized in [86] for the finite-dimensional case. Here we shall use this structure to characterize all admissible controllers.

First we give two lemmas to characterize all admissible controllers for a system that has the same structure as $\Sigma_{(((G_P)^\natural)_{\bar{P}})^\natural}$. Suppose that we have a Pritchard-Salamon system given by

$$\Sigma_{\bar{G}} = \Sigma(\bar{S}(\cdot), (\bar{B}_1 \ \ \bar{B}_2), \begin{pmatrix} \bar{C}_1 \\ \bar{C}_2 \end{pmatrix}, \begin{pmatrix} 0 & \bar{D}_{12} \\ \bar{D}_{21} & 0 \end{pmatrix})$$

and

$$\Sigma_{\bar{G}} \begin{cases} x(t) &= \bar{S}(t)x_0 + \int_0^t \bar{S}(t-s)(\bar{B}_1 w(s) + \bar{B}_2 u(s))ds \\ z(t) &= \bar{C}_1 x(t) + \bar{D}_{12} u(t) \\ y(t) &= \bar{C}_2 x(t) + \bar{D}_{21} w(t), \end{cases} \tag{5.59}$$

where $\bar{D}_{12}$ and $\bar{D}_{21}$ are both boundedly invertible. We also consider the description of the corresponding linear map for zero initial conditions, denoted by $\bar{G}$, as in (5.4):

$$\bar{G} : \begin{pmatrix} z(\cdot) \\ y(\cdot) \end{pmatrix} = \begin{pmatrix} \bar{G}_{11} & \bar{G}_{12} \\ \bar{G}_{21} & \bar{G}_{22} \end{pmatrix} \begin{pmatrix} w(\cdot) \\ u(\cdot) \end{pmatrix}, \tag{5.60}$$

where $\bar{G}_{ij}$ represent the corresponding linear maps. Comparing (5.60) with (2.111) we see that now the feedthrough operators of $\Sigma_{\bar{G}_{12}}$ and $\Sigma_{\bar{G}_{21}}$ are both invertible so that $\bar{G}_{12}$ and $\bar{G}_{21}$ are both invertible (see Lemma 2.26). As before, $\Sigma_{\bar{G}_{11}}$ and $\Sigma_{\bar{G}_{22}}$ have no feedthrough operator. The feedthrough operator of $\Sigma_{\bar{G}}$ given by

$$\bar{D} := \begin{pmatrix} 0 & \bar{D}_{12} \\ \bar{D}_{21} & 0 \end{pmatrix}$$

is boundedly invertible and there holds

$$\bar{D}^{-1} = \begin{pmatrix} 0 & (\bar{D}_{21})^{-1} \\ (\bar{D}_{12})^{-1} & 0 \end{pmatrix}.$$

According to Lemma 2.26, the linear map $\bar{G}^{-1}$ exists and it can be realized as the Pritchard-Salamon system

$$\Sigma_{\bar{G}^{-1}} =$$

$$\Sigma(\bar{S}_{-(\bar{B}_1 \;\; \bar{B}_2)\bar{D}^{-1}\begin{pmatrix} \bar{C}_1 \\ \bar{C}_2 \end{pmatrix}}(\cdot), (\bar{B}_1 \;\; \bar{B}_2)\bar{D}^{-1}, -\bar{D}^{-1}\begin{pmatrix} \bar{C}_1 \\ \bar{C}_2 \end{pmatrix}, \bar{D}^{-1}) =$$

$$\Sigma(\bar{S}_{-\bar{B}_1(\bar{D}_{21})^{-1}\bar{C}_2-\bar{B}_2(\bar{D}_{12})^{-1}\bar{C}_1}(\cdot), \left(\bar{B}_2(\bar{D}_{12})^{-1} \;\; \bar{B}_1(\bar{D}_{21})^{-1} \right),$$

$$\begin{pmatrix} -(\bar{D}_{21})^{-1}\bar{C}_2 \\ -(\bar{D}_{12})^{-1}\bar{C}_1 \end{pmatrix}, \begin{pmatrix} 0 & (\bar{D}_{21})^{-1} \\ (\bar{D}_{12})^{-1} & 0 \end{pmatrix}).$$

Next, we shall define a system $\Sigma_{\bar{G}_2}$ that is similar to $\Sigma_{\bar{G}^{-1}}$, the difference being that the roles of w and u and the roles of z and y are interchanged:

$$\Sigma_{\bar{G}_2} := \Sigma(\bar{S}_{-\bar{B}_1(\bar{D}_{21})^{-1}\bar{C}_2-\bar{B}_2(\bar{D}_{12})^{-1}\bar{C}_1}(\cdot), \left(\bar{B}_1(\bar{D}_{21})^{-1} \;\; \bar{B}_2(\bar{D}_{12})^{-1} \right),$$

$$\begin{pmatrix} -(\bar{D}_{12})^{-1}\bar{C}_1 \\ -(\bar{D}_{21})^{-1}\bar{C}_2 \end{pmatrix}, \begin{pmatrix} 0 & (\bar{D}_{12})^{-1} \\ (\bar{D}_{21})^{-1} & 0 \end{pmatrix}),$$

with the corresponding state-space description

$$\Sigma_{\bar{G}_2} \begin{cases} p_1(t) &= \bar{S}_{-\bar{B}_1(\bar{D}_{21})^{-1}\bar{C}_2-\bar{B}_2(\bar{D}_{12})^{-1}\bar{C}_1}(t)p_{10}+ \\ & \quad \int_0^t \bar{S}_{-\bar{B}_1(\bar{D}_{21})^{-1}\bar{C}_2-\bar{B}_2(\bar{D}_{12})^{-1}\bar{C}_1}(t-s) \\ & \quad (\bar{B}_1(\bar{D}_{21})^{-1}y_2(s) + \bar{B}_2(\bar{D}_{12})^{-1}v(s))ds \\ u_2(t) &= -(\bar{D}_{12})^{-1}\bar{C}_1 p_1(t) + (\bar{D}_{12})^{-1}v(t) \\ r(t) &= -(\bar{D}_{21})^{-1}\bar{C}_2 p_1(t) + (\bar{D}_{21})^{-1}y_2(t). \end{cases} \tag{5.61}$$

Denoting the corresponding linear map with zero initial conditions by

$$\bar{G}_2 : \begin{pmatrix} u_2(\cdot) \\ r(\cdot) \end{pmatrix} = \begin{pmatrix} \bar{G}_2^{11} & \bar{G}_2^{12} \\ \bar{G}_2^{21} & \bar{G}_2^{22} \end{pmatrix} \begin{pmatrix} y_2(\cdot) \\ v(\cdot) \end{pmatrix}, \tag{5.62}$$

it follows that

$$\begin{pmatrix} \bar{G}_{11} & \bar{G}_{12} \\ \bar{G}_{21} & \bar{G}_{22} \end{pmatrix}^{-1} = \begin{pmatrix} 0 & I \\ I & 0 \end{pmatrix} \begin{pmatrix} \bar{G}_2^{11} & \bar{G}_2^{12} \\ \bar{G}_2^{21} & \bar{G}_2^{22} \end{pmatrix} \begin{pmatrix} 0 & I \\ I & 0 \end{pmatrix}. \tag{5.63}$$

The reason for defining $\Sigma_{\bar{G}_2}$ will become clear in the sequel.
Next, let $\Sigma_{\bar{K}} = \Sigma(T(\cdot), N, L, R)$ be any controller of the form (5.3) for (5.59)

$$\Sigma_{\bar{K}} \left\{ \begin{array}{rcl} p(t) & = & T(t)p_0 + \int_0^t T(t-s)Ny_2(s)ds \\ u_2(t) & = & Lp(t) + Ry_2(t) \end{array} \right. \tag{5.64}$$

(as usual, the state-spaces are $\mathcal{W}_K \hookrightarrow \mathcal{V}_K$). The following holds.

Lemma 5.14
Let $\Sigma_{\bar{G}}$ be the Pritchard-Salamon system given by (5.59) such that $\bar{D}_{12}$ and $\bar{D}_{21}$ are both boundedly invertible and let $\Sigma_{\bar{K}}$ be a (Pritchard-Salamon) controller of the form (5.64). Furthermore, let $\bar{G}$ and $\bar{K}$ denote the corresponding linear maps for zero initial conditions. Finally, let $\Sigma_{\bar{G}_2}$ be the Pritchard-Salamon system defined by (5.61) with corresponding linear map given by (5.62). Then there exist separable Hilbert spaces $\mathcal{W}_\Lambda, \mathcal{V}_\Lambda$ and a Pritchard-Salamon system

$$\Sigma_\Lambda = \Sigma(S_\Lambda(\cdot), B_\Lambda, C_\Lambda, D_\Lambda) \tag{5.65}$$

with respect to $(\mathcal{W}_\Lambda, \mathcal{V}_\Lambda)$ given by

$$\Sigma_\Lambda \left\{ \begin{array}{rcl} \lambda(t) & = & S_\Lambda(t)\lambda_0 + \int_0^t S_\Lambda(t-s)B_\Lambda u_\Lambda(s)ds \\ y_\Lambda(t) & = & C_\Lambda\lambda(t) + D_\Lambda u_\Lambda(t) \end{array} \right. \tag{5.66}$$

with corresponding linear map Λ (for zero initial conditions) such that

$$\bar{K} = \bar{G}_2^{11} + \bar{G}_2^{12}\Lambda(I - \bar{G}_2^{22}\Lambda)^{-1}\bar{G}_2^{21} = \mathcal{F}(\bar{G}_2, \Lambda), \tag{5.67}$$

(note that $(I - \bar{G}_2^{22}\Lambda)^{-1}$ exists because there is no feedthrough term in $\bar{G}_2^{22}$).

The lemma implies that $\Sigma_{\bar{K}}$ can be seen as the interconnection of (5.61) and (5.66) with $u_\Lambda(\cdot) = r(\cdot)$ and $y_\Lambda(\cdot) = v(\cdot)$ as in Figure 5.4.

Proof
Since $\bar{G}_2^{21}$ and $\bar{G}_2^{12}$ are both invertible (the corresponding feedthrough operators $(\bar{D}_{12})^{-1}$ and $(\bar{D}_{21})^{-1}$ are invertible), we can define a linear map $\tilde{\Lambda}$ by

$$\tilde{\Lambda} := (\bar{G}_2^{12})^{-1}(\bar{K} - \bar{G}_2^{11})(\bar{G}_2^{21})^{-1} \tag{5.68}$$

and another linear map Λ by

$$\Lambda := (I + \tilde{\Lambda}\bar{G}_2^{22})^{-1}\tilde{\Lambda} \tag{5.69}$$

(note that $(I + \tilde{\Lambda}\bar{G}_2^{22})^{-1}$ exists because there is no feedthrough term in $\bar{G}_2^{22}$). Then $\tilde{\Lambda} = \Lambda(I - \bar{G}_2^{22}\Lambda)^{-1}$ and it is straightforward to show that (5.67) is

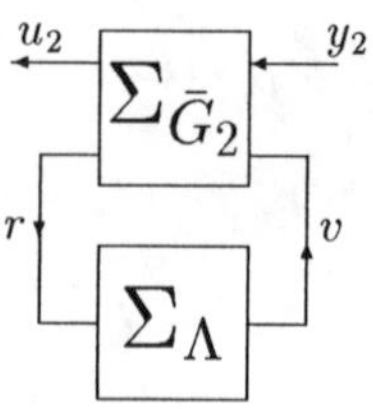

Figure 5.4: $\bar{K} = \mathcal{F}(\bar{G}_2, \Lambda)$.

satisfied. Furthermore, Λ can be considered as the linear map corresponding to a Pritchard-Salamon system of the form (5.66), because taking inverses, cascade interconnections and linear fractional transformations leaves the class of Pritchard-Salamon systems invariant, as noted at the beginning of this part **c)**. ∎

Now let $\Sigma_{\bar{G}_{zw}}$ denote the closed-loop system determined by $\Sigma_{\bar{G}}$ and $\Sigma_{\bar{K}}$ with $y_2(\cdot) = y(\cdot)$ and $u_2(\cdot) = u(\cdot)$. Denoting the corresponding linear map for zero initial conditions by $\bar{G}_{zw}$ we have

$$\bar{G}_{zw} = \bar{G}_{11} + \bar{G}_{12}\bar{K}(I - \bar{G}_{22}\bar{K})^{-1}\bar{G}_{21} = \mathcal{F}(\bar{G}, \bar{K}). \tag{5.70}$$

The following result may seem surprising, but in fact it follows from (5.63). It is reminiscent of the work of Redheffer [76], where linear fractional transformations are related to his so-called $*$-product and relationships between the $*$-product inverse and the ordinary inverse are established.

Lemma 5.15
Let $\bar{G}, \bar{G}_2$ and $\bar{K}$ be the linear maps corresponding to the Pritchard-Salamon systems (5.59), (5.61) *and* (5.64) *for zero initial conditions and denote $\bar{G}_{zw}$ as in* (5.70). *Furthermore, let Σ_Λ be any Pritchard-Salamon system of the form* (5.65) (*with corresponding linear map Λ*) *such that $\bar{K} = \mathcal{F}(\bar{G}_2, \Lambda)$. There holds*

$$\bar{G}_{zw} = \mathcal{F}(\bar{G}, \bar{K}) = \mathcal{F}(\bar{G}, \mathcal{F}(\bar{G}_2, \Lambda)) = \Lambda \tag{5.71}$$

and

$$\bar{K} = \mathcal{F}(\bar{G}_2, \Lambda) = \mathcal{F}(\bar{G}_2, \mathcal{F}(\bar{G}, \bar{K})). \tag{5.72}$$

Proof
Since $\bar{K} = \mathcal{F}(\bar{G}_2, \Lambda)$, the closed-loop system $\bar{G}_{zw}$ can be described as in Figure 5.5, where $\overset{i/o}{\Longleftrightarrow}$ means that the system on the right has the same input

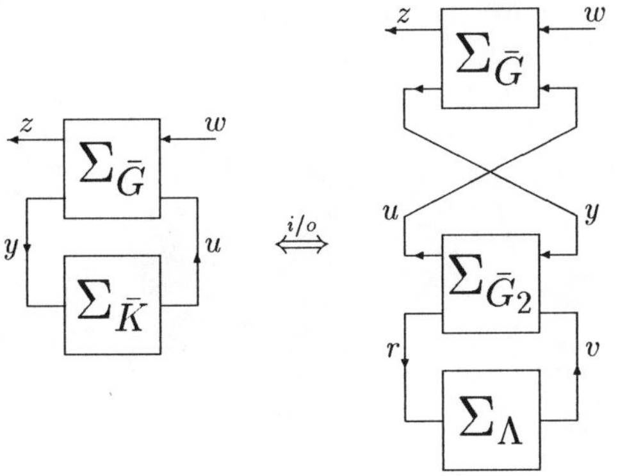

Figure 5.5: $\bar{G}_{zw}$

output map (zero initial conditions) as the system on the left. Now define $\Sigma_{\tilde{G}}$ as the interconnection of $\Sigma_{\bar{G}}$ given by (5.59) and $\Sigma_{\bar{G}_2}$ given by (5.61) with $y_2(\cdot) = y(\cdot)$ and $u_2(\cdot) = u(\cdot)$, as in Figure 5.6. Of course this type of feedback

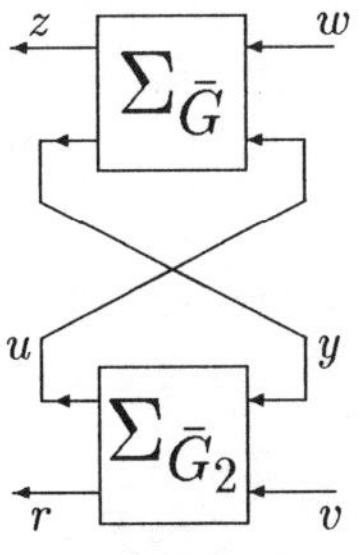

Figure 5.6: $\tilde{G}$

interconnection should be interpreted as in Section 2.7 on the state-spaces $\mathcal{W} \times \mathcal{W} \hookrightarrow \mathcal{V} \times \mathcal{V}$. It is straightforward to show that $x(\cdot), p_1(\cdot), z(\cdot)$ and $r(\cdot)$

of (5.59) and (5.61) now satisfy

$$\left\{\begin{array}{rcl} x(t)-p_1(t) &=& \bar{S}_{-\bar{B}_1(\bar{D}_{21})^{-1}\bar{C}_2}(t)(x_0-p_{10}) \\ x(t) &=& \bar{S}_{-\bar{B}_2(\bar{D}_{12})^{-1}\bar{C}_1}(t)x_0+\int_0^t \bar{S}_{-\bar{B}_2(\bar{D}_{12})^{-1}\bar{C}_1}(t-s) \\ && (\bar{B}_2(\bar{D}_{12})^{-1}\bar{C}_1(x(s)-p_1(s))+ \\ && \bar{B}_1 w(s)+\bar{B}_2(\bar{D}_{12})^{-1}v(s))ds \\ z(t) &=& \begin{pmatrix} \bar{C}_1 & -\bar{C}_1 \end{pmatrix}\begin{pmatrix} x(t) \\ p_1(t) \end{pmatrix}+v(t) \\ r(t) &=& \begin{pmatrix} (\bar{D}_{21})^{-1}\bar{C}_2 & -(\bar{D}_{21})^{-1}\bar{C}_2 \end{pmatrix}\begin{pmatrix} x(t) \\ p_1(t) \end{pmatrix}+w(t) \end{array}\right. \tag{5.73}$$

(formal differentiation shows that the above equations should hold, a rigorous proof can be given using Lemma 2.14, just as in the proof of (5.50) and (5.51)). Hence we may conclude that for $x(0)=0$ and $p_1(0)=0$, we have that $x(\cdot)-p_1(\cdot)=0$ and so

$$\begin{pmatrix} z(\cdot) \\ r(\cdot) \end{pmatrix} = \tilde{G}\begin{pmatrix} w(\cdot) \\ v(\cdot) \end{pmatrix} =$$

$$\begin{pmatrix} \tilde{G}_{11} & \tilde{G}_{12} \\ \tilde{G}_{21} & \tilde{G}_{22} \end{pmatrix}\begin{pmatrix} w(\cdot) \\ v(\cdot) \end{pmatrix} = \begin{pmatrix} 0 & I \\ I & 0 \end{pmatrix}\begin{pmatrix} w(\cdot) \\ v(\cdot) \end{pmatrix} \tag{5.74}$$

(in fact, this result is due to the special relationship between $\bar{G}$ and $\bar{G}_2$ in (5.63) and the work of Redheffer mentioned before). Now the fact that $\mathcal{F}(\bar{G},\mathcal{F}(\bar{G}_2,\Lambda))=\Lambda$ follows from Figure 5.5 and (5.74) and so we infer (5.71). Finally, (5.72) follows from (5.71). ∎

Using Lemmas 5.14, 5.15, formula (5.73) and some extra conditions, we can parametrize all admissible controllers of $\Sigma_{\bar{G}}$:

Lemma 5.16

Let $\Sigma_{\bar{G}}$ be the Pritchard-Salamon system given by (5.59) such that $\bar{D}_{12}$ and $\bar{D}_{21}$ are both boundedly invertible and let $\Sigma_{\bar{K}}$ be a (Pritchard-Salamon) controller of the form (5.64) such that $(T(\cdot),N)$ is admissibly stabilizable and $(L,T(\cdot))$ admissibly detectable. Furthermore, let $\Sigma_{\bar{G}_2}$ be the Pritchard-Salamon system defined by (5.61). Finally, suppose that $\bar{S}_{-\bar{B}_1(\bar{D}_{21})^{-1}\bar{C}_2}(\cdot)$ and $\bar{S}_{-\bar{B}_2(\bar{D}_{12})^{-1}\bar{C}_1}(\cdot)$ are both exponentially stable on $\mathcal{W}$ and $\mathcal{V}$. Then $\Sigma_{\bar{K}}$ is admissible for $\Sigma_{\bar{G}}$ if and only if there exists a Pritchard-Salamon system Σ_Λ of the form (5.65)-(5.66) (with corresponding linear map Λ), such that

$$\bar{K}=\mathcal{F}(\bar{G}_2,\Lambda),$$

$\|\Lambda\| < 1$ *and*

$S_\Lambda(\cdot)$ *is exponentially stable on* $\mathcal{W}_\Lambda$ *and* $\mathcal{V}_\Lambda$.

In this case, $\Sigma_{\mathcal{F}(\bar{G}_2,\Lambda)}$ *(with its natural realization on* $\mathcal{W} \times \mathcal{W}_\Lambda \hookrightarrow \mathcal{V} \times \mathcal{V}_\Lambda$*) is also itself admissible and we have* $\bar{G}_{zw} = \mathcal{F}(\bar{G}, \bar{K}) = \Lambda$.

Proof

(necessity) Suppose that $\Sigma_{\bar{K}}$ given by (5.64) is admissible for $\Sigma_{\bar{G}}$ given by (5.59).

First of all, Lemma 5.14 implies the existence of some Pritchard-Salamon system Σ_Λ (with corresponding linear map Λ) such that $\bar{K} = \mathcal{F}(\bar{G}_2, \Lambda)$. Furthermore, it follows from Lemma 5.15 that $\bar{G}_{zw} = \mathcal{F}(\bar{G}, \bar{K}) = \Lambda$ and since $\Sigma_{\bar{K}}$ is admissible for $\Sigma_{\bar{G}}$ we have $\|\Lambda\| = \|\bar{G}_{zw}\| < 1$. Finally, since $\Sigma_{\bar{K}}$ is an admissible controller for $\Sigma_{\bar{G}}$, the linear map $\bar{G}_{zw}$ has a realization $\Sigma_{\bar{G}_{zw}} = \Sigma(\mathcal{S}(\cdot), \mathcal{B}, \mathcal{C}, \mathcal{D})$ (as in (5.6)), where $\mathcal{S}(\cdot)$ is exponentially stable on $\mathcal{W} \times \mathcal{W}_K$ and $\mathcal{V} \times \mathcal{V}_K$. Since $\bar{G}_{zw} = \Lambda$, we can define a *new realization* of Λ by

$$\mathcal{W}_\Lambda := \mathcal{W} \times \mathcal{W}_K \quad ; \quad \mathcal{V}_\Lambda := \mathcal{V} \times \mathcal{V}_K$$

and

$$\Sigma(S_\Lambda(\cdot), B_\Lambda, C_\Lambda, D_\Lambda) := \Sigma(\mathcal{S}(\cdot), \mathcal{B}, \mathcal{C}, \mathcal{D}),$$

so that $S_\Lambda(\cdot)$ is exponentially stable on $\mathcal{W}_\Lambda$ and $\mathcal{V}_\Lambda$. This completes the proof of the necessity part.

(sufficiency) Suppose that the controller $\Sigma_{\bar{K}}$ is such that $\bar{K} = \mathcal{F}(\bar{G}_2, \Lambda)$ for some Pritchard-Salamon system Σ_Λ of the form (5.65)-(5.66), where $S_\Lambda(\cdot)$ is exponentially stable on $\mathcal{W}_\Lambda$ and $\mathcal{V}_\Lambda$ and $\|\Lambda\| < 1$.

First of all, we prove that $\Sigma_{\mathcal{F}(\bar{G}_2,\Lambda)}$ (with its natural realization on $\mathcal{W} \times \mathcal{W}_\Lambda \hookrightarrow \mathcal{V} \times \mathcal{V}_\Lambda$) is an admissible controller for $\Sigma_{\bar{G}}$. Hence, we have to show that the corresponding closed-loop system (see also the right hand side of Figure 5.5) is exponentially stable on $\mathcal{W} \times \mathcal{W} \times \mathcal{W}_\Lambda$ and $\mathcal{V} \times \mathcal{V} \times \mathcal{V}_\Lambda$ and that $\bar{G}_{zw} = \mathcal{F}(\bar{G}, \mathcal{F}(\bar{G}_2, \Lambda))$ satisfies $\|\bar{G}_{zw}\| < 1$. The idea is to apply Lemma 5.9 to the right hand side of Figure 5.5. First of all, we define the system $\Sigma_{\tilde{G}}$ as in the proof of Lemma 5.15 (see also Figure 5.6) and we recall that the equations of $\Sigma_{\tilde{G}}$ satisfy (5.73). We shall show that $\Sigma_{\tilde{G}}$ satisfies the assumptions (5.33), (5.34) and (5.35) of Lemma 5.9. Let the C_0-semigroup on $\mathcal{W} \times \mathcal{W}$ and $\mathcal{V} \times \mathcal{V}$ corresponding to $\Sigma_{\tilde{G}}$ be denoted by $\tilde{S}(\cdot)$, so that for zero inputs $w(\cdot)$ and $v(\cdot)$ in Figure 5.6 we have

$$\begin{pmatrix} x(t) \\ p_1(t) \end{pmatrix} = \tilde{S}(t) \begin{pmatrix} x_0 \\ p_{10} \end{pmatrix}.$$

We shall show that $\tilde{S}(\cdot)$ is exponentially stable on $\mathcal{W}\times\mathcal{W}$ and $\mathcal{V}\times\mathcal{V}$. Suppose that $(x_0, p_{10}) \in \mathcal{V}\times\mathcal{V}$ (or $(x_0, p_{10}) \in \mathcal{W}\times\mathcal{W}$, respectively). Consider (5.73) (with zero inputs $w(\cdot)$ and $v(\cdot)$) and recall the assumption that $\bar{S}_{-\bar{B}_1(\bar{D}_{21})^{-1}\bar{C}_2}(\cdot)$ and $\bar{S}_{-\bar{B}_2(\bar{D}_{12})^{-1}\bar{C}_1}(\cdot)$ are both exponentially stable on $\mathcal{W}$ and $\mathcal{V}$. It follows that $x(\cdot) - p_1(\cdot) \in L_2(0,\infty;\mathcal{V}\times\mathcal{V})$ (or $x(\cdot) - p_1(\cdot) \in L_2(0,\infty;\mathcal{W}\times\mathcal{W})$, respectively). Hence Lemma 2.24 implies that $x(\cdot) \in L_2(0,\infty;\mathcal{V}\times\mathcal{V})$ (or $x(\cdot) \in L_2(0,\infty;\mathcal{W}\times\mathcal{W})$, respectively). Thus it follows from Datko's result Lemma A.1 that $\tilde{S}(\cdot)$ is exponentially stable on on $\mathcal{V}\times\mathcal{V}$ and $\mathcal{W}\times\mathcal{W}$.

Furthermore, it follows trivially from (5.74) that $\Sigma_{\tilde{G}}$ is inner and that $(\tilde{G}_{21})^{-1}$ is bounded and so $\Sigma_{\tilde{G}}$ satisfies the assumptions (5.33), (5.34) and (5.35) of Lemma 5.9.

Now since $S_\Lambda(\cdot)$ is exponentially stable on $\mathcal{W}_\Lambda$ and $\mathcal{V}_\Lambda$ and $\|\Lambda\| < 1$, we can apply Lemma 5.9 to conclude that the closed-loop system $\Sigma_{\bar{G}_{zw}} = \Sigma_{\mathcal{F}(\bar{G},\mathcal{F}(\bar{G}_2,\Lambda))}$ is exponentially stable on $\mathcal{W}\times\mathcal{W}\times\mathcal{W}_\Lambda$ and $\mathcal{V}\times\mathcal{V}\times\mathcal{V}_\Lambda$ and $\|\bar{G}_{zw}\| < 1$. Therefore, $\Sigma_{\mathcal{F}(\bar{G}_2,\Lambda)}$ is an admissible controller for $\Sigma_{\bar{G}}$. We shall use this fact to show that $\Sigma_{\bar{K}}$ is also an admissible controller for $\Sigma_{\bar{G}}$: since $\bar{K} = \mathcal{F}(\bar{G}_2,\Lambda)$, it follows from Theorem 2.27 that

$$\begin{pmatrix} (I-\bar{K}\bar{G}_{22})^{-1} & (I-\bar{K}\bar{G}_{22})^{-1}\bar{K} \\ \bar{G}_{22}(I-\bar{K}\bar{G}_{22})^{-1} & (I-\bar{G}_{22}\bar{K})^{-1} \end{pmatrix}$$

is bounded (see also Remark 2.31). Now the idea is to use Theorem 2.27 once more. We have assumed that $(T(\cdot), N)$ is admissibly stabilizable and $(L, T(\cdot))$ is admissibly detectable and since the C_0-semigroups $\bar{S}_{-\bar{B}_1(\bar{D}_{21})^{-1}\bar{C}_2}(\cdot)$ and $\bar{S}_{-\bar{B}_2(\bar{D}_{12})^{-1}\bar{C}_1}(\cdot)$ are both exponentially stable on $\mathcal{W}$ and $\mathcal{V}$, we see that also $(\bar{S}(\cdot), \bar{B}_2)$ is admissibly stabilizable and $(\bar{C}_2, \bar{S}(\cdot))$ is admissibly detectable. Hence Theorem 2.27 implies that the closed-loop system determined by $\Sigma_{\bar{G}}$ and $\Sigma_{\bar{K}}$ is exponentially stable on $\mathcal{W}\times\mathcal{W}_K$ and $\mathcal{V}\times\mathcal{V}_K$ (again, see also Remark 2.31). Now since $\|\mathcal{F}(\bar{G},\bar{K})\| = \|\mathcal{F}(\bar{G},\mathcal{F}(\bar{G}_2,\Lambda))\| = \|\bar{G}_{zw}\| < 1$, we conclude that $\Sigma_{\bar{K}}$ given by (5.64) is an admissible controller for $\Sigma_{\bar{G}}$. ∎

Next we show that $\Sigma_{(((G_P)^\natural)_{\bar{P}})^\natural}$ satisfies the assumptions of Lemma 5.16 (recall that Σ_K is admissible for Σ_G if and only if Σ_K is admissible for $\Sigma_{(((G_P)^\natural)_{\bar{P}})^\natural}$). The system $\Sigma_{(((G_P)^\natural)_{\bar{P}})^\natural}$ is given by

$$\Sigma_{(((G_P)^\natural)_{\bar{P}})^\natural} \begin{cases} x(t) &= \bar{S}(t)x_0 + \int_0^t \bar{S}(t-s)(\bar{B}_1 w(s) + \bar{B}_2 u(s))ds \\ z(t) &= \bar{C}_1 x(t) + \bar{D}_{12} u(t) \\ y(t) &= \bar{C}_2 x(t) + \bar{D}_{21} w(t), \end{cases}$$

where $\bar{S}(\cdot), \bar{B}_1, \bar{B}_2, \bar{C}_1, \bar{C}_2, \bar{D}_{12}$ and $\bar{D}_{21}$ are defined by (5.20) with $\gamma = 1$ (the construction of $\Sigma_{((G_P)^\natural)_P}$ was explained at the beginning of this part **c)**). We see that $\Sigma_{(((G_P)^\natural)_{\bar{P}})^\natural}$ is a Pritchard-Salamon system of the form (5.59) (in particular, $\bar{D}_{12}$ and $\bar{D}_{21}$ are both invertible). Furthermore $\bar{S}_{-\bar{B}_1(\bar{D}_{21})^{-1}\bar{C}_2}(\cdot)$ and $\bar{S}_{-\bar{B}_2(\bar{D}_{12})^{-1}\bar{C}_1}(\cdot)$ are now given by $S_{\bar{P}}(\cdot)$ and $S_P(\cdot)$ and both C_0-semigroups are exponentially stable on $\mathcal{W}$ and $\mathcal{V}$ ($S_P(\cdot)$ was defined in Theorem 5.4 with $\gamma = 1$ and $S_{\bar{P}}(\cdot)$ is given by (5.58)). Hence we can apply Lemma 5.16 with $\Sigma_{\bar{G}}$ replaced by $\Sigma_{(((G_P)^\natural)_{\bar{P}})^\natural}$ and the parametrization result in Theorem 5.4 follows.

Part d): Sufficiency.
Suppose that there exist P and Q satisfying items $(i) - (iii)$ in Theorem 5.4. It follows from Appendix D, Lemma D.3 that $\bar{P}$ defined by

$$\bar{P} := Q(I - PQ)^{-1}$$

is nonnegative definite, satisfies $\bar{P} \in \mathcal{L}(\mathcal{W}', \mathcal{W})$, solves the Riccati equation (5.57) and is such that $S_{\bar{P}}(\cdot)$ defined by (5.58) is exponentially stable on $\mathcal{W}$ and $\mathcal{V}$. Hence we can define the system $\Sigma_{(((G_P)^\natural)_{\bar{P}})^\natural}$ just as we did in part **c)**. Using Lemmas 5.13 and 5.11 we can again conclude that Σ_K is admissible for Σ_G if and only if Σ_K is admissible for $\Sigma_{(((G_P)^\natural)_{\bar{P}})^\natural}$. As in part **c)**, $\Sigma_{(((G_P)^\natural)_{\bar{P}})^\natural}$ is of the form (5.59) and it satisfies the assumptions of Lemma 5.16. Defining the Pritchard-Salamon system $\Sigma_{\tilde{K}}$ as in (5.19) and $\tilde{K}$ as its corresponding linear map (as in Theorem 5.4; $\gamma = 1$), it follows from Lemma 5.16 that a particular admissible controller is given by $\Sigma_{\mathcal{F}(\tilde{K},0)}$ (this is sometimes called the *central controller*). This proves the sufficiency part of Theorem 5.4.

5.4 Relaxation of the a priori assumptions

In Section 5.1 we have given the complete solution to the $\mathcal{H}_\infty$-control problem with measurement-feedback for a system of the form (5.2). In the first part of this section we shall show how one can extend this result to nonzero D_{11} and D_{22}, using some loop-shifting ideas from [36, 79]. In the second part we show that also the assumptions (5.10)-(5.13) can be removed, however at the expense of introducing an extra, 'regularizing parameter'. As in Section 4.3 we assume that $\gamma = 1$ and we note that the general result can be obtained by scaling.

5.4.1 Including the feedthroughs

Consider the smooth Pritchard-Salamon system

$$\Sigma_G = \Sigma(S(\cdot), (B_1 \;\; B_2), \begin{pmatrix} C_1 \\ C_2 \end{pmatrix}, \begin{pmatrix} D_{11} & D_{12} \\ D_{21} & D_{22} \end{pmatrix})$$

given by

$$\Sigma_G \begin{cases} x(t) &= S(t)x_0 + \int_0^t S(t-s)(B_1 w(s) + B_2 u(s))ds \\ z(t) &= C_1 x(t) + D_{11} w(t) + D_{12} u(t) \\ y(t) &= C_2 x(t) + D_{21} w(t) + D_{22} u(t), \end{cases} \tag{5.75}$$

(see also (5.2)).
In Section 2.7 we have seen that for a controller $\Sigma_K = \Sigma(T(\cdot), N, L, R)$ of the form

$$\Sigma_K \begin{cases} p(t) &= T(t)p_0 + \int_0^t T(t-s)Nu_2(s)ds \\ y_2(t) &= Lp(t) + Ru_2(t), \end{cases} \tag{5.76}$$

the closed-loop system with $u_2(\cdot) = y(\cdot)$ and $y_2(\cdot) = u(\cdot)$ is well-posed if $I - D_{22}R$ (or, equivalently, $I - RD_{22}$) is boundedly invertible. Furthermore, the closed-loop system is given by (5.6). We shall show that Σ_K is admissible for Σ_G if and only if $\Sigma_{\hat{K}}$ is admissible for $\Sigma_{\hat{G}}$, where $\Sigma_{\hat{K}}$ and $\Sigma_{\hat{G}}$ are certain modifications of Σ_K and Σ_G and $\Sigma_{\hat{G}}$ has the property that its feedthrough operators $\hat{D}_{11}$ and $\hat{D}_{22}$ are zero. We shall use the loop-shifting ideas that were introduced in [36, 79] for the finite-dimensional case. First of all, we reduce the original problem to a problem with zero D_{11} and then we further reduce it to a problem in which also D_{22} is zero.

How to remove D_{11}.
If there exists an admissible controller Σ_K for Σ_G, it follows that

$$\|\tilde{G}_{zw}(\cdot)\|_\infty = \|\mathcal{C}(\cdot I - \mathcal{A}^{\tilde{\mathcal{V}}})^{-1}\mathcal{B} + \mathcal{D}\|_\infty < 1$$

(see Section 5.1). Hence, it follows from (2.23), that $\|\mathcal{D}\| < 1$. Considering the formula for $\mathcal{D}$ in (5.7), it follows that that there exists an operator $R_\infty \in \mathcal{L}(Y, U)$ such that $I - R_\infty D_{22}$ has a bounded inverse and

$$\bar{D}_{11} := D_{11} + D_{12}R_\infty(I - D_{22}R_\infty)^{-1}D_{21} \tag{5.77}$$

satisfies $\|\bar{D}_{11}\| < 1$. Assuming that such an R_∞ exists, we apply the preliminary output feedback

$$u(\cdot) := R_\infty y(\cdot) + \hat{u}(\cdot) \tag{5.78}$$

to Σ_G. Using the perturbation results from Section 2.4, it follows that the system equations of (5.75) can be reformulated as

$$\Sigma_{\bar{G}} \begin{cases} x(t) &= \bar{S}(t)x_0 + \int_0^t \bar{S}(t-s)(\bar{B}_1 w(s) + \bar{B}_2 \hat{u}(s))ds \\ z(t) &= \bar{C}_1 x(t) + \bar{D}_{11} w(t) + \bar{D}_{12} \hat{u}(t) \\ y(t) &= \bar{C}_2 x(t) + \bar{D}_{21} w(t) + \bar{D}_{22} \hat{u}(t), \end{cases} \tag{5.79}$$

where $\Sigma_{\bar{G}}$ is the smooth Pritchard-Salamon system

$$\Sigma_{\bar{G}} = \Sigma(\bar{S}(\cdot), (\bar{B}_1 \ \ \bar{B}_2), \begin{pmatrix} \bar{C}_1 \\ \bar{C}_2 \end{pmatrix}, \begin{pmatrix} \bar{D}_{11} & \bar{D}_{12} \\ \bar{D}_{21} & \bar{D}_{22} \end{pmatrix})$$

and

$$\begin{array}{rcl} \bar{B}_1 & := & B_1 + B_2 R_\infty (I - D_{22}R_\infty)^{-1} D_{21} \\ \bar{B}_2 & := & B_2(I - R_\infty D_{22})^{-1} \\ \bar{C}_1 & := & C_1 + D_{12}R_\infty(I - D_{22}R_\infty)^{-1}C_2 \\ \bar{C}_2 & := & (I - D_{22}R_\infty)^{-1}C_2 \\ \bar{D}_{12} & := & D_{12}(I - R_\infty D_{22})^{-1} \\ \bar{D}_{21} & := & (I - D_{22}R_\infty)^{-1}D_{21} \\ \bar{D}_{22} & := & (I - D_{22}R_\infty)^{-1}D_{22} \\ \bar{S}(\cdot) & := & S_{B_2R_\infty(I-D_{22}R_\infty)^{-1}C_2}(\cdot). \end{array}$$

The reason for this transformation is that the feedthrough operator from the disturbance input $w(\cdot)$ to the to-be-controlled output $z(\cdot)$, given by $\bar{D}_{11}$ in (5.77), now has norm smaller than one.
Furthermore, we define the controller $\Sigma_{\hat{K}} = \Sigma(T(\cdot), N, L, \hat{R})$ of the form

$$\Sigma_{\hat{K}} \begin{cases} p(t) & = & T(t)p_0 + \int_0^t T(t-s)Ny(s)ds \\ \hat{u}(t) & = & Lp(t) + \hat{R}y(t) \end{cases} \tag{5.80}$$

where $\hat{R} := R - R_\infty$ (of course, this transformed controller is related to Σ_K by (5.78)).
Next, we introduce the change of variables determined by

$$\begin{pmatrix} \hat{z}(\cdot) \\ w(\cdot) \end{pmatrix} = \Theta \begin{pmatrix} \hat{w}(\cdot) \\ z(\cdot) \end{pmatrix}, \tag{5.81}$$

where

$$\Theta := \begin{pmatrix} \bar{D}_{11} & (I - \bar{D}_{11}\bar{D}_{11}')^{\frac{1}{2}} \\ -(I - \bar{D}_{11}'\bar{D}_{11})^{\frac{1}{2}} & \bar{D}_{11}' \end{pmatrix}.$$

Σ_Θ can be considered as a simple, static Pritchard-Salamon system and schematically the transformation can be expressed as in Figure 5.7. Using the perturbation results of Section 2.4, it can be shown that with this change of variables, the system equations of $\Sigma_{\bar{G}}$ can be reformulated as

$$\Sigma_{\hat{G}} \begin{cases} x(t) & = & \hat{S}(t)x_0 + \int_0^t \hat{S}(t-s)(\hat{B}_1\hat{w}(s) + \hat{B}_2\hat{u}(s))ds \\ \hat{z}(t) & = & \hat{C}_1 x(t) + \hat{D}_{12}\hat{u}(t) \\ \hat{y}(t) = y(t) & = & \hat{C}_2 x(t) + \hat{D}_{21}\hat{w}(t) + \hat{D}_{22}\hat{u}(t), \end{cases} \tag{5.82}$$

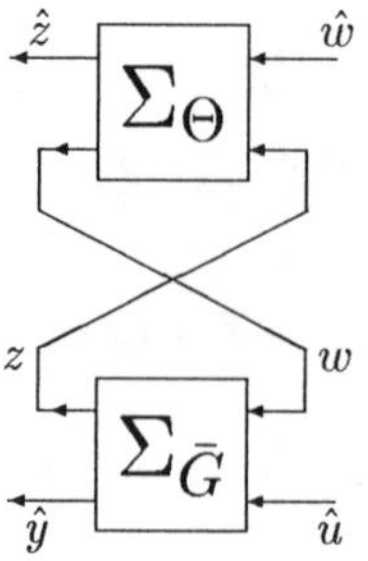

Figure 5.7: $\Sigma_{\hat{G}}$

where $\Sigma_{\hat{G}}$ is the smooth Pritchard-Salamon system

$$\Sigma_{\hat{G}} = \Sigma(\hat{S}(\cdot), (\hat{B}_1 \;\; \hat{B}_2), \begin{pmatrix} \hat{C}_1 \\ \hat{C}_2 \end{pmatrix}, \begin{pmatrix} 0 & \hat{D}_{12} \\ \hat{D}_{21} & \hat{D}_{22} \end{pmatrix})$$

and

$$\begin{aligned}
\hat{B}_1 &:= -\bar{B}_1(I - \bar{D}_{11}'\bar{D}_{11})^{-\frac{1}{2}} \\
\hat{B}_2 &:= \bar{B}_2 + \bar{B}_1\bar{D}_{11}'(I - \bar{D}_{11}\bar{D}_{11}')^{-1}\bar{D}_{12} \\
\hat{C}_1 &:= (I - \bar{D}_{11}\bar{D}_{11}')^{-\frac{1}{2}}\bar{C}_1 \\
\hat{C}_2 &:= \bar{C}_2 + \bar{D}_{21}\bar{D}_{11}'(I - \bar{D}_{11}\bar{D}_{11}')^{-1}\bar{C}_1 \\
\hat{D}_{12} &:= (I - \bar{D}_{11}\bar{D}_{11}')^{-\frac{1}{2}}\bar{D}_{12} \\
\hat{D}_{21} &:= -\bar{D}_{21}(I - \bar{D}_{11}'\bar{D}_{11})^{-\frac{1}{2}} \\
\hat{D}_{22} &:= \bar{D}_{22} + \bar{D}_{21}\bar{D}_{11}'(I - \bar{D}_{11}\bar{D}_{11}')^{-1}\bar{D}_{12} \\
\hat{S}(\cdot) &:= \bar{S}_{\bar{B}_1\bar{D}_{11}'(I-\bar{D}_{11}\bar{D}_{11}')^{-1}\bar{C}_1}(\cdot)
\end{aligned}$$

Note that the system $\Sigma_{\hat{G}}$ has no feedthrough operator from the disturbance to the to-be-controlled output. We have the following result.

Lemma 5.17
Suppose that there exists an R_∞ such that $\bar{D}_{11}$ defined by (5.77) satisfies $\|\bar{D}_{11}\| < 1$. Define the systems $\Sigma_{\bar{G}}$ by (5.79) and $\Sigma_{\hat{G}}$ by (5.82) and consider the three closed-loop systems of Figure 5.8. The controller Σ_K is admissible for Σ_G if and only if $\Sigma_{\hat{K}}$ is admissible for $\Sigma_{\bar{G}}$ if and only if $\Sigma_{\hat{K}}$ is admissible for $\Sigma_{\hat{G}}$.

Proof
In order to prove the first equivalence, we first note that the closed-loop system on the left in Figure 5.8 is well-posed if and only if the system in the middle is, because $I - \bar{D}_{22}(R - R_\infty) = I - (I - \bar{D}_{22}R_\infty)^{-1}D_{22}(R - R_\infty) = (I -$

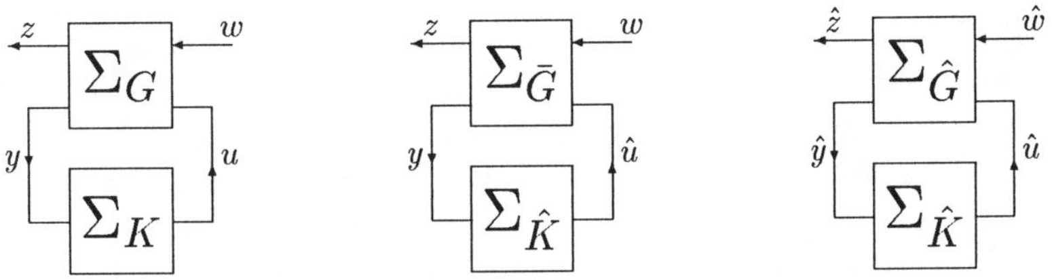

Figure 5.8: $\Sigma_{\mathcal{F}(G,K)}, \Sigma_{\mathcal{F}(\bar{G},\hat{K})}$ and $\Sigma_{\mathcal{F}(\hat{G},\hat{K})}$

$\bar{D}_{22}R_\infty)^{-1}(I - \bar{D}_{22}R)$. The rest of the proof of the first equivalence follows immediately from the definitions of $\Sigma_{\bar{G}}$ and $\Sigma_{\hat{K}}$ and the fact that $u(\cdot) = R_\infty y(\cdot) + \hat{u}(\cdot)$.

To prove the second equivalence of the lemma, we first deal with the well-posedness issue. Suppose that $\Sigma_{\hat{K}}$ is admissible for $\Sigma_{\bar{G}}$ (this means in particular that $I - \bar{D}_{22}\hat{R}$ has a bounded inverse). We have to prove that $I - \hat{D}_{22}\hat{R}$ also has a bounded inverse. There holds

$$I - \hat{D}_{22}\hat{R} = I - \bar{D}_{22}\hat{R} - \bar{D}_{21}\bar{D}'_{11}(I - \bar{D}_{11}\bar{D}'_{11})^{-1}\bar{D}_{12}\hat{R} =$$

$$\left(I - \bar{D}_{21}\bar{D}'_{11}(I - \bar{D}_{11}\bar{D}'_{11})^{-1}\bar{D}_{12}\hat{R}(I - \bar{D}_{22}\hat{R})^{-1}\right)(I - \bar{D}_{22}\hat{R}).$$

Using the fact that for bounded linear operators S and T we have $1 \in \rho(ST) \Leftrightarrow 1 \in \rho(TS)$ (see Lemma D.2, formula (D.1)) and that $I - \bar{D}_{22}\hat{R}$ has a bounded inverse, it follows that $I - \hat{D}_{22}\hat{R}$ has a bounded inverse if and only

$$I - \bar{D}_{12}\hat{R}(I - \bar{D}_{22}\hat{R})^{-1}\bar{D}_{21}\bar{D}'_{11}(I - \bar{D}_{11}\bar{D}'_{11})^{-1} =$$

$$(I - (\bar{D}_{11} + \bar{D}_{12}\hat{R}(I - \bar{D}_{22}\hat{R})^{-1}\bar{D}_{21})\bar{D}'_{11})(I - \bar{D}_{11}\bar{D}'_{11})^{-1} \tag{5.83}$$

has a bounded inverse. Since $\Sigma_{\hat{K}}$ is admissible for $\Sigma_{\bar{G}}$ it follows that the feedthrough operator of the closed-loop system satisfies

$$\|\bar{D}_{11} + \bar{D}_{12}\hat{R}(I - \bar{D}_{22}\hat{R})^{-1}\bar{D}_{21}\| < 1. \tag{5.84}$$

Furthermore, $\|\bar{D}'_{11}\| = \|\bar{D}_{11}\| < 1$ and so the result follows from (5.83) and (5.84).

Next, suppose that $\Sigma_{\hat{K}}$ is admissible for $\Sigma_{\hat{G}}$ (this means in particular that $I - \hat{D}_{22}\hat{R}$ has a bounded inverse). We want to prove that $I - \bar{D}_{22}\hat{R}$ also has a bounded inverse. It can be shown (using similar arguments as above) that $I - \bar{D}_{22}\hat{R}$ has a bounded inverse if and only $I - \hat{D}_{12}\hat{R}(I - \hat{D}_{22}\hat{R})^{-1}\hat{D}_{21}$ has. Since $\hat{D}_{12}\hat{R}(I - \hat{D}_{22}\hat{R})^{-1}\hat{D}_{21}$ is precisely the feedthrough operator of the closed-loop system $\Sigma_{\mathcal{F}(\hat{G},\hat{K})}$, which has norm smaller than one because $\Sigma_{\hat{K}}$ is admissible for $\Sigma_{\hat{G}}$, it follows that $I - \bar{D}_{22}\hat{R}$ has a bounded inverse.

To complete the proof of the second equivalence, we want to apply Lemma 5.9, using the structure of $\Sigma_{\hat{G}}$ in Figure 5.7. It is easy to see that Σ_Θ is inner and that Θ_{21} has a bounded inverse. Hence Lemma 5.9 with Σ_G replaced by Σ_Θ and Σ_K replaced by $\Sigma_{\mathcal{F}(\bar{G},\hat{K})}$ implies that $\Sigma_{\hat{K}}$ is admissible for $\Sigma_{\hat{G}}$ if and only if it is admissible for $\Sigma_{\bar{G}}$ and this completes the proof of the lemma. ∎

In the next lemma, it will be shown that if the regularity assumptions (5.10)-(5.13) are satisfied for Σ_G, then the corresponding assumptions for $\Sigma_{\hat{G}}$ are also satisfied. This will be useful for applying Theorem 5.4.

Lemma 5.18

Suppose that there exists an R_∞ such that $\bar{D}_{11}$ defined by (5.77) satisfies $\|\bar{D}_{11}\| < 1$ and define the systems $\Sigma_{\bar{G}}$ by (5.79) and $\Sigma_{\hat{G}}$ by (5.82). Furthermore, suppose that the pairs $(S(\cdot), B_2)$ and $(C_2, S(\cdot))$ are admissibly stabilizable and detectable, respectively. If Σ_G satisfies the regularity assumptions (5.10)-(5.13), we have that

(*i*) *there exists an $\epsilon > 0$ such that for all $(\omega, x, u) \in \mathbb{R} \times D(\hat{A}^V) \times U$ satisfying $i\omega x = \hat{A}^V x + \hat{B}_2 u$ there holds*

$$\left\|\hat{C}_1 x + \hat{D}_{12} u\right\|_Z^2 \geq \epsilon \|x\|_V^2,$$

$\hat{D}_{12}'\hat{D}_{12}$ is coercive,

(*ii*) *there exists an $\epsilon > 0$ such that for all $(\omega, x', y) \in \mathbb{R} \times D((\hat{A}^W)') \times Y$ satisfying $i\omega x' = (\hat{A}^W)'x' + \hat{C}_2' y$, there holds*

$$\left\|\hat{B}_1' x' + \hat{D}_{21}' y\right\|_W^2 \geq \epsilon \|x'\|_{W'}^2,$$

$\hat{D}_{21}\hat{D}_{21}'$ is coercive.

The proof below seems to be exactly the same as the finite-dimensional proof in [36]. However, we note that we really need the perturbation results of Section 2.4 and the results in Appendix C.1 in order to cope with the difficulties caused by the fact that we are dealing with Pritchard-Salamon systems here, rather than with finite-dimensional systems. Furthermore, some of the arguments that are used below are similar to the proof of (5.56) in Section 5.3.

Proof

The facts that $\hat{D}_{12}'\hat{D}_{12}$ and $\hat{D}_{21}\hat{D}_{21}'$ are coercive follow immediately from their definitions and the fact that $D_{12}'D_{12}$ and $D_{21}D_{21}'$ are coercive. To complete the proof of the first item, we note that for all $\omega \in \mathbb{R}$

$$\begin{pmatrix} i\omega I - \bar{A}^V & -\bar{B}_2 \\ \bar{C}_1 & \bar{D}_{12} \end{pmatrix} =$$

$$\begin{pmatrix} i\omega I - A^V & -B_2 \\ C_1 & D_{12} \end{pmatrix} \begin{pmatrix} I & 0 \\ 0 & (I - R_\infty D_{22})^{-1} \end{pmatrix} \begin{pmatrix} I & 0 \\ R_\infty C_2 & I \end{pmatrix}$$

and that

$$\begin{pmatrix} i\omega I - \hat{A}^V & -\hat{B}_2 \\ \hat{C}_1 & \hat{D}_{12} \end{pmatrix} =$$

$$\begin{pmatrix} I & -\bar{B}_1\bar{D}_{11}'(I - \bar{D}_{11}\bar{D}_{11}')^{-1} \\ 0 & (I - \bar{D}_{11}\bar{D}_{11}')^{-\frac{1}{2}} \end{pmatrix} \begin{pmatrix} i\omega I - \bar{A}^V & -\bar{B}_2 \\ \bar{C}_1 & \bar{D}_{12} \end{pmatrix}$$

(the perturbation results of Section 2.4 should be used here). Hence the proof of the first item follows from Appendix C.1. To complete the proof of the second item, we note that for all $\omega \in \mathbb{R}$

$$\begin{pmatrix} i\omega I - (\bar{A}^W)' & -\bar{C}_2' \\ \bar{B}_1' & \bar{D}_{21}' \end{pmatrix} =$$

$$\begin{pmatrix} i\omega I - (A^W)' & -C_2' \\ B_1' & D_{21}' \end{pmatrix} \begin{pmatrix} I & 0 \\ 0 & (I - R_\infty' D_{22}')^{-1} \end{pmatrix} \begin{pmatrix} I & 0 \\ R_\infty' B_2' & I \end{pmatrix}$$

and that

$$\begin{pmatrix} i\omega I - (\hat{A}^W)' & -\hat{C}_2' \\ \hat{B}_1' & \hat{D}_{21}' \end{pmatrix} =$$

$$\begin{pmatrix} I & -\bar{C}_1(I - \bar{D}_{11}\bar{D}_{11}')^{-1}\bar{D}_{11} \\ 0 & -(I - \bar{D}_{11}'\bar{D}_{11})^{-\frac{1}{2}} \end{pmatrix} \begin{pmatrix} i\omega I - (A^W)' & -C_2' \\ B_1' & D_{21}' \end{pmatrix}.$$

Appendix (C.1) again gives the result. ∎

With Lemmas 5.17 and 5.18 we have made the first step towards generalizing Theorem 5.4 to nonzero feedthrough operators: we have converted the problem for a system Σ_G with $D_{11} \neq 0$ to a problem for the system $\Sigma_{\hat{G}}$ that satisfies $\hat{D}_{11} = 0$. Next we show that also D_{22} can be removed.

How to remove D_{22}.
The idea that is used in this part is quite well known in finite-dimensional systems theory. Suppose that we have a smooth Pritchard-Salamon system

$$\Sigma_G = \Sigma(S(\cdot), (B_1 \;\; B_2), \begin{pmatrix} C_1 \\ C_2 \end{pmatrix}, \begin{pmatrix} 0 & D_{12} \\ D_{21} & D_{22} \end{pmatrix})$$

given by

$$\Sigma_G \begin{cases} x(t) &= S(t)x_0 + \int_0^t S(t-s)(B_1 w(s) + B_2 u(s))ds \\ z(t) &= C_1 x(t) + D_{12} u(t) \\ y(t) &= C_2 x(t) + D_{21} w(t) + D_{22} u(t), \end{cases} \tag{5.85}$$

(this is (5.2) with $D_{11} = 0$). Furthermore, suppose that we have a controller $\Sigma_K = \Sigma(T(\cdot), N, L, R)$ of the form (5.76) such that the closed-loop system is well-posed (hence $I - D_{22}R$ has a bounded inverse). With the change of variables

$$\tilde{y}(\cdot) := y(\cdot) - D_{22}u(\cdot)$$

the system equations of Σ_G can be reformulated as

$$\Sigma_{\tilde{G}} \begin{cases} x(t) &= S(t)x_0 + \int_0^t S(t-s)(B_1 w(s) + B_2 u(s))ds \\ z(t) &= C_1 x(t) + D_{12} u(t) \\ \tilde{y}(t) &= C_2 x(t) + D_{21} w(t), \end{cases} \tag{5.86}$$

where

$$\Sigma_{\tilde{G}} = \Sigma(S(\cdot), (B_1 \;\; B_2), \begin{pmatrix} C_1 \\ C_2 \end{pmatrix}, \begin{pmatrix} 0 & D_{12} \\ D_{21} & 0 \end{pmatrix}).$$

(note that $\Sigma_{\tilde{G}}$ has neither feedthrough from the disturbance to the to-be-controlled output nor feedthrough from the control input to the measured output). Of course the controller equations transform differently: using the perturbation results of Section 2.4, it is easy to see that with $y = \tilde{y} + D_{22}u$, the controller equations given by (5.76) can be reformulated as

$$\Sigma_{\tilde{K}} \begin{cases} p(t) &= \tilde{T}(t)p_0 + \int_0^t \tilde{T}(t-s)\tilde{N}\tilde{y}(s)ds \\ u(t) &= \tilde{L}p(t) + \tilde{R}\tilde{y}(t) \end{cases} \tag{5.87}$$

where $\Sigma_{\tilde{K}} = \Sigma(\tilde{T}(\cdot), \tilde{N}, \tilde{L}, \tilde{R})$ and

$$\begin{aligned} \tilde{N} &:= N(I - D_{22}R)^{-1} \\ \tilde{L} &:= (I - RD_{22})^{-1}L \\ \tilde{R} &:= R(I - D_{22}R)^{-1} \\ \tilde{T}(\cdot) &:= T_{N(I-D_{22}R)^{-1}D_{22}L}(\cdot). \end{aligned} \tag{5.88}$$

From the discussion above it follows that *if Σ_K is admissible for Σ_G, then $\Sigma_{\tilde{K}}$ determined by (5.88) is well-defined and an admissible controller for $\Sigma_{\tilde{G}}$.*

Before we try to derive a kind of converse of this statement, we show how the original controller can be obtained from $\Sigma_{\tilde{K}}$. Since $I - RD_{22}$ and $I - D_{22}R$ both have bounded inverses, there holds

$$\begin{aligned} I + \tilde{R}D_{22} &= I + (I - RD_{22})^{-1}RD_{22} = (I - RD_{22})^{-1} \\ I + D_{22}\tilde{R} &= I + D_{22}R(I - D_{22}R)^{-1} = (I - D_{22}R)^{-1} \end{aligned}$$

and so

$$\begin{aligned} N &= \tilde{N}(I + D_{22}\tilde{R})^{-1} \\ L &= (I + \tilde{R}D_{22})^{-1}\tilde{L} \\ R &= \tilde{R}(I + D_{22}\tilde{R})^{-1} \\ T(\cdot) &= \tilde{T}_{-\tilde{N}D_{22}(I+\tilde{R}D_{22})^{-1}\tilde{L}}(\cdot). \end{aligned} \tag{5.89}$$

Now suppose that we have a controller $\Sigma_{\tilde{K}} = \Sigma(\tilde{T}(\cdot), \tilde{N}, \tilde{L}, \tilde{R})$ such that $I + \tilde{R}D_{22}$ (or, equivalently, $I + D_{22}\tilde{R}$) has a bounded inverse. Define the controller Σ_K by (5.89). It is then easy to see that $I - RD_{22}$ and $I - D_{22}R$ both have bounded inverses and that (5.88) is satisfied. It follows from the discussion above that *if $\Sigma_{\tilde{K}}$ is an admissible controller for $\Sigma_{\tilde{G}}$ and $I + \tilde{R}D_{22}$ has a bounded inverse, then Σ_K defined by* (5.89) *is an admissible controller for Σ_G*. Hence, the following result requires no further proof.

Lemma 5.19
Σ_K is admissible for Σ_G given by (5.85) *if and only if $\Sigma_{\tilde{K}}$ is admissible for $\Sigma_{\tilde{G}}$ and $I + \tilde{R}D_{22}$ has a bounded inverse.*

Conclusion of subsection 5.4.1.
Using Lemmas 5.17, 5.18 and 5.19 we can generalize Theorem 5.4 to the case of nonzero feedthrough operators. We shall not state a formal theorem, but we explain the idea. Suppose that we have a smooth Pritchard-Salamon system Σ_G given by (5.75) that satisfies the regularity assumptions (5.10)-(5.13). If there exists an admissible controller for Σ_G we can define the transformed systems $\Sigma_{\hat{G}}$ and $\Sigma_{\tilde{\hat{G}}} =: \Sigma_{G_T}$. Then this controller can be transformed to obtain an admissible controller for Σ_{G_T}. We know that Σ_{G_T} satisfies $(D_T)_{11} = 0$, $(D_T)_{22} = 0$ and the corresponding regularity assumptions. Using Theorem 5.4 and the above lemmas it follows that there exists an admissible controller Σ_K for Σ_G if and only if there exist solutions to two coupled Riccati equations. Furthermore, these solutions can be used to find a parametrization of *all* admissible controllers for Σ_{G_T} of the form $\Sigma_{K_T} = \Sigma_{\mathcal{F}(K_0,\Lambda)}$. Finally, any admissible controller $\Sigma_{K_T} = \Sigma(T_T(\cdot), N_T, L_T, R_T)$ for Σ_{G_T} for which $I + R_T\hat{D}_{22}$ has a bounded inverse, can be transformed back to an admissible controller for Σ_G.

5.4.2 How to 'remove' the regularity assumptions

Here we show that the regularity assumptions (5.10)-(5.13) for the $\mathcal{H}_\infty$-control problem can be removed at the expense of an extra parameter (a similar procedure is followed in [88] for the bounded input/output case). Consider the smooth Pritchard-Salamon system of (5.75) and define the transformed system

$$\Sigma_{G_\epsilon} = \Sigma(S(\cdot), (B_{1\epsilon} \ \ B_2), \begin{pmatrix} C_{1\epsilon} \\ C_2 \end{pmatrix}, \begin{pmatrix} D_{11\epsilon} & D_{12\epsilon} \\ D_{21\epsilon} & D_{22} \end{pmatrix})$$

given by

$$\Sigma_{G_\epsilon} \begin{cases} x_\epsilon(t) &= S(t)x_0 + \int_0^t S(t-s)(B_{1\epsilon}w_\epsilon(s) + B_2 u_\epsilon(s))ds \\ z_\epsilon(t) &= C_{1\epsilon}x_\epsilon(t) + D_{11\epsilon}w_\epsilon(t) + D_{12\epsilon}u_\epsilon(t) \\ y_\epsilon(t) &= C_2 x_\epsilon(t) + D_{21\epsilon}w_\epsilon(t) + D_{22\epsilon}u_\epsilon(t), \end{cases} \tag{5.90}$$

where

$$C_{1\epsilon} := \begin{pmatrix} C_1 \\ \epsilon I_{\mathcal{V}} \\ 0 \end{pmatrix}, \ D_{11\epsilon} := \begin{pmatrix} D_{11} & 0 & 0 \\ 0 & 0 & 0 \\ 0 & 0 & 0 \end{pmatrix}, \ D_{12\epsilon} := \begin{pmatrix} D_{12} \\ 0 \\ \epsilon I_U \end{pmatrix},$$

$$B_{1\epsilon} := \begin{pmatrix} B_1 & \epsilon I_{\mathcal{W}} & 0 \end{pmatrix}, \ D_{21\epsilon} := \begin{pmatrix} D_{21} & 0 & \epsilon I_Y, \end{pmatrix}.$$

for some $\epsilon > 0$. Note that in the modified system Z is replaced by $Z \times \mathcal{V} \times U$ and W is replaced by $W \times \mathcal{W} \times Y$. It is easy to see that Σ_{G_ϵ} is again a smooth Pritchard-Salamon system. Furthermore, it satisfies the regularity assumptions (5.10)-(5.13) because

(*i*) for all $x \in \mathcal{W}$ and $u \in U$ we have $\|C_{1\epsilon}x + D_{12\epsilon}u\|^2_{Z\times\mathcal{V}\times U} \geq \epsilon^2 \|x\|^2_{\mathcal{V}}$,

(*ii*) $D'_{12\epsilon}D_{12\epsilon} \geq \epsilon^2 I_U$,

(*iii*) for all $x' \in \mathcal{V}'$ and $y \in Y$ we have $\|B'_{1\epsilon}x' + D'_{21\epsilon}y\|^2_{W\times\mathcal{W}'\times Y} \geq \epsilon^2 \|x'\|^2_{\mathcal{W}'}$,

(*iv*) $D_{21\epsilon}D'_{21\epsilon} \geq \epsilon^2 I_Y$.

Remark 5.20
We note that it is not necessary to use the same ϵ everywhere. On can think of introducing $\epsilon_1, \epsilon_2, \epsilon_3, \epsilon_4 \geq 0$, with possibly $\epsilon_i = 0$ for some i. This may be of help in the computational problem.

Let Σ_K given by (5.76) be a controller for Σ_G. If $I - D_{22}R$ has a bounded inverse, then the closed loop system is well-posed and given by (5.6) and the corresponding linear map $\mathcal{F}(G,K)$ is given by

$$z(\cdot) = \mathcal{F}(G,K)w(\cdot) = \mathcal{C}\int_0^t \mathcal{S}(t-s)\mathcal{B}w(s)ds + \mathcal{D}w(\cdot),$$

(see also (5.8)). It is easy to see that the closed-loop system $\Sigma_{\mathcal{F}(G,K)}$ is well-posed if and only the closed-loop system $\Sigma_{\mathcal{F}(G_\epsilon,K)}$ is (Σ_{G_ϵ} has the same D_{22}-operator). In this case, the closed-loop C_0-semigroup of $\Sigma_{\mathcal{F}(G_\epsilon,K)}$ is the same as that of $\Sigma_{\mathcal{F}(G,K)}$, because the B_2, C_2 and D_{22} operators of Σ_G and Σ_{G_ϵ} are the same (see (5.7)). Furthermore, the linear map $\mathcal{F}(G_\epsilon,K)$ is given by

$$z_\epsilon(\cdot) = \mathcal{F}(G_\epsilon,K)w_\epsilon(\cdot) = \mathcal{C}_\epsilon\int_0^t \mathcal{S}(t-s)\mathcal{B}_\epsilon w_\epsilon(s)ds + \mathcal{D}_\epsilon w_\epsilon(\cdot),$$

where

$$\mathcal{B}_\epsilon = \begin{pmatrix} \mathcal{B} & \epsilon I_{\mathcal{W}} & B_2R(I-D_{22}R)^{-1}\epsilon \\ & 0 & N(I-D_{22}R)^{-1}\epsilon \end{pmatrix}$$

$$\mathcal{C}_\epsilon = \begin{pmatrix} \mathcal{C} & \\ \epsilon I_{\mathcal{V}} & 0 \\ \epsilon R(I-D_{22}R)^{-1}C_2 & \epsilon(I-D_{22}R)^{-1}L \end{pmatrix}$$

$$\mathcal{D}_\epsilon = \begin{pmatrix} \mathcal{D} & 0 & \epsilon D_{12}R(I-D_{22}R)^{-1} \\ 0 & 0 & 0 \\ \epsilon R(I-D_{22}R)^{-1}D_{21} & 0 & \epsilon^2 R(I-D_{22}R)^{-1} \end{pmatrix}.$$

Now suppose that $I-D_{22}R$ is boundedly invertible and that the C_0-semigroup $\mathcal{S}(\cdot)$ (given in (5.7)) is exponentially stable on $\mathcal{W}\times\mathcal{W}_K$ and $\mathcal{V}\times\mathcal{V}_K$. According to the above discussion, this implies that both $\mathcal{F}(G_\epsilon,K)$ and $\mathcal{F}(G,K)$ are well-defined and bounded (from $L_2(0,\infty)$ to $L_2(0,\infty)$). Reformulating

$$z_\epsilon(\cdot) = \begin{pmatrix} z_1(\cdot) \\ z_2(\cdot) \\ z_3(\cdot) \end{pmatrix} \quad \text{and} \quad w_\epsilon(\cdot) = \begin{pmatrix} w_1(\cdot) \\ w_2(\cdot) \\ w_3(\cdot) \end{pmatrix},$$

we see that

$$\begin{pmatrix} z_1(\cdot) \\ z_2(\cdot) \\ z_3(\cdot) \end{pmatrix} = \mathcal{F}(G_\epsilon,K)\begin{pmatrix} w_1(\cdot) \\ w_2(\cdot) \\ w_3(\cdot) \end{pmatrix} =$$

$$\begin{pmatrix} \mathcal{F}(G,K) & T_{12}(\epsilon) & T_{13}(\epsilon) \\ T_{21}(\epsilon) & T_{22}(\epsilon) & T_{23}(\epsilon) \\ T_{31}(\epsilon) & T_{32}(\epsilon) & T_{33}(\epsilon) \end{pmatrix}\begin{pmatrix} w_1(\cdot) \\ w_2(\cdot) \\ w_3(\cdot) \end{pmatrix},$$

where the $T_{ij}(\epsilon)$ are bounded linear maps (from L_2 to L_2) such that $\|T_{ij}(\epsilon)\| \to 0$ as $\epsilon \to 0$ (this follows from the formulas for $\mathcal{B}_\epsilon, \mathcal{C}_\epsilon$ and $\mathcal{D}_\epsilon$, the fact that $\mathcal{S}(\cdot)$ is exponentially stable on $\mathcal{W} \times \mathcal{W}_K$ and $\mathcal{V} \times \mathcal{V}_K$ and Lemma 2.23). The following result requires no further proof.

Theorem 5.21
If the controller Σ_K is admissible for Σ_G, then there exists an $\epsilon_1 > 0$ such that for all $0 \le \epsilon \le \epsilon_1$, Σ_K is also admissible for Σ_{G_ϵ}. Conversely, if for some $\epsilon \ge 0$, Σ_K is admissible for Σ_{G_ϵ}, then Σ_K is also admissible for Σ_G.

Since Σ_{G_ϵ} satisfies the regularity assumptions, the results of subsection 5.4.1 can be applied and so we can restate the first part of Theorem 5.4, without using any a priori assumptions. We could state a result like the following: Let Σ_G be a Pritchard-Salamon system of the form (5.75). There exists a γ-admissible controller for Σ_G if and only if there exists a small enough ϵ and nonnegative definite stabilizing solutions P and Q to two Riccati equations (parametrized by ϵ) satisfying the coupling condition $r_\sigma^{\mathcal{W}'}(PQ) < \gamma^2$. Furthermore, P and Q can be used to construct a γ-admissible controller for Σ_G (in fact, we can construct infinitely many of such controllers). However, we see that this procedure leads to formulations depending on two parameters (γ and ϵ) instead of one and should therefore be avoided if possible. Finally, we note that for the finite-dimensional case different approaches have been considered to remove the a priori regularity conditions (see e.g. [86] and [83]), but it is not clear if these ideas can be generalized to the infinite-dimensional case.

Chapter 6

Examples and conclusions

In this last chapter we shall illustrate the theory developed in the previous chapters by means of some examples and we shall conclude the book with a discussion about the results obtained and possible topics for future research.

In Sections 6.1 and 6.2 we consider the class of delay systems that was introduced in Example 1.4 in more detail. This kind of delay system allows for a relatively simple realization as a Pritchard-Salamon system and so it is quite suitable for illustrating the theory that was developed in Chapters 2 to 5. The purpose is to give nothing more than an illustration; the theory applies to all the examples of Pritchard-Salamon systems given in (the references of) Chapter 1. Some of the aspects that will be considered are the construction of stabilizing state-feedbacks and stabilizing output-feedbacks, the solution to a simple linear quadratic control problem and a concretisation of the duality theory of Section 2.5. Furthermore, in Section 6.2 we shall consider a mixed-sensitivity problem for this class of delay systems. The solution to this $\mathcal{H}_\infty$-control problem is given in terms of the solvability of two coupled infinite-dimensional Riccati equations (using Theorem 5.4). Unfortunately, explicit solutions to these Riccati equations are not available, but this is not surprising for anyone familiar with infinite-dimensional Riccati equations. From standard LQ-theory it is known that solutions to infinite-dimensional Riccati equations can usually only be found approximately, by means of numerical approximation schemes. Hence, it would be interesting to look for approximation schemes for the Riccati equations corresponding to the mixed-sensitivity problem in particular, and for Pritchard-Salamon/$\mathcal{H}_\infty$-type Riccati equations in general. We shall discuss this and other practical issues in Section 6.3.

6.1 Delay systems in state-space

In this section we shall illustrate some of the preliminary results for Pritchard-Salamon systems in Chapters 2,3 by means of examples for a simple class of delay systems. In particular, we shall discuss the formal calculation of transfer functions, the existence and construction of dynamic output-feedback controllers, a linear quadratic control problem and duality for delay systems described by

$$\exp(-s)C_0(sI - A_0)^{-1}B_0, \tag{6.1}$$

where $A_0 \in \mathbb{R}^{n\times n}, B_0 \in \mathbb{R}^{n\times m}$ and $C_0 \in \mathbb{R}^{p\times n}$. In Example 1.4 we mentioned that a transfer function of the form (6.1) corresponds to the delay differential equation

$$\left\{ \begin{array}{rcl} \dot{z}(t) &=& A_0 z(t) + B_0 u(t), \\ y(t) &=& C_0 z(t-1), \end{array} \right. \tag{6.2}$$

where $z(t) \in \mathbb{R}^n, u(t) \in \mathbb{R}^m$ and $y(t) \in \mathbb{R}^p$. Furthermore, it follows from Examples 1.2,1.4 and 2.8 that this delay differential equation can be modelled as a smooth Pritchard-Salamon system, using the state

$$x(t) = (z(t), z_t) \in M_2 = \mathbb{R}^n \times L_2(-1,0\,;\mathbb{R}^n),$$

where

$$z_t(\tau) = z(t+\tau)\,; \quad -1 \leq \tau \leq 0\,;$$

(the past of $z(\cdot)$ up to 1 time unit). The 'larger' state-space is given by

$$\mathcal{V} = M_2 = \mathbb{R}^n \times L_2(-1,0\,;\mathbb{R}^n)$$

(with the obvious inner product). Furthermore, the operator $A^{\mathcal{V}}$ given by

$$\begin{array}{rcl} D(A^{\mathcal{V}}) &=& \{x = (\eta, \phi) \in \mathcal{V} \mid \phi \in W^{1,2}(-1,0\,;\mathbb{R}^n), \eta = \phi(0)\}, \\ A^{\mathcal{V}}x &=& (A_0\eta, \dot{\phi}). \end{array} \tag{6.3}$$

is the infinitesimal generator of a C_0-semigroup $S(\cdot)$ on $\mathcal{V}$ (we recall that $W^{1,2}(-1,0\,;\mathbb{R}^n)$ denotes the Sobolev space of absolute continuous functions in $L_2(-1,0\,;\mathbb{R}^n)$ whose derivative is in $L_2(-1,0\,;\mathbb{R}^n)$). The 'smaller' state-space is given by

$$\mathcal{W} = D(A^{\mathcal{V}}).$$

In order to avoid some technicalities regarding the definition of the inner product on $\mathcal{W} = D(A^{\mathcal{V}})$ we shall assume in the rest of this section that

$$0 \notin \sigma(A_0). \tag{6.4}$$

In Example 2.8 we used the inner product on $\mathcal{W}$ given by

$$< x, y >_{\mathcal{W}} = < \phi, \bar{\phi} >_{W^{1,2}}, \text{ for } x = (\eta, \phi)\;; y = (\bar{\eta}, \bar{\phi}),$$

but here we prefer to introduce the inner product defined by:

$$< x, y >_{\mathcal{W}} := < A^{\mathcal{V}}x, A^{\mathcal{V}}y >_{\mathcal{V}} = < A_0\eta, A_0\bar{\eta} >_{\mathbb{R}^n} + < \dot{\phi}, \dot{\bar{\phi}} >_{L_2(-1,0\,;\mathbb{R}^n)} .$$

Using assumption (6.4), it is easy to show that the corresponding norm $\|x\|_{\mathcal{W}} = (< A^{\mathcal{V}}x, A^{\mathcal{V}}x >_{\mathcal{V}})^{1/2}$ is equivalent to the norm $(< \phi, \phi >_{W^{1,2}})^{1/2}$. Hence it follows that the C_0-semigroup $S(\cdot)$ restricts to a C_0-semigroup on $\mathcal{W}$. Furthermore, it follows from Example 2.8 that the operator $B : \mathbb{R}^m \to \mathcal{V}$ given by

$$Bu = (B_0 u, 0) \tag{6.5}$$

satisfies $B \in \mathcal{L}(\mathbb{R}^m, \mathcal{V})$ and that B is an admissible input operator and that the operator $C : \mathcal{W} \to \mathbb{R}^p$ given by

$$Cx = C_0\phi(-1) \quad \text{for} \quad x = (\eta, \phi) \in \mathcal{W} \tag{6.6}$$

satisfies $C \in \mathcal{L}(\mathcal{W}, \mathbb{R}^p)$ and that C is an admissible output operator. Hence, the system given by

$$\begin{cases} x(t) &= S(t)x_0 + \int_0^t S(t-s)Bu(s)ds \\ y(t) &= Cx(t), \end{cases} \tag{6.7}$$

is a smooth Pritchard-Salamon system. In the following example we shall show that the transfer function of this system given by $C(sI - A^{\mathcal{V}})^{-1}B$ (see Section 2.3) is indeed equal to $\exp(-s)C_0(sI - A_0)^{-1}B_0$.

Example 6.1 (*calculation of the transfer function*)
First of all, we show that the spectrum of $A^{\mathcal{V}}$ is a pure point spectrum and that $\sigma(A^{\mathcal{V}}) = \sigma(A_0)$. Suppose first that

$$(\lambda I - A^{\mathcal{V}})x = 0 \quad \text{for some nonzero } x = (\eta, \phi) \in D(A^{\mathcal{V}}). \tag{6.8}$$

Then (η, ϕ) satisfies

$$\begin{cases} (\lambda I - A_0)\eta &= 0 \\ \lambda\phi - \dot{\phi} &= 0 \\ \phi(0) &= \eta \end{cases}$$

and it follows that

$$\begin{cases} (\lambda I - A_0)\eta &= 0 \\ \phi(\theta) &= \exp(\lambda\theta)\eta. \end{cases}$$

Hence, there exists a nonzero (η, ϕ) satisfying (6.8) if and only if $\lambda \in \sigma(A_0)$ and so the point spectrum of $A^{\mathcal{V}}$ is equal to $\sigma(A_0)$. Next, let $\lambda \notin \sigma(A_0)$. We shall show that $\lambda \in \rho(A^{\mathcal{V}})$. Consider solving

$$(\lambda I - A^{\mathcal{V}})x = \bar{x} = (\bar{\eta}, \bar{\phi}) \text{ for } x = (\eta, \phi) \in D(A^{\mathcal{V}}),$$

or, equivalently,

$$\begin{cases} (\lambda I - A_0)\eta &= \bar{\eta} \\ \lambda\phi - \dot{\phi} &= \bar{\phi} \\ \phi(0) &= \eta. \end{cases}$$

It follows that $(\eta, \phi) = (\lambda I - A^{\mathcal{V}})^{-1}(\bar{\eta}, \bar{\phi})$ is given by

$$\begin{cases} \eta &= (\lambda I - A_0)^{-1}\bar{\eta} \\ \phi(\theta) &= \exp(\lambda\theta)\eta - \int_0^\theta \exp(\lambda(\theta - \sigma))\bar{\phi}(\sigma)d\sigma \end{cases}$$

and it is easy to see that $(\lambda I - A^{\mathcal{V}})^{-1} \in \mathcal{L}(\mathcal{V})$; thus $\lambda \in \rho(A^{\mathcal{V}})$.
The calculation of $C(sI - A^{\mathcal{V}})^{-1}B$ is now straightforward: for any $s \in \rho(A^{\mathcal{V}})$ and $u \in \mathbb{R}^m$ we have

$$(sI - A^{\mathcal{V}})^{-1}Bu = (sI - A^{\mathcal{V}})^{-1}(B_0u, 0) =$$

$$((sI - A_0)^{-1}B_0u, \exp(s\theta)(sI - A_0)^{-1}B_0u)$$

and since $C(\eta, \phi) = C_0\phi(-1)$ for $(\eta, \phi) \in D(A^{\mathcal{V}})$, it follows that

$$C(sI - A^{\mathcal{V}})^{-1}Bu = \exp(-s)C_0(sI - A_0)^{-1}B_0u.$$

□

6.1.1 Dynamic controllers for delay systems

In Section 2.7 we discussed the mathematical treatment of dynamic output-feedback for Pritchard-Salamon systems. One of the important goals that may be attained using dynamic controllers is stability. Theorem 2.30 shows that we can always find a stabilizing controller for a Pritchard-Salamon system if the pairs $(S(\cdot), B)$ and $(C, S(\cdot))$ are admissibly stabilizable and admissibly detectable, respectively (it follows from Theorem 2.27 that this last condition is also necessary). In this subsection, we shall discuss dynamic output-feedback for the delay system given by (6.3),(6.5), (6.6) and (6.7). We shall assume that the pairs (A_0, B_0) and (C_0, A_0) are stabilizable and detectable, respectively.

In the literature one can find many results about stability, stabilizability and detectability for this delay system (see e.g. [72, 80], where in fact the more general case of Example 2.8 is treated). We note that because of the fact that $D(A^{\mathcal{V}}) = \mathcal{W}$, stability of $S(\cdot)$ on $\mathcal{V}$ is equivalent to stability of $S(\cdot)$ on $\mathcal{W}$. Moreover, any admissible perturbation $S_{HF}(\cdot)$ of $S(\cdot)$ (i.e. H and F are admissible) is stable on $\mathcal{W}$ if and only if it is stable on $\mathcal{V}$ (use Theorem 2.20). Furthermore, it follows from [72, 80] that the pairs $(S(\cdot), B)$ and $(C, S(\cdot))$ are admissibly stabilizable and admissibly detectable (we assume that the pairs (A_0, B_0) and (C_0, A_0) are stabilizable and detectable, respectively).

Stability of $S(\cdot)$ is determined by the spectrum of A_0: $S(\cdot)$ is exponentially stable on $\mathcal{W}$ and $\mathcal{V}$ if and only if $\sigma(A_0) \subset \mathbb{C}^-$ (this is of course not very surprising if one keeps the representation (6.2) in mind). Furthermore, if one chooses the output operator $F : \mathcal{W} \to \mathbb{R}^m$ given by

$$Fx = F(\eta, \phi) = F_0\eta, \tag{6.9}$$

where F_0 is such that $\sigma(A_0 + B_0F_0) \in \mathbb{C}^-$, it follows that $F \in \mathcal{L}(\mathcal{W}, \mathbb{R}^m)$ is an admissible output operator such that $S_{BF}(\cdot)$ is exponentially stable on $\mathcal{W}$ and $\mathcal{V}$ (stability is now completely determined by $\sigma(A_0 + B_0F_0)$). Here we note that admissibility follows easily since in fact $F \in \mathcal{L}(\mathcal{V}, \mathbb{R}^m)$.

Finding an admissible input operator $H \in \mathcal{L}(\mathbb{R}^p, \mathcal{V})$ such that $S_{HC}(\cdot)$ is exponentially stable on $\mathcal{W}$ and $\mathcal{V}$ is not so straightforward. One general method would be to use the technique of splitting up the state-space $\mathcal{V}$ in a stable and an unstable subspace (as in [92, 17]) and to partition the input and output operators accordingly. Then only the unstable part needs to be stabilized. In [80] a different method is considered.

In the following simple example we shall give explicit formulas for an admissible output operator F and an admissible input operator H such that $S_{BF}(\cdot)$ and $S_{HC}(\cdot)$ are exponentially stable on $\mathcal{W}$ and $\mathcal{V}$. Furthermore, we shall give a stabilizing dynamic controller for this system, both in state-space and in the frequency domain.

Example 6.2 (*a stabilizing controller for* $\exp(-s)/(s-1)$)
Let A_0, B_0 and C_0 be given by

$$\begin{cases} A_0 &= 1 \\ B_0 &= 1 \\ C_0 &= 1. \end{cases}$$

Then the corresponding Pritchard-Salamon system given by (6.3),(6.5),(6.6) and (6.7) has the transfer function $\exp(-s)/(s-1)$. The operator $F : \mathcal{W} \to \mathbb{R}$ given by

$$Fx = F(\eta, \phi) = -2\eta,$$

is an admissible output operator in $\mathcal{L}(\mathcal{W}, \mathbb{R})$ such that $S_{BF}(\cdot)$ is exponentially stable on $\mathcal{W}$ and $\mathcal{V}$ (this follows from the above). In order to find a stabilizing input operator H we note that the 'unstable eigenvector' of $A^{\mathcal{V}}$ corresponding to the eigenvalue 1 is given by

$$c(1, \exp(\theta)), \quad \text{for } c \neq 0$$

(see Example 6.1). Defining $H \in \mathcal{L}(\mathbb{R}, \mathcal{W})$ by

$$Hu = -2\exp(1)u(1, \exp(\theta)),$$

it can be shown that $S_{HC}(\cdot)$ is exponentially stable on $\mathcal{W}$ and $\mathcal{V}$. One could prove this result using the theory of stable and unstable subspaces mentioned above (this is what motivated the particular choice of H), but we shall briefly discuss another proof. Using the fact that for all $s \in \mathbb{C}$

$$(sI - A^{\mathcal{V}} - HC)Hu = (sH - A^{\mathcal{V}}H - HCH)u = (s+1)Hu,$$

it follows that

$$(sI - A^{\mathcal{V}} - HC)^{-1}Hu = (1/(s+1))Hu$$

and so

$$C(sI - A^{\mathcal{V}} - HC)^{-1}H = (1/(s+1))CH = -2/(s+1)\ ;$$

$$C(sI - A^{\mathcal{V}} - HC)^{-1}H \in \mathcal{H}_\infty.$$

It is not difficult to show that the pairs $(S_{HC}(\cdot), H)$ and $(C, S_{HC}(\cdot))$ are admissibly stabilizable and detectable, respectively, and so the exponential stability of $S_{HC}(\cdot)$ follows from Lemma 2.23.

Using the above defined operators F and H, it follows from Theorem 2.30 that the Pritchard-Salamon controller given by

$$\Sigma_K \begin{cases} p(t) &= S_{BF+HC}(t)p_0 - \int_0^t S_{BF+HC}(t-s)Hy(s)ds \\ \\ u(t) &= Fp(t)(t) \end{cases}$$

is admissibly stabilizing. We shall conclude the example by showing that the transfer function of this controller is given by

$$-F(sI - A^{\mathcal{V}} - BF - HC)^{-1}H = \frac{-4\exp(1)(s-1)}{s^2+2s+1-4\exp(1-s)}. \tag{6.10}$$

We only have to solve

$$(sI - A^{\mathcal{V}} - BF - HC)(\eta, \phi) = H = -2\exp(1)(1, \exp(\theta)) \tag{6.11}$$

for $(\eta, \phi) \in D(A^{\mathcal{V}})$; the transfer function is then given by $-F(\eta, \phi) = 2\eta$. Reformulation of the above equation gives

$$\begin{cases} (s+1)\eta + 2\exp(1)\phi(-1) &= -2\exp(1) \\ s\phi - \dot{\phi} + 2\exp(1+\theta)\phi(-1) &= -2\exp(1+\theta) \\ \phi(0) &= \eta \end{cases}$$

Hence, $\phi(-1)$ can be expressed in η:

$$\phi(-1) = -1 - \frac{\eta(1+s)}{2\exp(1)} \tag{6.12}$$

and so the differential equation for ϕ becomes

$$\dot{\phi} = s\phi - \eta(1+s)\exp(\theta).$$

Using $\phi(0) = \eta$ it follows that

$$\phi(\theta) = \exp(s\theta)\eta - \frac{\eta(1+s)\exp(s\theta)(\exp(\theta(1-s)) - 1)}{1-s}.$$

The combination of this last equation with (6.12) gives an equation for η, which in turn implies that

$$\eta = \frac{-2\exp(1)(s-1)}{s^2+2s+1-4\exp(1-s)}$$

and so (6.10) follows. □

It is interesting to note that for the simple delay system above one can find an explicit coprime factorization of the transfer function, including the Bezout factors, i.e. one can find transfer functions $M(s), N(s), X(s)$ and $Y(s)$ in $\mathcal{H}_\infty$ such that

$$\exp(-s)/(s-1) = N(s)M^{-1}(s) \; ; \; M(s)X(s) - N(s)Y(s) = 1.$$

Using the above defined stabilizing F and H it follows that one can take

$$M(s) = F(sI - A^{\mathcal{V}} - BF)^{-1}B + I = (s-1)/(s+1)$$

$$N(s) = C(sI - A^{\mathcal{V}} - BF)^{-1}B = \exp(-s)/(s+1)$$

$$Y(s) = -F(sI - A^{\mathcal{V}} - HC)^{-1}H = -4\exp(1)/(s+1)$$

$$X(s) = -F(sI - A^{\mathcal{V}} - HC)^{-1}B + I =$$

$$(s^2 + 2s + 1 - 4\exp(1-s))/(s^2 - 1).$$

We note that the above formulas in terms of the system parameters and the stabilizing feedbacks are the same as for the finite-dimensional case (see e.g. [13, 60]); the calculation of the respective transfer functions is straightforward. Furthermore, it is known that the system given by its transfer function $X^{-1}(s)Y(s)$ is a stabilizing controller for $\exp(-s)/(s-1) = N(s)M^{-1}(s)$. This controller is the same as the one of Example 6.2:

$$X^{-1}(s)Y(s) = -F(sI - A^{\mathcal{V}} - BF - HC)^{-1}H =$$

$$-4\exp(1)(s-1)/(s^2 + 2s + 1 - 4\exp(1-s)).$$

This method of finding coprime factorizations (including Bezout factors) can be considered as an alternative to the complicated method in Kamen et al. [44] and references therein.

Finally, we note that most of the manipulations above are feasible for transfer functions of the form $\exp(-\tau s)C_0(sI - A_0)^{-1}B_0$. As explained above, it is simple to find a stabilizing feedback F and the problem is to find an admissible $H \in \mathcal{L}(\mathbb{R}^p, \mathcal{W})$ such that $S_{HC}(\cdot)$ is exponentially stable on $\mathcal{W}$ and $\mathcal{V}$. Now, H should be constructed using the (generalized) unstable eigenvectors of $A^{\mathcal{V}}$. For more general delay systems and PDE-systems in the Pritchard-Salamon class F and H may be obtained using the same principles, but instead of explicit formulas one obtains numerical approximations.

6.1.2 A linear quadratic control problem

In general, it is impossible to find explicit solutions to (infinite-dimensional) Riccati equations. Therefore, one usually considers (numerical) approximations (see Banks and Burns [4], Ito and Tran [42], Kappel [45], Gibson [35], Lasiecka and Triggiani [57] and references therein for LQ-type Riccati equations). In this subsection, however, we shall give an explicit solution to a special infinite-dimensional LQ-type Riccati equation. Consider again the Pritchard-Salamon system

$$\begin{cases} x(t) &= S(t)x_0 + \int_0^t S(t-s)Bu(s)ds \\ y(t) &= Cx(t), \end{cases}$$

where $S(\cdot)$ is the C_0-semigroup generated by $A^{\mathcal{V}}$ (given by (6.3)), and B and C are given by (6.5) and (6.6), respectively. As before, we shall assume that the pairs (A_0, B_0) and (C_0, A_0) are stabilizable and detectable, respectively, which, in turn, implies that the pairs $(S(\cdot), B)$ and $(C, S(\cdot))$ are admissibly stabilizable and detectable (see also subsection 6.1.1). To this system we associate the quadratic cost function

$$J(x_0, u(\cdot)) := \int_0^\infty (\|y(t)\|_Y^2 + \|u(t)\|_U^2)dt = \int_0^\infty (\|Cx(t)\|_Y^2 + \|u(t)\|_U^2)dt$$

for all $u(\cdot)$ in U_{adm} (recall that U_{adm} is the set of all L_2-inputs $u(\cdot)$ such that the state $x(\cdot)$ is in $L_2(0, \infty; \mathcal{V})$). This corresponds to the cost function given by (3.57) in Section 3.3, where

$$\mathcal{F}(x, u) :=< Cx, Cx >_Y + < u, u >_U$$

(this is the so-called 'standard problem', which was already considered by Pritchard and Salamon in [72]). Since $(C, S(\cdot))$ is admissibly detectable it follows from Appendix C that there exists an ϵ such that for all $(\omega, x, u) \in \mathbb{R} \times D(A^{\mathcal{V}}) \times U$ satisfying $i\omega x = A^{\mathcal{V}}x + Bu$ there holds

$$\mathcal{F}(x, u) = \|Cx\|_Y^2 + \|u\|_U^2 \geq \epsilon\|x\|_{\mathcal{V}}^2.$$

In fact, this frequency domain condition can still be proved if the assumption that (C_0, A_0) is detectable is replaced by the assumption that A_0 has no undetectable modes on the imaginary axis; see also Remark 3.13. Hence, we can apply Theorem 3.10:

- there exists a unique $\bar{u}(\cdot) \in L_2(0, \infty, U)$ such that

$$\inf_{u(\cdot)\in U_{adm}} J(x_0, u(\cdot)) = \min_{u(\cdot)\in U_{adm}} J(x_0, u(\cdot)) = J(x_0, \bar{u}(\cdot)),$$

- there exists a (unique) self-adjoint $P \in \mathcal{L}(\mathcal{V})$ such that for all $x, y \in D(A^{\mathcal{V}})$

$$< A^{\mathcal{V}}x, Py >_{\mathcal{V}} + < Px, A^{\mathcal{V}}y >_{\mathcal{V}} -$$

$$< PBB^*Px, y >_{\mathcal{V}} + < Cx, Cy >_Y = 0 \tag{6.13}$$

  and $A^{\mathcal{V}} - BB^*P$ is the generator of an exponentially stable C_0-semigroup on $\mathcal{V}$,

- the minimizing $\bar{u}(\cdot)$ can be given in feedback form:

$$\bar{u}(\cdot) = -B^* P x(\cdot),$$

and

$$\inf_{u(\cdot)\in U_{adm}} J(x_0, u(\cdot)) = J(x_0, \bar{u}(\cdot)) =< x_0, P x_0 >_{\mathcal{V}} .$$

Due to the special structure of the specific problem that we consider here, we can give an explicit solution to the Riccati equation: the operator $P \in \mathcal{L}(\mathcal{V})$ satisfying (6.13) is given by

$$P = \begin{pmatrix} P_0 & 0 \\ 0 & C_0^T C_0\, Id \end{pmatrix} \quad \text{i.e.} \quad P(\eta, \phi) = (P_0 \eta, C_0^T C_0 \phi),$$

where P_0 is the solution to the finite-dimensional Riccati equation

$$A_0^T P_0 + P_0 A_0 - P_0 B_0 B_0^T P_0 + C_0^T C_0 = 0$$

such that $\sigma(A_0 - B_0 B_0^T P_0) \in \mathbb{C}^-$ (the existence and uniqueness of such P_0 follows from the assumptions on A_0, B_0, C_0). Before giving a simple intuitive explanation of this expression for P, we shall give the proof that this P is indeed the solution. First of all, we note that $B^* \in \mathcal{L}(\mathcal{V}, U)$ is given by

$$B^* x = B_0^T \eta \quad \text{for} \quad x = (\eta, \phi) \in \mathcal{V},$$

so that

$$B^* P x = B^*(P_0 \eta, C_0^T C_0 \phi) = B_0^T P_0 \eta.$$

For all $x = (\eta, \phi) \in D(A^{\mathcal{V}})$ there holds

$$< A^{\mathcal{V}} x, P x >_{\mathcal{V}} + < P x, A^{\mathcal{V}} x >_{\mathcal{V}} - < P B B^* P x, x >_{\mathcal{V}} + < C x, C x >_Y =$$

$$2 < \begin{pmatrix} A_0 \eta \\ \dot{\phi} \end{pmatrix}, \begin{pmatrix} P_0 \eta \\ C_0^T C_0 \phi \end{pmatrix} >_{\mathcal{V}} -$$

$$< B^* P x, B^* P x >_U + < C_0 \phi(-1), C_0 \phi(-1) >_Y =$$

$$< (A_0^T P_0 + P_0 A_0 - P_0 B_0 B_0^T P_0) \eta, \eta >_{\mathbb{R}^n} +$$

$$\int_{-1}^{0} \frac{d}{d\sigma} < C_0 \phi(\sigma), C_0 \phi(\sigma) > d\sigma + < C_0 \phi(-1), C_0 \phi(-1) >_Y = 0,$$

where in the last step we used the fact that $\phi(0) = \eta$ for $(\eta, \phi) \in D(A^{\mathcal{V}})$ and the Riccati equation for P_0. The closed-loop generator $A^{\mathcal{V}} - B B^* P$ is given by

$$(A^{\mathcal{V}} - B B^* P)(\eta, \phi) = ((A_0 - B_0 B_0^T P_0) \eta, \dot{\phi})$$

and since $\sigma(A_0 - B_0B_0^TP_0) \in \mathbb{C}^-$ it follows that $A^{\mathcal{V}} - BB^*P$ is the generator of an exponentially stable C_0-semigroup on $\mathcal{V}$ (see also the remark after (6.9)).

Next, we shall give an intuitive explanation for the particular expression for P. Recall that the Pritchard-Salamon system (6.7) represents the delay differential system

$$\begin{cases} \dot{z}(t) &= A_0z(t) + B_0u(t), \\ y(t) &= C_0z(t-1), \end{cases}$$

using the state

$$x(t) = (z(t), z(t+\cdot)) \in \mathbb{R}^n \times L_2(-1,0\,;\mathbb{R}^n)$$

(the initial condition for the differential equation for $z(\cdot)$ is given by $z(0) = \eta_0$, $z(\tau) = \phi_0(\tau)$ for $\tau \in (-1,0)$ and $(\eta_0, \phi_0) = x_0$). Thus, the cost criterion can be reformulated as

$$J(x_0, u(\cdot)) = \int_0^\infty (\|y(t)\|_Y^2 + \|u(t)\|_U^2)dt =$$

$$\int_0^\infty (\|C_0z(t-1)\|_Y^2 + \|u(t)\|_U^2)dt =$$

$$\int_0^1 \|C_0z(t-1)\|_Y^2 dt + \int_1^\infty \|C_0z(t-1)\|_Y^2 dt + \int_0^\infty \|u(t)\|_U^2 dt =$$

$$\int_{-1}^0 \|C_0z(t)\|_Y^2 dt + \int_0^\infty (\|C_0z(t)\|_Y^2 + \|u(t)\|_U^2)dt =$$

$$\int_{-1}^0 \|C_0\phi_0(t)\|_Y^2 dt + < P_0\eta_0, \eta_0 >_{\mathbb{R}^n} + \int_0^\infty \|u(t) + B_0^TP_0z(t)\|_U^2 dt,$$

where in the last step we used the Riccati equation for P_0, the differential equation for $z(\cdot)$ and a standard finite-dimensional completion of the squares argument. The minimizing control is given by

$$\bar{u}(\cdot) = -B_0^TP_0z(\cdot) = -B^*Px(\cdot)$$

and the optimal cost is given by

$$\int_{-1}^0 \|C_0\phi_0(\tau)\|_Y^2 dt + < P_0\eta_0, \eta_0 >_{\mathbb{R}^n} = < Px_0, x_0 >_{\mathcal{V}}$$

and so the expression for P is not so surprising after all.

Remark 6.3

In Delfour et al. [21], the authors consider the LQ-problem for another class of delay systems (hereditary differential systems). These systems can also be modelled as Pritchard-Salamon systems using $\mathcal{W} = \mathcal{V} = \mathbb{R}^n \times L_2(-1,0\,;\mathbb{R}^n)$ as the state-space (hence bounded inputs and outputs). They give a method to reduce the corresponding operator-valued Riccati equation to a system of partial differential equations, which may then be solved using numerical techniques. It would be interesting to try to extend this method to other classes of delay systems, such as the class considered in this section. However, the solutions to their Riccati equations have different properties. In general, a bounded selfadjoint operator P from $\mathbb{R}^n \times L_2(-1,0\,;\mathbb{R}^n)$ to $\mathbb{R}^n \times L_2(-1,0\,;\mathbb{R}^n)$ can be represented in the following way:

$$P = \begin{pmatrix} P_{11} & P_{12} \\ P_{21} & P_{22} \end{pmatrix},$$

where $P_{11} \in \mathbb{R}^{nxn}$, $P_{12} \in \mathcal{L}(L_2(-1,0\,;\mathbb{R}^n), \mathbb{R}^n)$, $P_{21} = P_{12}^*$ and $P_{22} \in \mathcal{L}(L_2(-1,0;\mathbb{R}^n), L_2(-1,0\,;\mathbb{R}^n))$. The point is that in their case P_{22} can be represented as

$$(P_{22}\phi)(\tau) = \int_{-1}^{0} K(\tau,\sigma)\phi(\sigma)ds,$$

for some square integrable $K(\cdot,\cdot)$ (in fact, they are able to derive a partial differential equation for $K(\cdot,\cdot)$). This represents an important difference with our example, because we have

$$(P_{22}\phi)(\tau) = C_0^T C_0 \phi(\tau) = \int_{-1}^{0} \delta(\tau-\sigma) C_0^T C_0 \phi(\sigma)ds$$

(our P_{22} is not compact).

Remark 6.4

In the previous remark we saw that the unboundedness of C can lead to a 'different kind of solution' to a Riccati equation. We can show that such a solution P can be written as $P = P_1 + P_2$, where P_1 is the solution to a (Pritchard-Salamon-type) Lyapunov equation and P_2 is a solution to a Riccati equation of the *bounded* type. The properties of P and the difficulties of calculating P due to the unboundedness of C are then captured by the solution to the Lyapunov equation. Let $\Sigma(S(\cdot), B, C, 0)$ be a smooth Pritchard-Salamon system and let $R \in \mathcal{L}(\mathcal{V})$ be some self-adjoint operator. Suppose that we are looking for a solution $P \in \mathcal{L}(\mathcal{V})$ of

$$< A^{\mathcal{V}}x, Py >_{\mathcal{V}} + < Px, A^{\mathcal{V}}y >_{\mathcal{V}} + < PRPx, y >_{\mathcal{V}} + < Cx, Cy >_{Y} = 0, \quad (6.14)$$

such that $A^{\mathcal{V}} + RP$ generates an exponentially stable C_0-semigroup. Furthermore, suppose that $F \in \mathcal{L}(\mathcal{V}, U)$ is such that $A^{\mathcal{V}} + BF$ generates an exponentially stable C_0-group on $\mathcal{V}$. Now, let $P_1 \in \mathcal{L}(\mathcal{V})$ be the (unique) solution to the Lyapunov equation

$$< (A^{\mathcal{V}} + BF)x, P_1 y >_{\mathcal{V}} + < P_1 x, (A^{\mathcal{V}} + BF)y >_{\mathcal{V}} = - < Cx, Cy >_Y \quad (6.15)$$

(we note that P_1 can be found using Lemma 3.1). Substitution of this expression for $< Cx, Cy >_Y$ in (6.14) gives the following Riccati equation for $P_2 := P - P_1$:

$$< (A^{\mathcal{V}} + RP_1)x, P_2 y >_{\mathcal{V}} + < P_2 x, (A^{\mathcal{V}} + RP_1)y >_{\mathcal{V}} + < P_2 R P_2 x, y >_{\mathcal{V}} =$$

$$- < (P_1 R P_1 - P_1 BF - F^* B^* P_1)x, y >_{\mathcal{V}} . \quad (6.16)$$

We note that this last equation is of the bounded type: all operators (except $A^{\mathcal{V}}$) are bounded on (or to, respectively) $\mathcal{V}$. Furthermore, P is a stabilizing solution to (6.14) if and only if $P = P_1 + P_2$, where P_1 is the solution to (6.15) and P_2 is the solution to (6.16) such that $(A^{\mathcal{V}} + RP_1) + RP_2$ generates an exponentially stable C_0-group on $\mathcal{V}$. Finally, we note that the right-hand side of the Riccati equation (6.16) is not necessarily semi-definite: this is the price that must be paid in order to get rid of the unbounded term.

6.1.3 Duality

In Section 2.5 we extensively discussed the issue of duality for Pritchard-Salamon systems. We devoted particular attention to how the duals of $\mathcal{W}$ and $\mathcal{V}$ should/can be represented. As before, we shall consider the Pritchard-Salamon system $\Sigma(S(\cdot), B, C, 0)$, where $S(\cdot)$ is the C_0-semigroup generated by $A^{\mathcal{V}}$ (given by (6.3)), and B and C are given by (6.5) and (6.6), respectively. We recall that the 'larger space' $\mathcal{V}$ is given by

$$\mathcal{V} = \mathbb{R}^n \times L_2(-1, 0\, ; \mathbb{R}^n)$$

with inner product given by

$$< x, y >_{\mathcal{V}} = < (\eta, \phi), (\bar{\eta}, \bar{\phi}) >_{\mathcal{V}} = < \eta, \bar{\eta} >_{\mathbb{R}^n} + < \phi, \bar{\phi} >_{L_2(-1,0\,;\mathbb{R}^n)}$$

and the 'smaller space' $\mathcal{W}$ is given by

$$\mathcal{W} = D(A^{\mathcal{V}})$$

with inner product given by

$$< x, y >_{\mathcal{W}} = < A^{\mathcal{V}} x, A^{\mathcal{V}} y >_{\mathcal{V}}$$

(we made sure that $0 \in \rho(A^{\mathcal{V}})$). We shall calculate the dual of $\mathcal{V}$ (denoted by $\mathcal{V}'$) for the case that the dual of $\mathcal{W}$ is identified with itself, according to the procedure of Section 2.5 ($\mathcal{W}$ is used as a pivot space; we take $\mathcal{W} = \mathcal{H}$ in (2.69)). This means that the representation of $\mathcal{L}(\mathcal{W}, \mathbb{R})$ is chosen to be $\{\mathcal{W}, (i_{\mathcal{W}})^{-1}\}$, where $i_{\mathcal{W}}$ is the canonical isomorphism from $\mathcal{L}(\mathcal{W}, \mathbb{R})$ to $\mathcal{W}$. Furthermore, denoting the restriction of $i_{\mathcal{W}}$ to $\mathcal{L}(\mathcal{V}, \mathbb{R})$ by $i_{\mathcal{W}} |_{\mathcal{L}(\mathcal{V},\mathbb{R})}$, it follows that $\mathcal{V}'$ is the image of this map:

$$\begin{array}{ccc} \mathcal{L}(\mathcal{V}, \mathbb{R}) & \hookrightarrow & \mathcal{L}(\mathcal{W}, \mathbb{R}) \\ \\ \downarrow i_{\mathcal{W}} |_{\mathcal{L}(\mathcal{V},\mathbb{R})} & & \downarrow i_{\mathcal{W}} \\ \\ \mathcal{V}' & \hookrightarrow & \mathcal{W} = \mathcal{W}'. \end{array}$$

The representation of $\mathcal{L}(\mathcal{V}, \mathbb{R})$ is given by $\{\mathcal{V}', (i_{\mathcal{W}} |_{\mathcal{L}(\mathcal{V},\mathbb{R})})^{-1}\}$ and the inner product on $\mathcal{V}'$ and the duality pairing $< \cdot, \cdot >_{<\mathcal{V}',\mathcal{V}>}$ are defined accordingly (compare with (2.55) and (2.56)). All this seems to be quite abstract, but we can obtain some explicit formulas. First of all, since $\mathcal{L}(\mathcal{V}, \mathbb{R}) \hookrightarrow \mathcal{L}(\mathcal{W}, \mathbb{R})$, we have for all $f \in \mathcal{L}(\mathcal{V}, \mathbb{R})$ and all $v \in \mathcal{W}$

$$\begin{cases} f(v) = < i_{\mathcal{V}} f, v >_{\mathcal{V}} \\ \\ f(v) = < i_{\mathcal{W}} f, v >_{\mathcal{W}} = < A^{\mathcal{V}} i_{\mathcal{W}} f, A^{\mathcal{V}} v >_{\mathcal{V}} . \end{cases}$$

Hence, $A^{\mathcal{V}} i_{\mathcal{W}} f \in D((A^{\mathcal{V}})^*)$ (note that $(A^{\mathcal{V}})^*$ represents the *Hilbert* adjoint of $A^{\mathcal{V}}$) and we have

$$< i_{\mathcal{V}} f, v >_{\mathcal{V}} = < A^{\mathcal{V}} i_{\mathcal{W}} f, A^{\mathcal{V}} v >_{\mathcal{V}} = < (A^{\mathcal{V}})^* A^{\mathcal{V}} i_{\mathcal{W}} f, v >_{\mathcal{V}}$$

which implies that

$$i_{\mathcal{V}} = (A^{\mathcal{V}})^* A^{\mathcal{V}} i_{\mathcal{W}} |_{\mathcal{L}(\mathcal{V},\mathbb{R})} \quad \text{or} \quad (A^{\mathcal{V}})^{-1}((A^{\mathcal{V}})^*)^{-1} i_{\mathcal{V}} = i_{\mathcal{W}} |_{\mathcal{L}(\mathcal{V},\mathbb{R})} .$$

It follows that for $v_1', v_2', v' \in \mathcal{V}$ and $v \in \mathcal{V}$

$$\begin{cases} \mathcal{V}' = (A^{\mathcal{V}})^{-1}((A^{\mathcal{V}})^*)^{-1} i_{\mathcal{V}} \mathcal{L}(\mathcal{V}, \mathbb{R}) = (A^{\mathcal{V}})^{-1}((A^{\mathcal{V}})^*)^{-1} \mathcal{V} \\ \\ < v_1', v_2' >_{\mathcal{V}'} = < (A^{\mathcal{V}})^* A^{\mathcal{V}} v_1', (A^{\mathcal{V}})^* A^{\mathcal{V}} v_2' >_{\mathcal{V}} \\ \\ < v', v >_{<\mathcal{V}',\mathcal{V}>} = < (A^{\mathcal{V}})^* A^{\mathcal{V}} v', v >_{\mathcal{V}} . \end{cases}$$

Using the fact that $\mathcal{V}' = (A^{\mathcal{V}})^{-1}((A^{\mathcal{V}})^*)^{-1}\mathcal{V}$ and the definition of $(A^{\mathcal{W}})^*$ (the Hilbert adjoint of $A^{\mathcal{W}}$) it is straightforward to show that in fact

$$\begin{cases} D((A^{\mathcal{W}})^*) = \mathcal{V}' = (A^{\mathcal{V}})^{-1}((A^{\mathcal{V}})^*)^{-1}\mathcal{V} \\ \\ (A^{\mathcal{W}})^* = (A^{\mathcal{V}})^{-1}(A^{\mathcal{V}})^* A^{\mathcal{V}}. \end{cases} \tag{6.17}$$

We note that since $\mathcal{W} = \mathcal{W}'$, there holds

$$\begin{cases} D((A^{\mathcal{W}})') = D((A^{\mathcal{W}})^*) \\ (A^{\mathcal{W}})' = (A^{\mathcal{W}})^*. \end{cases} \tag{6.18}$$

Furthermore, using formula (2.61) and the fact that $(A^{\mathcal{V}})^* A^{\mathcal{V}}$ is the isometry from $\mathcal{V}'$ to $\mathcal{V}$, we can relate $(A^{\mathcal{V}})'$ to $(A^{\mathcal{V}})^*$ as follows

$$\begin{cases} D((A^{\mathcal{V}})') = (A^{\mathcal{V}})^{-1}((A^{\mathcal{V}})^*)^{-1} D((A^{\mathcal{V}})^*) \\ (A^{\mathcal{V}})' = (A^{\mathcal{V}})^{-1}(A^{\mathcal{V}})^* A^{\mathcal{V}}. \end{cases} \tag{6.19}$$

It follows from Section 2.5 that $S'(\cdot)$ is a C_0-semigroup on $\mathcal{W}'$ and $\mathcal{V}'$, with infinitesimal generators $(A^{\mathcal{W}})'$ and $(A^{\mathcal{V}})'$, respectively, and that

$$(A^{\mathcal{V}})' = (A^{\mathcal{W}})' \quad \text{on} \quad D((A^{\mathcal{V}})'), \tag{6.20}$$

and we see that (6.20) is confirmed by (6.17), (6.18) and (6.19).

We note that the above results do not depend on the fact that we are considering delay systems: we only used that $\mathcal{W} = D(A^{\mathcal{V}})$ and that the inner product on $\mathcal{W}$ is given by $< x, y >_{\mathcal{W}} = < A^{\mathcal{V}} x, A^{\mathcal{V}} y >_{\mathcal{V}}$.

Next, we shall give the expressions for $D((A^{\mathcal{V}})^*)$ and $(A^{\mathcal{V}})^*$ for the particular delay system that is considered throughout this section. It follows from a result in [17] that

$$D((A^{\mathcal{V}})^*) = \{(\eta, \phi) \in \mathcal{V} \mid \phi \in W^{1,2}(-1, 0\,; \mathbb{R}^n), \phi(-1) = 0\}$$

and

$$(A^{\mathcal{V}})^*(\eta, \phi) = (A_0^T \eta + \phi(0), -\dot{\phi}).$$

We shall not prove this here, but in order to make it plausible we calculate for $(\eta, \phi) \in D(A^{\mathcal{V}})$ and $(\bar{\eta}, \bar{\phi}) \in D((A^{\mathcal{V}})^*)$

$$< A^{\mathcal{V}}(\eta, \phi), (\bar{\eta}, \bar{\phi}) >_{\mathcal{V}} = < (A_0 \eta, \dot{\phi}), (\bar{\eta}, \bar{\phi}) >_{\mathcal{V}} =$$

$$< A_0 \eta, \bar{\eta} >_{\mathbb{R}^n} + \int_{-1}^{0} < \dot{\phi}(\sigma), \bar{\phi}(\sigma) >_{\mathbb{R}^n} d\sigma = < \eta, A_0^T \bar{\eta} >_{\mathbb{R}^n} +$$

$$< \phi(\sigma), \bar{\phi}(\sigma) >_{\mathbb{R}^n} \Big|_{\sigma=-1}^{\sigma=0} - \int_{-1}^{0} < \phi(\sigma), \dot{\bar{\phi}}(\sigma) >_{\mathbb{R}^n} d\sigma =$$

$$< \eta, A_0^T \bar{\eta} >_{\mathbb{R}^n} + < \eta, \bar{\phi}(0) >_{\mathbb{R}^n} - \int_{-1}^{0} < \phi(\sigma), \dot{\bar{\phi}}(\sigma) >_{\mathbb{R}^n} d\sigma =$$

$$< (\eta, \phi), (A^{\mathcal{V}})^*(\bar{\eta}, \bar{\phi}) >_{\mathcal{V}} .$$

One can of course also consider the case where $\mathcal{V}$ is the pivot space and then calculate $\mathcal{W}'$ according to the procedure in Section 2.5 (see the derivation of (2.52)). We shall not go into details, but we note that $\mathcal{W}'$ is the completion of $\mathcal{V}$ with the inner product $< ((A^{\mathcal{V}})^*)^{-1}v_1, ((A^{\mathcal{V}})^*)^{-1}v_2 >_{\mathcal{V}}$.

If one only considers the state-feedback for the $\mathcal{H}_\infty$-control problem (or for the LQ-control problem), the issue of duality is not so important. The formulation for the state-feedback Riccati equation (4.9) can be stated without introducing a $\mathcal{V}'$ which is not equal to $\mathcal{V}$ itself (see also Remark 4.6). The measurement case, however, does ask for a rigorous treatment of duality (see Theorem 5.4 and the 'dual' Riccati equation in (5.16)).

6.2 The mixed-sensitivity problem for delay systems

6.2.1 Introduction and statement of the problem

In this section, we shall consider an important example of the standard $\mathcal{H}_\infty$-control problem: the mixed-sensitivity minimization problem (see also Section 1.1). We shall treat the problem for the class of delay systems that was introduced in Section 6.1. We have seen that this class of delay systems can be modelled by smooth Pritchard-Salamon systems and so it will be possible to use the main results of the book: Theorem 4.4 and Theorem 5.4.

The sensitivity minimization problem for delay systems is a difficult mathematical problem and for a single delay $\exp(-s\tau)$ this was solved in [32, 29]. Since then, more general problems have been considered in [30, 67, 68] (an introductory survey can be found in [65]). These references propose a frequency domain approach to the problem. In [65, 67, 68], the idea is to transform the continuous time problem into a discrete-time one which is then solved to yield the optimal infinite-dimensional controller. This solution is then transformed back into continuous time, where it is not only often infinite-dimensional, but improper as well. Consequently, approximations need to be made at this stage to yield an implementable (sub-optimal) controller. While this seems to yield good results in practice (see [58, 68] and [32]), a general theory for these approximations is still lacking (there does exist some approximation theory for specific problems). The difficulty with designing finite-dimensional controllers is that it is not sufficient to design good approximations to the infinite-dimensional optimal controller. One also needs to achieve a certain performance level at the same time. In [66] an approximation for certain scalar plants is proposed. However, one of the assumptions is that $(1-s)\frac{dg}{ds} \in H_1$, which eliminates a lot of interesting plants, including $\exp(-s\tau)$

and $\exp(-s\tau)/(a+s)$. Despite this, the method did work satisfactorily on the (example) plant $\exp(-0.01s)/(s-1)$, even though it did not satisfy the theoretical assumptions. A different approach is taken in [77]; one first approximates the infinite-dimensional system and then designs an optimal controller for the approximating finite-dimensional plant. In order to obtain the approximation one needs to first obtain an inner-outer factorization. In general, this step has to be performed numerically which is a difficult problem in itself, even for scalar plants (see [28]). We note that in both of the above frequency domain approaches one usually needs explicit inner-outer factorizations of infinite-dimensional systems at a certain stage. While this is straightforward for simple systems (such as $\exp(-s\tau)P(s)$ with $P(s)$ rational and scalar), this is an unsolved problem for general irrational multivariable transfer functions (in [28] some numerical results are obtained). With the exception of [58], all the mixed-sensitivity control designs have been for scalar plants. The main problem is that of obtaining an explicit inner-outer factorization via a good numerical algorithm. Summarizing, we can say that although the frequency domain approach provides existence results for $\mathcal{H}_\infty$-control problems for general delay systems, a completely satisfactory methodology for real controller design for these systems is still lacking. This is why we propose the state-space approach to the problem.

The mixed sensitivity minimization problem is to find a stabilizing controller with transfer matrix $K(s)$ which minimizes the $\mathcal{H}_\infty$-norm

$$\left\| \begin{pmatrix} W_1(\cdot)(I-P(\cdot)K(\cdot))^{-1} \\ W_2(\cdot)K(\cdot)(I-P(\cdot)K(\cdot))^{-1} \end{pmatrix} \right\|_\infty ,$$

where in our case $P(s) = \exp(-\tau s)C_0(sI-A_0)^{-1}B_0$ is the original plant, $W_1(s)$ and $W_2(s)$ are certain weighting transfer matrices given by

$$W_1(s) = D_1 + C_1(sI-A_1)^{-1}B_1 \text{ and } W_2(s) = D_2 + C_2(sI-A_2)^{-1}B_2$$

and $A_0, A_1, A_2, B_0, B_1, B_2, C_0, C_1, C_2, D_1, D_2$ are real matrices of compatible dimensions. We note that $K(\cdot)$ is called a *stabilizing* controller for $P(\cdot)$ if the transfer function

$$\begin{pmatrix} (I-P(\cdot)K(\cdot))^{-1} & K(\cdot)(I-P(\cdot)K(\cdot))^{-1} \\ P(\cdot)(I-K(\cdot)P(\cdot))^{-1} & (I-K(\cdot)P(\cdot))^{-1} \end{pmatrix}$$

has all its entries in $\mathcal{H}_\infty$. In the sequel we shall have (Pritchard-Salamon) state-space realizations of $K(\cdot)$ and $P(\cdot)$ and it follows from Theorem 2.27 that if these realizations are admissibly stabilizable/detectable, then $K(\cdot)$ is stabilizing for $P(\cdot)$ if and only if the corresponding closed-loop system is exponentially stable on the extended state-spaces.

For the motivation for this problem we refer to the introduction (Section 1.1). We also mentioned there that the choice of the weights $W_1(\cdot), W_2(\cdot)$

is an important issue and that it typically depends on the kind of application. Here we shall assume that $W_1(\cdot)$ is strictly proper ($D_1 = 0$) and that $W_2(\cdot), (W_2)^{-1}(\cdot) \in \mathbb{R}\mathcal{H}_\infty$. We also mentioned that in some papers the mixed sensitivity problem is defined differently: the second component $W_2(\cdot)K(\cdot)(I-P(\cdot)K(\cdot))^{-1}$ may be replaced by $W_2(\cdot)P(\cdot)K(\cdot)(I - P(\cdot)K(\cdot))^{-1}$. This last choice corresponds to robustness optimization with respect to *multiplicative* perturbations. We prefer the first formulation because it corresponds to a *regular* $\mathcal{H}_\infty$-control problem, whereas the second choice leads to a *singular* $\mathcal{H}_\infty$-control problem.

In the next subsection we shall show how the results of Chapter 5 can be applied to the mixed-sensitivity problem. Inherent to the state-space approach is the fact that we consider the *sub-optimal* problem; we are looking for controllers that achieve a certain performance bound, rather than those that minimize the performance criterion. Thus, the problem that is actually considered here is the sub-optimal mixed-sensitivity problem. We stress that this represents hardly any loss of generality: in the above mentioned frequency domain approaches one usually has to perform numerical approximations at a certain stage, in order to find implementable (rational) controllers, and this leads automatically to sub-optimal controllers.

6.2.2 Main result

First of all, we show how the mixed-sensitivity problem for delay systems can be reformulated as a standard $\mathcal{H}_\infty$-control problem. Then we shall show that, under some conditions, the assumptions of Theorem 5.4 are satisfied, so that we can apply the theorem.

Let the transfer functions $P(s) = \exp(-\tau s)C_0(sI - A_0)^{-1}B_0$, $W_1(s) = C_1(sI - A_1)^{-1}B_1$ and $W_2(s) = D_2 + C_2(sI - A_2)^{-1}B_2$ be given as above, with $A_i \in \mathbb{R}^{n_i \times n_i}$, $B_i \in \mathbb{R}^{n_i \times m_i}$, $C_i \in \mathbb{R}^{p_i \times n_i}$ ($n_i, m_i, p_i \in \mathbb{N}$ for $i = 0, 1, 2$), $D_2 \in \mathbb{R}^{p_2 \times m_2}$, $m_1 = p_0$ and $p_2 = m_2 = m_0$. We assume (without loss of generality) that the pairs (A_i, B_i) ((C_i, A_i), respectively) are controllable (observable, respectively) for $i = 0, 1, 2$. Furthermore, we assume that $W_2(\cdot)$ is invertible and that $W_1(\cdot), W_2(\cdot), (W_2)^{-1}(\cdot) \in \mathcal{H}_\infty$. Finally, we assume that A_0 has no poles on the imaginary axis:

$$\sigma(A_0) \cap i\mathbb{R} = \emptyset. \tag{6.21}$$

Defining the transfer function $G(\cdot)$ by

$$G(\cdot) := \left(\begin{array}{c|c} W_1(\cdot) & W_1(\cdot)P(\cdot) \\ 0 & W_2(\cdot) \\ \hline I & P(\cdot) \end{array} \right), \tag{6.22}$$

it follows that the transfer function of the mixed-sensitivity criterion can be considered as a linear fractional transformation:

$$\begin{pmatrix} W_1(\cdot)(I - P(\cdot)K(\cdot))^{-1} \\ W_2(\cdot)K(\cdot)(I - P(\cdot)K(\cdot))^{-1} \end{pmatrix} = \mathcal{F}(G(\cdot), K(\cdot)), \tag{6.23}$$

where $\mathcal{F}$ is defined as in (2.119):

$$\mathcal{F}(G, K) = G_{11} + G_{12}K(I - G_{22}K)^{-1}G_{21}.$$

We note that $\mathcal{F}(G(\cdot), K(\cdot))$ corresponds to the transfer function of a closed-loop system as in Figure 2.3. Therefore, in order to formulate the mixed-sensitivity problem as a standard $\mathcal{H}_\infty$-control problem, we shall define a Pritchard-Salamon system Σ_G such that its transfer function is given by (6.22). First of all, we define the input and output spaces $W := \mathbb{R}^{m_1}, U := \mathbb{R}^{m_2}, Z := \mathbb{R}^{p_1} \times \mathbb{R}^{p_2}, Y := \mathbb{R}^{p_0}$ and the 'larger' state space $\mathcal{V} := \mathbb{R}^{n_1} \times \mathbb{R}^{n_2} \times \mathbb{R}^{n_0} \times L_2(-\tau, 0; \mathbb{R}^{n_0})$ (with the obvious inner product). Furthermore, we define

$$\mathcal{B}_1 := \begin{pmatrix} B_1 \\ 0 \\ 0 \\ 0 \end{pmatrix}, \mathcal{B}_2 := \begin{pmatrix} 0 \\ B_2 \\ B_0 \\ 0 \end{pmatrix}, \mathcal{C}_1 := \begin{pmatrix} C_1 & 0 & 0 & 0 \\ 0 & C_2 & 0 & 0 \end{pmatrix},$$

$$\mathcal{C}_2 := \begin{pmatrix} 0 & 0 & 0 & C_0 < \delta_{-\tau}, \cdot > \end{pmatrix}, \mathcal{D}_{12} = \begin{pmatrix} 0 \\ D_2 \end{pmatrix}, \quad \mathcal{D}_{21} = I_{p_0 \times p_0},$$

$$D(\mathcal{A}^{\mathcal{V}}) = \{(x_1, x_2, \eta, \phi) \in \mathcal{V} \mid \phi \in W^{1,2}(-\tau, 0\,; \mathbb{R}^n), \eta = \phi(0)\}$$

$$\mathcal{A}^{\mathcal{V}} \begin{pmatrix} x_1 \\ x_2 \\ \eta \\ \phi \end{pmatrix} = \begin{pmatrix} A_1 & 0 & 0 & B_1 C_0 < \delta_{-\tau}, \cdot > \\ 0 & A_2 & 0 & 0 \\ 0 & 0 & A_0 & 0 \\ 0 & 0 & 0 & \frac{d}{d\theta} \end{pmatrix} \begin{pmatrix} x_1 \\ x_2 \\ \eta \\ \phi \end{pmatrix}$$

Finally, we define the 'smaller' state space $\mathcal{W} := D(\mathcal{A}^{\mathcal{V}})$ with inner product given by

$$< \begin{pmatrix} x_1^1 \\ x_2^1 \\ \eta^1 \\ \phi^1 \end{pmatrix}, \begin{pmatrix} x_1^2 \\ x_2^2 \\ \eta^2 \\ \phi^2 \end{pmatrix} >_{\mathcal{W}} := < \mathcal{A}^{\mathcal{V}} \begin{pmatrix} x_1^1 \\ x_2^1 \\ \eta^1 \\ \phi^1 \end{pmatrix}, \mathcal{A}^{\mathcal{V}} \begin{pmatrix} x_1^2 \\ x_2^2 \\ \eta^2 \\ \phi^2 \end{pmatrix} >_{\mathcal{V}}$$

(we note that our assumptions imply that $0 \in \rho(\mathcal{A}^{\mathcal{V}})$). It follows from Section 6.1 and the perturbation results of Section 2.4 that $\mathcal{A}^{\mathcal{V}}$ generates a C_0-semigroup $\mathcal{S}(\cdot)$ on $\mathcal{V}$, which restricts to a C_0-semigroup on $\mathcal{W}$, and that

$$\Sigma_G = \Sigma(\mathcal{S}(\cdot), (\mathcal{B}_1 \;\; \mathcal{B}_2), \begin{pmatrix} \mathcal{C}_1 \\ \mathcal{C}_2 \end{pmatrix}, \begin{pmatrix} 0 & \mathcal{D}_{12} \\ \mathcal{D}_{21} & 0 \end{pmatrix}) \tag{6.24}$$

is a smooth Pritchard-Salamon system as defined in Definition 2.3. We note that because of $\mathcal{W} = D(\mathcal{A}^{\mathcal{V}})$, any admissible perturbation of $\mathcal{S}(\cdot)$ is exponentially stable on $\mathcal{V}$ if and only if it is exponentially stable on $\mathcal{W}$ (see Theorem 2.20).

The dual spaces $\mathcal{V}'$ and $\mathcal{W}'$ can be chosen as in Chapter 5, Section 5.1. We recall that the actual choice is not so important as long as the conditions (2.64), (2.65) and (2.66) are satisfied. We recall that the case where the pivot space is chosen to be equal to $\mathcal{W}$ is extensively treated in subsection 6.1.3.

Next, we shall show that the transfer function of Σ_G is given by

$$\begin{pmatrix} \mathcal{C}_1 \\ \mathcal{C}_2 \end{pmatrix} (sI - \mathcal{A}^{\mathcal{V}})^{-1}(\mathcal{B}_1 \;\; \mathcal{B}_2) + \begin{pmatrix} 0 & \mathcal{D}_{12} \\ \mathcal{D}_{21} & 0 \end{pmatrix} =$$

$$\left(\begin{array}{c|c} W_1(s) & W_1(s)P(s) \\ 0 & W_2(s) \\ \hline I & P(s) \end{array}\right) = G(s), \tag{6.25}$$

for all $s \notin \sigma(A_0) \cup \sigma(A_1) \cup \sigma(A_2)$. Indeed, for such s and for a given vector $(\bar{x}_1, \bar{x}_2, \bar{\eta}, \bar{\phi}) \in \mathcal{V}$ the equation

$$(sI - \mathcal{A}^{\mathcal{V}}) \begin{pmatrix} x_1 \\ x_2 \\ \eta \\ \phi \end{pmatrix} = \begin{pmatrix} (sI - A_1)x_1 - B_1C_0\phi(-\tau) \\ (sI - A_2)x_2 \\ (sI - A_0)\eta \\ (s - \frac{d}{d\theta})\phi \end{pmatrix} = \begin{pmatrix} \bar{x}_1 \\ \bar{x}_2 \\ \bar{\eta} \\ \bar{\phi} \end{pmatrix} \tag{6.26}$$

has a unique solution $(x_1, x_2, \eta, \phi) \in D(\mathcal{A}^{\mathcal{V}})$ given by

$$\begin{pmatrix} x_1 \\ x_2 \\ \eta \\ \phi(\cdot) \end{pmatrix} = (sI - \mathcal{A}^{\mathcal{V}})^{-1} \begin{pmatrix} \bar{x}_1 \\ \bar{x}_2 \\ \bar{\eta} \\ \bar{\phi} \end{pmatrix} =$$

$$\begin{pmatrix} (sI - A_1)^{-1}[B_1C_0(\exp(-\tau s)(sI - A_0)^{-1}\bar{\eta} - \\ \qquad \int_0^{-\tau} \exp(-s(\tau + \sigma))\bar{\phi}(\sigma)d\sigma) + \bar{x}_1] \\ \\ (sI - A_2)^{-1}\bar{x}_2 \\ \\ (sI - A_0)^{-1}\bar{\eta} \\ \\ \exp(s\cdot)(sI - A_0)^{-1}\bar{\eta} - \int_0^{\cdot} \exp(s(\cdot - \sigma))\bar{\phi}(\sigma)d\sigma \end{pmatrix}. \tag{6.27}$$

Using (6.27), it follows that $(sI - \mathcal{A}^{\mathcal{V}})^{-1} \in \mathcal{L}(\mathcal{V})$ and so $s \in \rho(\mathcal{A}^{\mathcal{V}})$. The proof of (6.25) is now straightforward and left to the reader. Hence, for any

Pritchard-Salamon controller Σ_K of the form (5.3), with transfer function $K(\cdot)$, the closed-loop transfer function is given by

$$\mathcal{F}(G(\cdot), K(\cdot)) = \begin{pmatrix} W_1(\cdot)(I - P(\cdot)K(\cdot))^{-1} \\ W_2(\cdot)K(\cdot)(I - P(\cdot)K(\cdot))^{-1} \end{pmatrix}$$

(see also (6.23)). Thus, the mixed-sensitivity problem has been reformulated as a standard $\mathcal{H}_\infty$-control problem for the Pritchard-Salamon system (6.24), where the operators $\mathcal{A}^{\mathcal{V}}, \mathcal{B}_1, \mathcal{B}_2, \mathcal{C}_1, \mathcal{C}_2, \mathcal{D}_{12}$ and $\mathcal{D}_{21}$ are given as above.

Next, we shall show that the assumptions (5.10)-(5.13) of Theorem (5.4) are satisfied. First of all, $\mathcal{D}_{21}\mathcal{D}_{21}' = I_{p_0 \times p_0}$ and $\mathcal{D}_{12}'\mathcal{D}_{12} = D_2'D_2$, so that (5.11) and (5.13) are satisfied (recall that $W_2(\cdot)$ is assumed to be invertible, so that D_2 is also invertible).

In order to discuss the frequency domain assumptions (5.10) and (5.12), we recall that if $s \notin \sigma(A_0) \cup \sigma(A_1) \cup \sigma(A_2)$, then $s \in \rho(\mathcal{A}^{\mathcal{V}})$ and $(sI - \mathcal{A}^{\mathcal{V}})^{-1}$ is given by (6.27). We note that since $W_1(\cdot)$ and $W_2(\cdot)$ are assumed to have minimal realizations, we have

$$\sigma(A_1) \cap i\mathbb{R} = \emptyset \quad \text{and} \quad \sigma(A_2) \cap i\mathbb{R} = \emptyset, \tag{6.28}$$

(recall that $W_1(\cdot), W_2(\cdot) \in \mathbb{R}\mathcal{H}_\infty$). Furthermore, we recall the assumption that A_0 has no poles on the imaginary axis (assumption (6.21)). It follows that

$$\sigma(\mathcal{A}^{\mathcal{V}}) \cap i\mathbb{R} = \emptyset \tag{6.29}$$

and, in fact, we have

$$\sup_{\omega \in \mathbb{R}} \|(i\omega I - \mathcal{A}^{\mathcal{V}})^{-1}\|_{\mathcal{L}(\mathcal{V})} < \infty \tag{6.30}$$

(this follows from (6.27)). Using (6.29), it is easy to see that $(\omega, x, u) \in \mathbb{R} \times D(\mathcal{A}^{\mathcal{V}}) \times U$ satisfies $i\omega x = \mathcal{A}^{\mathcal{V}}x + \mathcal{B}_2 u$ if and only if $x = (i\omega I - \mathcal{A}^{\mathcal{V}})^{-1}\mathcal{B}_2 u$ and using (6.30) it follows that there exists a $c_1 > 0$ such that for all $\omega \in \mathbb{R}$

$$\|x\|_{\mathcal{V}} = \|(i\omega I - \mathcal{A}^{\mathcal{V}})^{-1}\mathcal{B}_2 u\|_{\mathcal{V}} \leq c_1\|u\|_U. \tag{6.31}$$

This means that (5.10) follows, if we can prove that

$$\|(\mathcal{C}_1(i\omega I - \mathcal{A}^{\mathcal{V}})^{-1}\mathcal{B}_2 + \mathcal{D}_{12})u\|_Z \geq c_2\|u\|_U, \tag{6.32}$$

for some $c_2 > 0$. Now, since

$$(\mathcal{C}_1(i\omega I - \mathcal{A}^{\mathcal{V}})^{-1}\mathcal{B}_2 + \mathcal{D}_{12})u = \begin{pmatrix} W_1(i\omega)P(i\omega)u \\ W_2(i\omega)u \end{pmatrix}$$

(see (6.25)), (6.32) follows from the fact that $W_2(\cdot)^{-1} \in \mathcal{H}_\infty$. Hence, the frequency domain condition (5.10) is satisfied.

Next, we shall show that the second frequency domain condition (5.12) is also satisfied. First of all, we have $\mathcal{W} = D(A^{\mathcal{V}})$ and so there exists an easy relationship between $\mathcal{A}^{\mathcal{W}}$ and $\mathcal{A}^{\mathcal{V}}$ (this is quite well-known, see e.g. [73]):
Let $\lambda \in \mathbb{R} \cap \rho(\mathcal{A}^{\mathcal{V}})$ and define $S : \mathcal{V} \to \mathcal{W}$; $Sv := (\lambda I - \mathcal{A}^{\mathcal{V}})^{-1}v$. It follows that $D(\mathcal{A}^{\mathcal{W}}) = SD(\mathcal{A}^{\mathcal{V}})$ and $\mathcal{A}^{\mathcal{W}}w = \mathcal{A}^{\mathcal{V}}w$ for all $w \in D(\mathcal{A}^{\mathcal{W}})$. Furthermore, $\rho(\mathcal{A}^{\mathcal{W}}) = \rho(\mathcal{A}^{\mathcal{V}})$ and for all $s \in \rho(\mathcal{A}^{\mathcal{W}})$ there holds $(sI - \mathcal{A}^{\mathcal{W}})^{-1} = S(sI - \mathcal{A}^{\mathcal{V}})^{-1}S^{-1} = (sI - \mathcal{A}^{\mathcal{V}})^{-1} \mid_{\mathcal{W}}$. Hence,

$$\|(sI - \mathcal{A}^{\mathcal{W}})^{-1}\|_{\mathcal{L}(\mathcal{W})} \le \|S\|_{\mathcal{L}(\mathcal{V},\mathcal{W})}\|(sI - \mathcal{A}^{\mathcal{V}})^{-1}\|_{\mathcal{L}(\mathcal{V})}\|S^{-1}\|_{\mathcal{L}(\mathcal{W},\mathcal{V})} \quad (6.33)$$

for all $s \in \rho(\mathcal{A}^{\mathcal{W}})$. Recalling (6.30), it follows from (6.33) that

$$\sup_{\omega\in\mathbb{R}} \|(i\omega I - \mathcal{A}^{\mathcal{W}})^{-1}\|_{\mathcal{L}(\mathcal{W})} = \sup_{\omega\in\mathbb{R}} \|(i\omega I - (\mathcal{A}^{\mathcal{W}})')^{-1}\|_{\mathcal{L}(\mathcal{W}')} < \infty. \quad (6.34)$$

As in the proof of (5.10), $(\omega, x^*, y) \in \mathbb{R} \times D((\mathcal{A}^{\mathcal{W}})') \times Y$ satisfies $i\omega x^* = (\mathcal{A}^{\mathcal{W}})'x^* + \mathcal{C}_2'y$ if and only if $x^* = (i\omega I - (\mathcal{A}^{\mathcal{W}})')^{-1}\mathcal{C}_2'y$, and using (6.34) it follows that there exists a $c_1 > 0$ such that for all $\omega \in \mathbb{R}$

$$\|x^*\|_{\mathcal{W}'} = \|(i\omega I - (\mathcal{A}^{\mathcal{W}})')^{-1}\mathcal{C}_2'y\|_{\mathcal{W}'} \le c_1\|y\|_Y. \quad (6.35)$$

Hence, (5.12) follows, if we can prove that

$$\|(\mathcal{B}_1'(i\omega I - (\mathcal{A}^{\mathcal{W}})')^{-1}\mathcal{C}_2' + \mathcal{D}_{21}')y\|_W \ge c_2\|y\|_Y, \quad (6.36)$$

for some $c_2 > 0$. But since $\mathcal{D}_{21} = I_{p_0 \times p_0}$ and $\mathcal{C}_2(i\omega I - \mathcal{A}^{\mathcal{V}})^{-1}\mathcal{B}_1 = 0$ (see (6.25)), it easily follows that (6.36) is satisfied. Hence, the frequency domain condition (5.12) is satisfied. The following theorem is now an immediate consequence of Theorem 5.4.

Theorem 6.5
Let $W_1(\cdot), W_2(\cdot)$ and $P(\cdot)$ be given as in the beginning of this subsection. Define the corresponding Pritchard-Salamon system Σ_G as in (6.24) and let $\gamma > 0$. Then there exists a stabilizing Pritchard-Salamon controller Σ_K such that

$$\left\| \begin{pmatrix} W_1(\cdot)(I - P(\cdot)K(\cdot))^{-1} \\ W_2(\cdot)K(\cdot)(I - P(\cdot)K(\cdot))^{-1} \end{pmatrix} \right\|_{\mathcal{H}_\infty} < \gamma,$$

if and only if the following conditions hold:
(i) there exists a $P \in \mathcal{L}(\mathcal{V}, \mathcal{V}')$ with $P = P' \ge 0$ satisfying

$$< Px, (\mathcal{A}^{\mathcal{V}} - \mathcal{B}_2(\mathcal{D}_{12}'\mathcal{D}_{12})^{-1}\mathcal{D}_{12}'\mathcal{C}_1)y >_{<\mathcal{V}',\mathcal{V}>} +$$

$$< (\mathcal{A}^{\mathcal{V}} - \mathcal{B}_2(\mathcal{D}_{12}'\mathcal{D}_{12})^{-1}\mathcal{D}_{12}'\mathcal{C}_1)x, Py >_{<\mathcal{V},\mathcal{V}'>} +$$

$$< P(\gamma^{-2}\mathcal{B}_1\mathcal{B}_1' - \mathcal{B}_2(\mathcal{D}_{12}'\mathcal{D}_{12})^{-1}\mathcal{B}_2')Px, y >_{<\mathcal{V}',\mathcal{V}>} +$$

$$< (I - \mathcal{D}_{12}(\mathcal{D}_{12}'\mathcal{D}_{12})^{-1}\mathcal{D}_{12}')\mathcal{C}_1x, (I - \mathcal{D}_{12}(\mathcal{D}_{12}'\mathcal{D}_{12})^{-1}\mathcal{D}_{12}')\mathcal{C}_1y >_Z = 0 \quad (6.37)$$

for all $x, y \in D(A^{\mathcal{V}})$, *such that the* C_0*-semigroup* $\mathcal{S}_P(\cdot)$ *defined by*

$$\mathcal{S}_P(\cdot) = \mathcal{S}_{\begin{pmatrix} \mathcal{B}_1 & \mathcal{B}_2 \end{pmatrix}\begin{pmatrix} \gamma^{-2}\mathcal{B}_1' P \\ -(\mathcal{D}_{12}'\mathcal{D}_{12})^{-1}(\mathcal{B}_2' P + \mathcal{D}_{12}'\mathcal{C}_1) \end{pmatrix}}(\cdot), \tag{6.38}$$

is exponentially stable on $\mathcal{W}$ *and* $\mathcal{V}$,
(*ii*) *there exists a* $Q \in \mathcal{L}(\mathcal{W}', \mathcal{W})$ *with* $Q = Q' \geq 0$ *satisfying*

$$\begin{aligned} &< Qx, ((A^{\mathcal{W}})' - \mathcal{C}_2'(\mathcal{D}_{21}\mathcal{D}_{21}')^{-1}\mathcal{D}_{21}\mathcal{B}_1')y >_{<\mathcal{W},\mathcal{W}'>} + \\ &< ((A^{\mathcal{W}})' - \mathcal{C}_2'(\mathcal{D}_{21}\mathcal{D}_{21}')^{-1}\mathcal{D}_{21}\mathcal{B}_1')x, Qy >_{<\mathcal{W}',\mathcal{W}>} + \\ &< Q(\gamma^{-2}\mathcal{C}_1'\mathcal{C}_1 - \mathcal{C}_2'(\mathcal{D}_{21}\mathcal{D}_{21}')^{-1}\mathcal{C}_2)Qx, y >_{<\mathcal{W},\mathcal{W}'>} + \\ &< (I - \mathcal{D}_{21}'(\mathcal{D}_{21}\mathcal{D}_{21}')^{-1}\mathcal{D}_{21})\mathcal{B}_1'x, (I - \mathcal{D}_{21}'(\mathcal{D}_{21}\mathcal{D}_{21}')^{-1}\mathcal{D}_{21})\mathcal{B}_1'y >_{\mathcal{W}} = 0 \end{aligned} \tag{6.39}$$

for all $x, y \in D((A^{\mathcal{W}})')$, *such that the* C_0*-semigroup* $\mathcal{S}_Q(\cdot)$ *defined via its adjoint as*

$$\mathcal{S}_Q'(\cdot) = \mathcal{S}'_{\begin{pmatrix} \mathcal{C}_1' & \mathcal{C}_2' \end{pmatrix}\begin{pmatrix} \gamma^{-2}\mathcal{C}_1 Q \\ -(\mathcal{D}_{21}\mathcal{D}_{21}')^{-1}(\mathcal{C}_2 Q + \mathcal{D}_{21}\mathcal{B}_1') \end{pmatrix}}(\cdot), \tag{6.40}$$

is exponentially stable on $\mathcal{W}$ *and* $\mathcal{V}$,
(*iii*)

$$r_\sigma^{\mathcal{W}'}(PQ) = r_\sigma^{\mathcal{V}}(QP) < \gamma^2. \tag{6.41}$$

Furthermore, in this case the controller given by (5.3) *with*

$$\begin{aligned} T(\cdot) &:= \bar{\mathcal{S}}_{-\bar{\mathcal{B}}_1(\mathcal{D}_{21}\mathcal{D}_{21}')^{-1/2}\bar{\mathcal{C}}_2 - \bar{\mathcal{B}}_2(\mathcal{D}_{12}'\mathcal{D}_{12})^{-1/2}\bar{\mathcal{C}}_1}(\cdot), \\ N &:= \bar{\mathcal{B}}_1(\mathcal{D}_{21}\mathcal{D}_{21}')^{-1/2}, \\ L &:= -(\mathcal{D}_{12}'\mathcal{D}_{12})^{-1/2}\bar{\mathcal{C}}_1, \qquad R := 0, \end{aligned} \tag{6.42}$$

where

$$\begin{aligned} \bar{\mathcal{C}}_1 &:= (\mathcal{D}_{12}'\mathcal{D}_{12})^{-1/2}(\mathcal{B}_2'P + \mathcal{D}_{12}'\mathcal{C}_1) \\ \bar{\mathcal{C}}_2 &:= \mathcal{C}_2 + \gamma^{-2}\mathcal{D}_{21}\mathcal{B}_1'P \\ \bar{\mathcal{B}}_1 &:= (\bar{P}(\mathcal{C}_2' + \gamma^{-2}P\mathcal{B}_1\mathcal{D}_{21}') + \mathcal{B}_1\mathcal{D}_{21}')(\mathcal{D}_{21}\mathcal{D}_{21}')^{-1/2} \\ \bar{\mathcal{B}}_2 &:= \mathcal{B}_2 + \gamma^{-2}\bar{P}(P\mathcal{B}_2 + \mathcal{C}_1'\mathcal{D}_{12}) \\ \bar{\mathcal{S}}(\cdot) &:= \mathcal{S}_{\gamma^{-2}\mathcal{B}_1\mathcal{B}_1'P + \gamma^{-2}\bar{P}(P\mathcal{B}_2 + \mathcal{C}_1'\mathcal{D}_{12})(\mathcal{D}_{12}'\mathcal{D}_{12})^{-1}(\mathcal{B}_2'P + \mathcal{D}_{12}'\mathcal{C}_1)}(\cdot) \\ \bar{P} &:= Q(I - \gamma^{-2}PQ)^{-1}, \end{aligned} \tag{6.43}$$

is γ*-admissible.*

Remark 6.6
At the beginning of this section we assumed that the pairs (A_0, B_0) and (C_0, A_0) are controllable and observable, respectively. Using the results in [72], it can therefore be shown that the pairs $(\mathcal{S}(\cdot), \mathcal{B}_2)$ and $(\mathcal{C}_2, \mathcal{S}(\cdot))$ of the Pritchard-Salamon system Σ_G in (6.24) are admissibly stabilizable and admissibly detectable, respectively (see also subsection 6.1.1). Hence, it follows from Theorem 2.30 that there exists a stabilizing Pritchard-Salamon controller Σ_{K_0} for Σ_G. Thus, for γ sufficiently large there exist operators P and Q satisfying items $(i), (ii)$ and (iii) of Theorem 6.5. (see also Remark 5.5).

We note that both Riccati equations are of the *bounded* type: In the first Riccati equation the output operator

$$(I - \mathcal{D}_{12}(\mathcal{D}_{12}'\mathcal{D}_{12})^{-1}\mathcal{D}_{12}')\mathcal{C}_1$$

is bounded on $\mathcal{V}$ because $\mathcal{C}_1$ is and in the second Riccati equation the output operator

$$(I - \mathcal{D}_{21}'(\mathcal{D}_{21}\mathcal{D}_{21}')^{-1}\mathcal{D}_{21})\mathcal{B}_1'$$

is bounded on $\mathcal{W}'$ because $\mathcal{B}_1'$ is.

Of course there remain two difficult problems: how can we find solutions of the operator-valued Riccati equations in Theorem 6.5 and how can we (approximately) implement an infinite-dimensional controller of the form (5.3), (6.42)? We have seen in subsection 6.1.2 that in some cases it is possible to find explicit solutions to infinite-dimensional Riccati equations, but in general some approximation method would have to be used. In the next section we shall discuss these issues in some detail, but basically they are left as topics for future research.

6.3 Conclusions and directions for future research

In this final section we list the main contributions of the book and we indicate some directions for future research.

We have extended the finite-dimensional results for the regular sub-optimal $\mathcal{H}_\infty$-control problem with measurement-feedback of Doyle et al. [26] and Glover and Doyle [37] to the Prichard-Salamon class. In particular, given a Pritchard-Salamon system satisfying certain a priori assumptions, we have shown that the existence of a (Pritchard-Salamon) controller that solves the sub-optimal $\mathcal{H}_\infty$-control problem is equivalent to the solvability of two coupled Riccati equations. In the case that the problem is solvable, we have given a parametrization of all (Pritchard-Salamon) controllers that solve the problem. As for the finite-dimensional case, the solution to the *optimal* $\mathcal{H}_\infty$-control problem can in principle be found (approximately) using a procedure called γ-iteration.

As far as the a priori assumptions are concerned, we have shown that for Pritchard-Salamon systems all the finite-dimensional regularity conditions of Glover and Doyle [37] can be replaced by their natural infinite-dimensional counterparts. Furthermore, these regularity conditions can even be removed, however, at the expense of introducing an extra 'regularizing parameter'. We note that for the finite-dimensional case the regularity conditions can be removed in a more elegant way (see Stoorvogel [86] and Scherer [83]), but it is not clear if this would be possible for infinite-dimensional systems. One of the problems in this respect is the fact that it is not feasible to restrict an infinite-dimensional system to a controllability subspace.

In the process of generalizing the finite-dimensional results on $\mathcal{H}_\infty$-theory, we have developed interesting system theoretic results for the Pritchard-Salamon class. In particular, the book shows the nice duality properties of Pritchard-Salamon systems and the fact that the class is closed under dynamic output-feedback. Furthermore, using the results of [15], we have developed an appropriate stability/stabilizability theory for the Pritchard-Salamon class.

One of the by-products of our approach to the $\mathcal{H}_\infty$-control problem is the solution to the linear quadratic control problem (LQ-problem) in Chapter 3. We have shown that the solvability of the LQ-problem is equivalent to a frequency domain inequality and the solvability of an LQ-type Riccati equation. A special case of the frequency domain inequality emerges in the solution to the $\mathcal{H}_\infty$-control problem (invariant zeros condition). Incidentally, we claim that the so-called $\mathcal{H}_2$-problem (see [26] for the finite-dimensional case) can be solved for the Pritchard-Salamon class using exactly the same techniques that were used in the solution to the $\mathcal{H}_\infty$-control problem. In fact, by letting γ tend to ∞, the corresponding Riccati equations can be obtained from the $\mathcal{H}_\infty$-Riccati equations and the $\mathcal{H}_2$-optimal controller can be obtained from the central controller.

Due to the duality properties mentioned above, the Pritchard-Salamon class is such that a system Σ_G belongs to the class if and only $\Sigma_{G^\natural}$ does, and $\Sigma_{(G^\natural)^\natural} = \Sigma_G$ (here $\natural$ is used to denote the transpose, or 'dual' of a system, see (2.79)). Furthermore, a controller Σ_K is γ-admissible for Σ_G if and only if $\Sigma_{K^\natural}$ is γ-admissible for $\Sigma_{G^\natural}$. These facts are crucial in the derivation of the second ('dual') Riccati equation and the coupling condition. The classes of infinite-dimensional systems that are considered in [8, 57, 59] and the many references therein, do not have such nice duality properties. Moreover, the perturbation theory for these classes is not as rich as for the Pritchard-Salamon class. In some classes the notion of stabilizing output feedback may not even be defined. Thus, it seems that the Pritchard-Salamon class of infinite-dimensional systems is the only one that is suitable for the state-space approach to the $\mathcal{H}_\infty$-control problem with measurement-feedback, whenever one wants to allow for unbounded input and output operators simultaneously.

We note that the frequency domain (skew Toeplitz) approach of Tannenbaum and Özbay et al. [90, 65, 67] (and references therein) can deal with plants outside the Pritchard-Salamon class (e.g. a pure delay $\exp(-s)$), but certain assumptions in their work are rather restrictive. For instance, in the skew Toeplitz approach one usually needs explicit inner/outer factorizations of the plants and these are (in general) very hard to find, as explained in [65]. For some examples they are able to find optimal (infinite-dimensional) $\mathcal{H}_\infty$-controllers, which then need to be approximated (see also [58, 66, 68] and subsection 6.2.1).

In the introduction we mentioned that several interesting problems regarding robustness optimization, performance optimization and combinations thereof fit into the general $\mathcal{H}_\infty$-control framework (see Francis [34], Curtain [12]), Tannenbaum [90] and the many references therein). In Section 6.2 we have examined one example: the mixed-sensitivity problem for delay systems. In principle, these problems may now be solved for any system in the Pritchard-Salamon class, using Theorem 5.4 and the γ-iteration procedure. However, two kinds of practical difficulties immediately arise: how can we solve operator-valued Riccati equations and how can one implement (or approximate) infinite-dimensional controllers? (The γ-admissible central controller of Theorem 5.4 has the same state-space as the original system). Naturally, similar questions arose in the infinite-dimensional $LQ(G)$-theory and there exists some theory for the corresponding Riccati equations.

As far as the first kind of practical difficulty is concerned, for particular examples of the operators A, B, C etc. it may be possible to reduce a Riccati equation to an equation that can be solved using an ad hoc (numerical) method (see e.g. Delfour [20] and references therein and Remark 6.3). In particular, Riccati equations corresponding to delay equations as in Examples 1.2,1.4 may allow for such ad hoc solutions, due to the particular form of the corresponding state-space realizations (see also subsection 6.1.2). It would be interesting to consider these direct methods for $\mathcal{H}_\infty$-type Riccati equations, especially because $\mathcal{H}_\infty$-control of systems described by delay equations is such an important topic.

In the case that one cannot obtain an exact solution of an infinite-dimensional Riccati equation (the 'generic' case) it may be feasible to try to find approximating solutions, using general approximation schemes. At the moment there exists a rather well developed approximation theory for infinite-dimensional LQ-type Riccati equations (see Banks and Burns [4], Ito and Tran [42], Kappel [45] Gibson [35], Lasiecka and Triggiani [57] and references therein) and so it would be interesting to try to extend this theory to $\mathcal{H}_\infty$-type Riccati equations. The idea is usually to construct a finite-dimensional (state-space) approximation of the plant, to solve the corresponding finite-dimensional problem (leading to a sequence of solutions to finite-dimensional Riccati equations) and then to show that the finite-dimensional solutions converge to the infinite-dimensional ones. This procedure is rather indirect and so it would also be interesting to find (and compare with) more direct methods. In this respect we should mention the spectral factorization method of Callier and Winkin [10] for the LQ-problem and the Hamiltonian theory for infinite-dimensional systems, which is being developed by Kuiper and Zwart [54]. Finally, we have seen in Remark 6.4 that the problem of solving Riccati equations of the Pritchard-Salamon type with unbounded terms like $< Cx, Cx >_Y$ can be reduced to solving a (Pritchard-Salamon type) Lyapunov equation and a Riccati equation of the bounded type. Hence, an approximation theory for general Riccati equations of the bounded type would suffice.

As far as the second kind of practical difficulty is concerned, implementing infinite-dimensional controllers is only possible in very special cases and so one usually has to consider finite-dimensional approximations. For the (sub-optimal) $\mathcal{H}_\infty$-control problem this amounts to the following question: if there exists a stabilizing (Pritchard-Salamon) controller Σ_K for Σ_G such that $\|\mathcal{F}(G,K)\| < \gamma$ (i.e. Σ_K is γ-admissible), does there exist a finite-dimensional controller Σ_{K_n} that approximates Σ_K such that Σ_{K_n} stabilizes Σ_G and $\|\mathcal{F}(G,K_n)\| < \gamma$? In other words, is it possible to find an approximating controller Σ_{K_n} such that the closed-loop performance does not deteriorate too much? In principle, one way to obtain Σ_{K_n} would be to try to approximate the infinite-dimensional central controller $\Sigma_{\mathcal{F}(\tilde{K},0)}$ from Theorem 5.4 by using the finite-dimensional approximations of the solutions to the Riccati equations mentioned above. It is an interesting open problem to further develop this approximation procedure to construct finite-dimensional γ-admissible controllers.

For special $\mathcal{H}_\infty$-control problems there already exists some approximation theory for finding finite-dimensional sub-optimal controllers (see e.g. Curtain [12], Özbay [65, 66] and Rodriguez and Dahleh [77]). One approach (sometimes called the *direct* approach) is to try to approximate the transfer function K of the (optimal) controller Σ_K by K_n in a suitable way (e.g. L_∞, gap metric, etc.), so that the closed loop performance $\|\mathcal{F}(G,K_n)\|$ converges to $\|\mathcal{F}(G,K)\|$. The choice of the approximation should of course depend on the form of the generalized plant Σ_G corresponding to the particular $\mathcal{H}_\infty$-control problem. This idea was used in [65] for a particular mixed-sensitivity problem. We note that this idea can be combined with the procedure proposed in the previous paragraph. Another approximation scheme is the following: find a finite-dimensional approximation Σ_{G_n} of the infinite-dimensional plant Σ_G, design a finite-dimensional controller Σ_{K_n} for Σ_{G_n} using finite-dimensional $\mathcal{H}_\infty$-theory, apply Σ_{K_n} to Σ_G and hope that Σ_{K_n} stabilizes Σ_G for n large enough and that $\|\mathcal{F}(G, K_n)\|$ converges to 'the optimal value' (this is sometimes called the *indirect* approach). Using this scheme, it seems that all the theory about infinite-dimensional Riccati equations is not necessary if one is only interested in sub-optimal controllers. However, in general it is not clear if Σ_{K_n} will stabilize Σ_G for n large enough, because this depends critically on the kind of approximation of Σ_G and the kind of $\mathcal{H}_\infty$-control problem. Furthermore, it is not at all clear how $\|\mathcal{F}(G, K_n)\|$ is going to behave. In this respect, we note that Smith and Engelbert [85, 27] have obtained some positive and some negative results regarding the continuous dependence of the optimal performance $\inf_K \|\mathcal{F}(G, K)\|$ on the plant G for special $\mathcal{H}_\infty$-control problems. On the other hand, the above described indirect approach was used successfully in [12] for two robust control problems and in [77] for a sensitivity minimization problem. It is our opinion that the indirect approach may work if the approximations are chosen very carefully, depending on the particular $\mathcal{H}_\infty$-control problem and that general convergence results may be obtained by using the existence of solutions to the infinite-dimensional Riccati equations.

Concluding, in this book we have set the first step towards solving $\mathcal{H}_\infty$-control problems for infinite-dimensional systems using state-space techniques, but there are still many challenging problems left to solve.

Appendix A

Stability theory

In this section we quote some well-known results about the theory of stability and stabilizability for semigroup control systems. We note that (as usual) a semigroup $T(\cdot)$ on a Hilbert space X is called *exponentially stable* if there exists some $\alpha > 0$ and some $M > 0$ such that $\|T(t)\| \leq M\exp(-\alpha t)$ for all $t \geq 0$.

A.1

Here we shall quote two very useful results that were obtained by Datko in [18] and [19]. The first result gives a necessary and sufficient condition for a C_0-semigroup to be exponentially stable and is taken from [18].

Lemma A.1
Let $T(\cdot)$ be a C_0-semigroup on a separable Hilbert space X. Then $T(\cdot)$ is exponentially stable if and only if $T(\cdot)x \in L_2(0,\infty;X)$ for all $x \in X$.

Datko's second result is proved in [19] and relates exponential stabilizability to the solvability of the standard LQ-problem for systems with a bounded input operator.

Lemma A.2
Suppose that $T(\cdot)$ is a C_0-semigroup on a separable Hilbert space X with infinitesimal generator A and let $B \in \mathcal{L}(U,X)$, where U is another separable Hilbert space. Consider the system

$$x(t) = T(t)x_0 + \int_0^t T(t-s)Bu(s)ds$$

and suppose that for all $x_0 \in X$ there exists a $u(\cdot) \in L_2(0,\infty;U)$ such that $x(\cdot) \in L_2(0,\infty;X)$. The following hold:

(*i*) *There exists a unique* $\bar{u}(\cdot) \in L_2(0,\infty;U)$ *that minimizes the cost function*

$$J = \int_0^\infty (\|x(t)\|_X^2 + \|u(t)\|_U^2)dt$$

and $\bar{u}(\cdot)$ *is of feedback form:* $\bar{u}(\cdot) = Kx(\cdot)$ *for some* $K \in \mathcal{L}(X,U)$.

(*ii*) *The pair* (A,B) *is exponentially stabilizable (this a consequence of the first item).*

A.2

Here we state a useful result that follows imediately from Zabczyk [102, Lemma 3]. It generalizes a Lyapunov type argument to infinite dimensions.

Lemma A.3
Let $T(\cdot)$ *be a* C_0*-semigroup on a Hilbert space* X *with infinitesimal generator* A. *Suppose that there exist operators* $C \in \mathcal{L}(X,Y)$ *and* $G \in \mathcal{L}(Y,X)$ *(where* Y *is some Hilbert space) such that* $A + GC$ *is the infinitesimal generator of an exponentially stable semigroup (or, in other words, the pair* (C,A) *is exponentially detectable). If there exists an operator* $P \in \mathcal{L}(X)$ *such that* $P = P^* \geq 0$ *and for all* $x, y \in D(A)$

$$< Ax, Py >_X + < Px, Ay >_X \leq - < Cx, Cy >_Y,$$

then $T(\cdot)$ *is exponentially stable.*

Appendix B

Differentiability and some convergence results

In this section we shall give several results regarding differentiablity of systems equations and convergence theory. Some of these results are well-known and some will be proved here.

B.1

Suppose that we have an exponentially stable semigroup $T(\cdot)$ *on a separable Hilbert space* X *and consider*

$$x(t) = T(t)x_0 + \int_0^t T(t-\tau)f(\tau)d\tau \tag{B.1}$$

for some $f(\cdot) \in L_2(0,\infty;X)$ *and some* $x_0 \in X$*. Then*

$$x(\cdot) \in L_2(0,\infty;X), \tag{B.2}$$

$$x(\cdot) \text{ is strongly continuous} \tag{B.3}$$

and

$$x(t) \xrightarrow{X} 0 \text{ as } t \to \infty. \tag{B.4}$$

If $x_0 = 0$*, the mapping* $\mathcal{S}$ *from* $f(\cdot)$ *to* $x(\cdot)$ *satisfies*

$$\mathcal{S} \in \mathcal{L}(L_2(0,\infty;X), L_2(0,\infty;X)). \tag{B.5}$$

Proof:
Since $T(\cdot)$ is exponentially stable, $T(t)x_0 \in L_2(0,\infty;X)$. Furthermore,

$$\|\int_0^{\cdot} T(\cdot-\tau)f(\tau)d\tau\|_2 \le \|\int_0^{\cdot} \|T(\cdot-\tau)\| \cdot \|f(\tau)\|_X d\tau\|_2$$

$$\leq \|T(\cdot)\|_1 \|f(\cdot)\|_2, \tag{B.6}$$

using the property of convolutions $\|(f * g)(\cdot)\|_2 \leq \|f(\cdot)\|_1 \|g(\cdot)\|_2$ and the fact that $T(\cdot)$ is exponentially stable. Hence we infer (B.2) and (B.5). Furthermore, (B.3) follows from [16, section 2.2]. In order to prove (B.4) we define

$$\| \int_0^t T(t-\tau) f(\tau) d\tau \|_X \leq \int_0^t \|T(t-\tau)\| \cdot \|f(\tau)\|_X d\tau =: g(t)$$

and since $\|T(\cdot)\|$ and $\|f(\cdot)\|_X$ are both in $L_2(0,\infty;\mathbb{R})$ we may conclude that $g(t) \rightarrow 0$ as $t \rightarrow \infty$ (see e.g. [22] for another result about convolutions). Finally, since $T(t)x_0 \stackrel{X}{\rightarrow} 0$ as $t \rightarrow \infty$, property (B.4) follows.

B.2

Suppose that A is the infinitesimal generator of a semigroup $T(\cdot)$ on a separable Hilbert space X and $B \in \mathcal{L}(U,X)$, where U is also a separable Hilbert space and suppose that (A,B) is exponentially stabilizable (this means that there exists an $F \in \mathcal{L}(X,U)$ such that $A + BF$ generates an exponentially stable C_0-semigroup). Consider the controlled system

$$x(t) = T(t)x_0 + \int_0^t T(t-\tau)(Bu(\tau) + f(\tau))d\tau, \tag{B.7}$$

where $u(\cdot) \in L_2(0,\infty;U)$ and $f(\cdot) \in L_2(0,\infty;X)$. If $x(\cdot) \in L_2(0,\infty;X)$, there holds:

$$x(\cdot) \text{ is strongly continuous,} \tag{B.8}$$

$$x(t) \stackrel{X}{\rightarrow} 0 \text{ as } t \rightarrow \infty. \tag{B.9}$$

Proof:
(B.8) is immediate from [16, section 2.2].
Define $u_1(\cdot) := u(\cdot) - Fx(\cdot)$, where F is such that $A + BF$ generates an exponentially stable semigroup $T_{BF}(\cdot)$. Then $u_1(\cdot) \in L_2(0,\infty;U)$ and

$$x(t) = T_{BF}(t)x_0 + \int_0^t T_{BF}(t-\tau)(Bu_1(\tau) + f(\tau))d\tau. \tag{B.10}$$

(B.9) now follows from Appendix B.1.

B.3

Let $T(\cdot)$ be a C_0-semigroup on a separable Hilbert space X with infinitesimal generator A, let $f \in CD(0,\infty;X)$ (continuously differentiable) and suppose that $x_0 \in D(A)$. Then $x(\cdot)$ defined by

$$x(t) = T(t)x_0 + \int_0^t T(t-\tau)f(\tau)d\tau, \tag{B.11}$$

is also continuously differentiable and we have

$$\dot{x} = Ax + f; \quad x(t) \in D(A) \text{ for all } t \geq 0.$$

(This is a well known result: see e.g. [16, section 2.2]).

B.4

Let $T(\cdot)$ be an exponentially stable C_0-semigroup on a separable X with infinitesimal generator A, let $f \in L_2(-\infty,0;X)$ and consider

$$s(t) := \int_{-\infty}^t T(t-\tau)f(\tau)d\tau \text{ for } t \leq 0.$$

Then

$$s(\cdot) \in L_2(-\infty,0;X),$$

$s(\cdot)$ *is strongly continuous*

$$s(t) \xrightarrow{X} 0 \text{ as } t \to -\infty.$$

the mapping $\mathcal{S}$ from $f(\cdot)$ to $s(\cdot)$ satisfies

$$\mathcal{S} \in \mathcal{L}(L_2(-\infty,0;X), L_2(-\infty,0;X)).$$

Proof:
The proof is similar to the proof of Appendix B.1 and therefore deleted.

B.5

Consider

$$s(t) := \int_{-\infty}^t T(t-\tau)f(\tau)d\tau \text{ for } t \leq 0,$$

where $T(\cdot)$ *is an exponentially stable semigroup with infinitesimal generator* A *and* $f(\cdot) \in L_2(-\infty, 0; X)$ *is continuously differentiable such that* $\dot{f}(\cdot) \in L_1(-\infty, 0; X)$*. Then*

$$s(\cdot) \textit{ is continuously differentiable} \tag{B.12}$$

and

$$\dot{s} = As + f \textit{ and } s(t) \in D(A) \textit{ for all } t \leq 0. \tag{B.13}$$

Proof:
First we note that these results are small but necessary extensions of the results in Appendix B.3.

Since $T(\cdot)$ is exponentially stable and $\dot{f}(\cdot) \in L_1(-\infty, 0; X)$, we have the following fact:

$$\int_{-\infty}^{t} \|T(t-\tau)\dot{f}(\tau)\| d\tau < \infty \tag{B.14}$$

(the proof of this fact is similar to the proof of (B.2),(B.5); here the property $\|(f * g)(\cdot)\|_1 \leq \|f(\cdot)\|_1 \|g(\cdot)\|_1$ should be used).
Proceeding as in [16, section 2.2], there holds

$$s(t) = \int_{-\infty}^{t} T(t-\tau)\{\int_{-\infty}^{\tau} \dot{f}(\alpha) d\alpha\} d\tau = \int_{-\infty}^{t} \{\int_{\alpha}^{t} T(t-\tau)\dot{f}(\alpha) d\tau\} d\alpha,$$

using Fubini's Theorem (see e.g. [38, Theorem 3.7.13]). Furthermore, we know that for all $x \in X$

$$\int_{\alpha}^{t} T(t-\tau)x d\tau \in D(A) \text{ and } A\int_{\alpha}^{t} T(t-\tau)x d\tau = T(t-\alpha)x - x, \tag{B.15}$$

so that

$$\int_{-\infty}^{t} \|A\int_{\alpha}^{t} T(t-\tau)\dot{f}(\alpha) d\tau\| d\alpha = \int_{-\infty}^{t} \|T(t-\alpha)\dot{f}(\alpha) - \dot{f}(\alpha)\| d\alpha < \infty,$$

using (B.14) and the fact that $\dot{f}(\cdot) \in L_1(-\infty, 0; X)$. Since A is closed we can therefore use a result from [16, section 6.2] to infer that for all $t \geq 0$, $s(t) = \int_{-\infty}^{t}\{\int_{\alpha}^{t} T(t-\tau)\dot{f}(\alpha) d\tau\} d\alpha \in D(A)$ and

$$As(t) = A\int_{-\infty}^{t} \{\int_{\alpha}^{t} T(t-\tau)\dot{f}(\alpha) d\tau\} d\alpha = \int_{-\infty}^{t} A\{\int_{\alpha}^{t} T(t-\tau)\dot{f}(\alpha) d\tau\} d\alpha$$

$$= \int_{-\infty}^{t} T(t-\alpha)\dot{f}(\alpha) d\alpha - f(t), \tag{B.16}$$

where in the last step we have used (B.15).
On the other hand,

$$s(t) = \int_{-\infty}^{t} T(t-\tau)f(\tau)d\tau = \int_{\infty}^{0} T(\alpha)f(t-\alpha)d\alpha.$$

Now since $\frac{d}{dt}T(\alpha)f(t-\alpha) = T(\alpha)\dot{f}(t-\alpha)$ is integrable as a function of α (use a result like (B.14)), we can conclude that $s(\cdot)$ is continuously differentiable and

$$\dot{s}(t) = \int_{\infty}^{0} T(\alpha)\dot{f}(t-\alpha)d\alpha = \int_{-\infty}^{t} T(t-\tau)\dot{f}(\tau)d\tau.$$

The combination of this last equation and (B.16) gives $\dot{s}(t) = As(t) + f(t)$.

B.6

Consider a Pritchard-Salamon system $\Sigma_G = \Sigma(S(\cdot), B, C, D)$ *of the form*

$$\Sigma_G \begin{cases} x(t) &= S(t)x_0 + \int_0^t S(t-s)Bu(s)ds \\ \\ y(t) &= Cx(t) + Du(t), \end{cases}$$

(just as in (2.7)) and let $T > 0$ *be arbitrary. Furthermore, let* $u_n(\cdot) \in CD(0,\infty;U)$ *be such that* $\|u_n(\cdot) - u(\cdot)\|_{L_2(0,T;U)} \to 0$ *and* $x_{0n} \in D(A^{\mathcal{V}})$ *be such that* $x_{0n} \xrightarrow{\mathcal{V}} x_0$ *as* $n \to \infty$.
Let $x_n(\cdot)$ *be given by*

$$x_n(t) = S(t)x_{0n} + \int_0^t S(t-s)Bu_n(s)ds.$$

It follows that $x_n(\cdot)$ *is continuously differentiable,* $x_n(t) \in D(A^{\mathcal{V}})$ *for all* $t \geq 0$,

$$\dot{x}_n(t) = A^{\mathcal{V}}x_n(t) + Bu_n(t), \ \textit{for all}\ t \geq 0, \tag{B.17}$$

$$\|x_n(\cdot) - x(\cdot)\|_{L_2(0,T;\mathcal{V})} \to 0 \ \textit{as}\ n \to \infty, \tag{B.18}$$

$$x_n(T) \xrightarrow{\mathcal{V}} x(T) \ \textit{as}\ n \to \infty \tag{B.19}$$

and

$$\|Cx_n(\cdot) - Cx(\cdot)\|_{L_2(0,T;Y)} \to 0 \ \textit{as}\ n \to \infty. \tag{B.20}$$

Proof:
Property (B.17) follows from Appendix B.3. Furthermore,

$$\|x_n(\cdot) - x(\cdot)\|_{L_2(0,T;\mathcal{V})} \leq \|S(\cdot)(x_{0n} - x_0)\|_{L_2(0,T;\mathcal{V})} +$$

$$\left\| \int_0^{\cdot} S(\cdot - s)(Bu_n(s) - Bu(s))ds \right\|_{L_2(0,T;\mathcal{V})} \leq \|S(\cdot)(x_{0n} - x_0)\|_{L_2(0,T;\mathcal{V})} +$$

$$\left\| \int_0^{\cdot} \|S(\cdot - s)\|_{\mathcal{L}(\mathcal{V})} \|B\|_{\mathcal{L}(U,\mathcal{V})} \|u_n(s) - u(s)\|_U ds \right\|_{L_2(0,T)}$$

and using the property of convolutions $\|(f * g)(\cdot)\|_2 \leq \|f(\cdot)\|_1 \|g(\cdot)\|_2$, we may deduce (B.18).

To prove (B.19), we estimate

$$\|x_n(T) - x(T)\|_{\mathcal{V}} \leq \|S(T)x_{0n} - S(T)x_0\|_{\mathcal{V}} +$$

$$\| \int_0^T S(T - s)(Bu_n(s) - Bu(s))ds\|_{\mathcal{V}} \leq \|S(T)\|_{\mathcal{L}(\mathcal{V})} \|x_{0n} - x_0\|_{\mathcal{V}} +$$

$$\int_0^T \|S(T - s)\|_{\mathcal{L}(\mathcal{V})} \|B\|_{\mathcal{L}(U,\mathcal{V})} \|u_n(s) - u(s)\|_U ds$$

and using Cauchy-Schwarz this implies the desired result.

Finally, we prove (B.20). Here we have to be careful with the fact that $C \in \mathcal{L}(\mathcal{W}, Y)$ is not bounded on $\mathcal{V}$. We have

$$\|Cx_n(\cdot) - Cx(\cdot)\|_{L_2(0,T;Y)} \leq \left\|\overline{CS}(\cdot)(x_{0n} - x_0)\right\|_{L_2(0,T;Y)} +$$

$$\left\| C \int_0^{\cdot} S(\cdot - s)B(u_n(s) - u(s))ds \right\|_{L_2(0,T;Y)} \leq \left\|\overline{CS}(\cdot)(x_{0n} - x_0)\right\|_{L_2(0,T;Y)}$$

$$+ \|C\|_{\mathcal{L}(\mathcal{W},Y)} \left\| \int_0^{\cdot} S(\cdot - s)B(u_n(s) - u(s))ds \right\|_{L_2(0,T;\mathcal{W})} \tag{B.21}$$

Since C is admissible, we have

$$\left\|\overline{CS}(\cdot)(x_{0n} - x_0)\right\|_{L_2(0,T;Y)} \to 0 \text{ as } n \to \infty$$

and the fact that B is admissible implies that

$$\left\| \int_0^{\cdot} S(\cdot - s)B(u_n(s) - u(s))ds \right\|_{L_2(0,T;\mathcal{W})} \to 0 \text{ as } n \to \infty.$$

Hence, (B.20) follows from (B.21).

B.7

Consider

$$r(t) = \int_t^\infty T(s-t)f(s)ds \quad t \geq 0; r_n(t) = \int_t^\infty T(s-t)f_n(s)ds \quad t \geq 0,$$

where $T(\cdot)$ is an exponentially stable semigroup with infinitesimal generator A on X, $f(\cdot) \in L_2(0,\infty;X)$ and $f_n(\cdot) \in CCD(0,\infty;X)$ converges to $f(\cdot)$ (in the L_2-norm) as $n \to \infty$. Furthermore, suppose that $T > 0$. It follows from Appendix B.4 *that $r(\cdot), r_n(\cdot) \in L_2(0,\infty;X)$. Furthermore,*

$$\|r_n(\cdot) - r(\cdot)\|_{L_2(0,\infty;X)} \to 0 \text{ and } r_n(T) \xrightarrow{X} r(T) \text{ as } n \to \infty.$$

The proof of this analogous to the proof of Appendix B.6 and therefore omitted.

Appendix C

The invariant zeros condition

C.1

Consider the smooth Pritchard-Salamon system $\Sigma(S(\cdot), B_2, C_1, D_{12})$ and recall the 'invariant zeros condition' (4.4):

$$\text{there exists an } \epsilon > 0 \text{ such that for all } (\omega, x, u) \in \mathbb{R} \times D(A^{\mathcal{V}}) \times U$$

$$\text{satisfying } i\omega x = A^{\mathcal{V}}x + B_2 u, \text{ there holds}$$

$$\|C_1 x + D_{12}u\|_Z^2 \geq \epsilon \|x\|_{\mathcal{V}}^2 . \tag{4.4}$$

(recall that smoothness means that $D(A^{\mathcal{V}}) \hookrightarrow \mathcal{W}$ so that $C_1 x$ is well-defined for all $x \in D(A^{\mathcal{V}})$). Furthermore, we suppose that $D_{12}'D_{12}$ is coercive.
The following statements hold:

(*i*) *The following condition is equivalent to* (4.4)*:*

$$\text{there exists an } \epsilon > 0 \text{ such that for all } (\omega, x, u) \in \mathbb{R} \times D(A^{\mathcal{V}}) \times U$$

$$\text{satisfying } i\omega x = A^{\mathcal{V}}x + B_2 u, \text{ there holds}$$

$$\|C_1 x + D_{12}u\|_Z^2 \geq \epsilon(\|x\|_{\mathcal{V}}^2 + \|u\|_U^2). \tag{C.1}$$

(*ii*) *If assumption* (4.12) *is satisfied* (*i.e.* $D_{12}'[C_1 \;\; D_{12}] = [0 \;\; I]$), *then* (4.4) *is implied by the condition that there exists an admissible input operator* $H \in \mathcal{L}(Z, \mathcal{V})$ *such that* $S_{HC_1}(\cdot)$ *is exponentially stable on* $\mathcal{V}$.

(*iii*) *Let* $F \in \mathcal{L}(\mathcal{W}, U)$ *be an admissible output operator. For all* $x \in D(A^{\mathcal{V}})$, $u \in U$ *and* $\omega \in \mathbb{R}$ *there holds*

$$\begin{pmatrix} i\omega I - A_{B_2F}^{\mathcal{V}} & -B_2 \\ C_1 + D_{12}F & D_{12} \end{pmatrix} \begin{pmatrix} x \\ u \end{pmatrix} =$$

$$\begin{pmatrix} i\omega I - A^{\mathcal{V}} & -B_2 \\ C_1 & D_{12} \end{pmatrix} \begin{pmatrix} I & 0 \\ F & I \end{pmatrix} \begin{pmatrix} x \\ u \end{pmatrix}$$

and the following condition is equivalent to (4.4)*:*

there exists a $\delta > 0$ *such that for all* $(\omega, x, v) \in \mathbb{R} \times D(A^{\mathcal{V}}) \times U$

satisfying $i\omega x = A^{\mathcal{V}}_{B_2F}x + B_2v$*, there holds*

$$\|(C_1 + D_{12}F)x + D_{12}v\|_Z^2 \geq \delta \|x\|_{\mathcal{V}}^2 . \tag{C.2}$$

(*iv*) *If* $S(\cdot)$ *is exponentially stable on* $\mathcal{V}$*, the following condition is equivalent to* (4.4)*:*

there is an $\epsilon > 0$ *such that for all* $\omega \in \mathbb{R}$ *and* $u \in U$

$$\|(C_1(i\omega I - A^{\mathcal{V}})^{-1}B_2 + D_{12})u\|_Z^2 \geq \epsilon \|u\|_U^2. \tag{C.3}$$

(*v*) *If the pair* $(A^{\mathcal{V}}, B)$ *is exponentially stabilizable on* $\mathcal{V}$*, the following condition is equivalent to* (4.4)*:*

there is an $\epsilon > 0$ *such that for all* $(\omega, x, u) \in \mathbb{R} \times D(A^{\mathcal{V}}) \times U$

$$\left\| \begin{matrix} i\omega x - A^{\mathcal{V}}x - B_2u \\ C_1x + D_{12}u \end{matrix} \right\|_{\mathcal{V}\times Z}^2 \geq \epsilon(\|x\|_{\mathcal{V}}^2 + \|u\|_U^2). \tag{C.4}$$

Before we prove these results we have to give a lemma that shall be used to prove item (*v*). The lemma partially extends Lemma 3.1. One of the difficulties in the proof is caused by the fact that we only assume stability on the 'larger space' $\mathcal{V}$.

Lemma C.1

Let $\Sigma(S(\cdot), B, C, D)$ *be a smooth Pritchard-Salamon system of the form* (2.7) *such that* $S(\cdot)$ *is exponentially stable on* $\mathcal{V}$*. Then the operator* $\mathcal{C}^\infty$ *defined by*

$$\mathcal{C}^\infty x := CS(\cdot)x, \text{ for } x \in \mathcal{W}, \tag{C.5}$$

has a unique bounded extension on $\mathcal{V}$ *(denoted by the same symbol), so that*

$$\mathcal{C}^\infty \in \mathcal{L}(\mathcal{V}, L_2(0,\infty;Y)) \cap \mathcal{L}(\mathcal{V}, L_1(0,\infty;Y)) \tag{C.6}$$

Furthermore, for all $x \in \mathcal{V}$*, the Laplace transform of* $\mathcal{C}^\infty x$ *is given by*

$$\widehat{\mathcal{C}^\infty x}(\cdot) = C(\cdot I - A^{\mathcal{V}})^{-1}x \in \mathcal{H}_2(Y) \cap \mathcal{H}_\infty(Y) \tag{C.7}$$

and there holds

$$C(\cdot I - A^{\mathcal{V}})^{-1} \in \mathcal{H}_\infty(\mathcal{L}(\mathcal{V}, Y)). \tag{C.8}$$

Proof

It follows from [72, Remark 3.5 (i)], that there exists some $c > 0$ such that for all $x \in \mathcal{W}$,

$$\|CS(\cdot)x\|_{L_2(0,\infty;Y)} \leq c\|x\|_{\mathcal{V}} \tag{C.9}$$

We claim that there exists another constant $c_1 > 0$ such that for all $x \in \mathcal{W}$,

$$\|CS(\cdot)x\|_{L_1(0,\infty;Y)} \leq c_1\|x\|_{\mathcal{V}}. \tag{C.10}$$

The proof of this claim is similar to the proof in [72, Remark 3.5 (i)] of (C.9). The fact that $S(\cdot)$ is exponentially stable on $\mathcal{V}$ implies the existence of some $T > 0$ such that $\|S(t)\|_{\mathcal{L}(\mathcal{V})} < 1$ for all $t \geq T$. Furthermore, the admissibility of C implies the existence of some c_T such that for all $x \in \mathcal{W}$

$$\|CS(\cdot)x\|_{L_2(0,T;Y)} \leq c_T\|x\|_{\mathcal{V}}$$

but this implies that for some $c_{T1} > 0$

$$\|CS(\cdot)x\|_{L_1(0,T;Y)} \leq c_{T1}\|x\|_{\mathcal{V}} \tag{C.11}$$

(using Cauchy-Schwarz). Now, we can estimate

$$\int_0^\infty \|CS(t)x\|_Y dt = \sum_{k=0}^\infty \int_{kT}^{(k+1)T} \|CS(t)x\|_Y dt =$$

$$\sum_{k=0}^\infty \int_0^T \|CS(t+kT)x\|_Y dt = \sum_{k=0}^\infty \|CS(\cdot)S(kT)x\|_{L_1(0,T;Y)} \leq$$

$$\sum_{k=0}^\infty c_{T1}\|S(kT)x\|_{\mathcal{V}} \qquad \text{using (C.11)}$$

$$\leq \sum_{k=0}^\infty c_{T1}\|S(T)\|^k_{\mathcal{L}(\mathcal{V})}\|x\|_{\mathcal{V}} \leq c_1\|x\|_{\mathcal{V}},$$

for some $c_1 > 0$, using the fact that $\|S(T)\|_{\mathcal{L}(\mathcal{V})} < 1$. This proves (C.10). Since $\mathcal{W} \hookrightarrow \mathcal{V}$, (C.9) and (C.10) imply that $\mathcal{C}^\infty$ defined by (C.5) satisfies

$$\mathcal{C}^\infty \in \mathcal{L}(\mathcal{W}, L_1(0,\infty;Y)) \cap \mathcal{L}(\mathcal{W}, L_2(0,\infty;Y))$$

and that $\mathcal{C}^\infty$ has bounded extensions $\mathcal{C}_1^\infty \in \mathcal{L}(\mathcal{V}, L_1(0,\infty;Y))$ and $\mathcal{C}_2^\infty \in \mathcal{L}(\mathcal{V}, L_2(0,\ \infty;Y))$. Furthermore, it is straightforward to show that these extensions must be the same because for all $T > 0$, $L_2(0,T;Y) \subset L_1(0,T;Y)$. Denoting both extensions by $\mathcal{C}^\infty$, this proves (C.6).

It follows that for $x \in \mathcal{V}$ we have $\mathcal{C}^\infty x \in L_1(0,\infty;Y) \cap L_2(0,\infty;Y)$ and so we infer that

$$\widehat{\mathcal{C}^\infty x}(\cdot) \in \mathcal{H}_2(Y) \cap \mathcal{H}_\infty(Y)$$

(this follows from the Paley-Wiener Theorem and the fact that the Laplace transform of an $L_1(0,\infty)$ function is in $\mathcal{H}_\infty$). In order to complete the proof of (C.7), we have to show that for all $x \in \mathcal{V}$

$$\widehat{\mathcal{C}^\infty x}(s) = C(sI - A^{\mathcal{V}})^{-1}x \text{ for all } s \in \mathbb{C}^+. \tag{C.12}$$

First of all we note that $C(sI - A^{\mathcal{V}})^{-1}x$ is well-defined for $s \in \mathbb{C}^+$ because $(sI - A^{\mathcal{V}})^{-1} \in \mathcal{L}(\mathcal{V}, \mathcal{W})$ and moreover that

$$C(sI - A^{\mathcal{V}})^{-1} \text{ is holomorphic w.r.t. the topology of } \mathcal{L}(\mathcal{V}, Y) \text{ on } \mathbb{C}^+$$

(this follows as in the proof of Lemma 3.4). Since $\mathcal{W} \hookrightarrow \mathcal{V}$ there exists a sequence $x_n \in \mathcal{W}$ such that $\|x_n - x\|_{\mathcal{V}} \to 0$ as $n \to \infty$. Now let $s \in \mathbb{C}^+$ be such that Re$(s) >$max$(\omega^{\mathcal{W}}_{S(\cdot)}, \omega^{\mathcal{V}}_{S(\cdot)})$. There holds

$$\widehat{\mathcal{C}^\infty x_n}(s) = \int_0^\infty \exp(-st)(\mathcal{C}^\infty x_n)(t)dt = \int_0^\infty \exp(-st)CS(t)x_n dt =$$

$$C\int_0^\infty \exp(-st)S(t)x_n(t)dt = C(sI - A^{\mathcal{W}})^{-1}x_n = C(sI - A^{\mathcal{V}})^{-1}x_n, \tag{C.13}$$

(in the last step we have used that $(sI - A^{\mathcal{V}})^{-1} \mid_{\mathcal{W}} = (sI - A^{\mathcal{W}})^{-1}$, see Lemma 2.12). Since $\mathcal{C}^\infty \in \mathcal{L}(\mathcal{V}, L_1(0,\infty;Y))$ it is easy to see that for $s \in \mathbb{C}^+$

$$\widehat{\mathcal{C}^\infty x_n}(s) \xrightarrow{Y} \widehat{\mathcal{C}^\infty x}(s) \text{ as } n \to \infty,$$

because

$$\|\widehat{\mathcal{C}^\infty x_n}(s) - \widehat{\mathcal{C}^\infty x}(s)\|_Y \leq \int_0^\infty \exp(-\text{Re}(s)t)\|(\mathcal{C}^\infty x_n)(t) - (\mathcal{C}^\infty x)(t)\|_Y dt.$$

On the other hand, since $C(sI - A^{\mathcal{V}})^{-1} \in \mathcal{L}(\mathcal{V}, Y)$, there holds

$$C(sI - A^{\mathcal{V}})^{-1}x_n \xrightarrow{Y} C(sI - A^{\mathcal{V}})^{-1}x \text{ as } n \to \infty.$$

Therefore, (C.13) implies that

$$\widehat{\mathcal{C}^\infty x}(s) = C(sI - A^{\mathcal{V}})^{-1}x.$$

Since $\widehat{\mathcal{C}^\infty x}(\cdot)$ and $C(\cdot I - A^{\mathcal{V}})^{-1}x$ are both holomorphic on $\mathbb{C}^+$, (C.12) follows and this, in turn, completes the proof of (C.7)

Finally, we prove (C.8). We have already seen that $C(sI - A^{\mathcal{V}})^{-1}$ is holomorphic w.r.t. the topology of $\mathcal{L}(\mathcal{V}, Y)$ on $\mathbb{C}^+$, so that we only have to prove boundedness. For all $x \in \mathcal{V}$ and $s \in \mathbb{C}^+$ there holds

$$\|C(sI - A^{\mathcal{V}})^{-1}x\|_Y = \|\widehat{\mathcal{C}^\infty x}(s)\|_Y \text{ (using (C.12))}$$

$$= \| \int_0^\infty \exp(-st)(\mathcal{C}^\infty x)(t)dt\|_Y \le \int_0^\infty \|(\mathcal{C}^\infty x)(t)\|dt \le c\|x\|_\mathcal{V},$$

where c is independent of s and x (the last step follows from (C.6)). Hence

$$\sup_{s\in\mathbb{C}^+} \|C(sI - A^\mathcal{V})^{-1}\|_{\mathcal{L}(\mathcal{V},Y)} = \sup_{s\in\mathbb{C}^+} \sup_{\|x\|_\mathcal{V}=1} \|C(sI - A^\mathcal{V})^{-1}x\|_Y < \infty$$

and this concludes the proof of Lemma C.1. ∎

Next, we prove the statements about the invariant zeros condition.
Proof of (i):
This result is in fact a special case of the result that is treated in the proof of Theorem 3.10 about the LQ-problem (the proof of $(iii) \Rightarrow (i)$). The frequency domain inequality related to the LQ-problem has now the following special form: for all $(\omega, x, u) \in \mathbb{R} \times D(A^\mathcal{V}) \times U$ satisfying $i\omega x = A^\mathcal{V}x + B_2u$, there holds

$$\mathcal{F}(x, u) = \|C_1x + D_{12}u\|_Z^2 .$$

Proof of (ii):
Suppose that $D_{12}'[C_1 \;\; D_{12}] = [0 \;\; I]$ and that there exists an admissible input operator $H \in \mathcal{L}(Z, \mathcal{V})$ such that $S_{HC_1}(\cdot)$ is exponentially stable on $\mathcal{V}$.
Now if (4.4) were not satisfied, there would exist a sequence $(\omega_n, x_n, u_n) \in \mathbb{R} \times D(A^\mathcal{V}) \times U$ with $\|x_n\|_\mathcal{V}^2 = 1$ such that $i\omega_n x_n - A^\mathcal{V}x_n - B_2u_n = 0$ and

$$\|C_1x_n + D_{12}u_n\|_Z^2 = \|C_1x_n\|_Z^2 + \|u_n\|_U^2 \to 0.$$

Hence $u_n \overset{U}{\to} 0$, $i\omega_n x_n - A^\mathcal{V}x_n \overset{\mathcal{V}}{\to} 0$ and $C_1x_n \overset{Z}{\to} 0$ as n tends to infinity. It follows that $(i\omega_n I - (A^\mathcal{V} + HC_1))x_n \overset{\mathcal{V}}{\to} 0$. On the other hand, $S_{HC_1}(\cdot)$ is exponentially stable on $\mathcal{V}$, so that its generator on $\mathcal{V}$ given by $A^\mathcal{V} + HC_1$ satisfies $\|(i\omega I - (A^\mathcal{V} + HC_1))^{-1}\|_{\mathcal{L}(\mathcal{V})} \le const$ for all $\omega \in \mathbb{R}$. Hence we conclude that $x_n \overset{\mathcal{V}}{\to} 0$ as n tends to infinity, but now this contradicts the fact that $\|x_n\|_\mathcal{V}^2 = 1$.
Proof of (iii):
The first statement follows immediately from the perturbation results of Lemma 2.13: $D(A^\mathcal{V}) = D(A^\mathcal{V}_{B_2F})$ and $A^\mathcal{V}_{B_2F} = A^\mathcal{V} + B_2F$. Next, we prove that (4.4) implies (C.2). Suppose that $i\omega x = A^\mathcal{V}_{B_2F}x + B_2v$ for $(\omega, x, v) \in \mathbb{R} \times D(A^\mathcal{V}) \times U$. Since $i\omega x = A^\mathcal{V}x + B_2(Fx + v)$, it follows from (4.4) that

$$\|(C_1 + D_{12}F)x + D_{12}v\|_Z^2 = \|C_1x + D_{12}(Fx + v)\|_Z^2 \ge \epsilon\,\|x\|_\mathcal{V}^2 ,$$

which proves (C.2).
To prove that (C.2) implies (4.4), we can simply use the result above with $A^\mathcal{V}$ replaced by $A^\mathcal{V}_{B_2F}$, C_1 replaced by $C_1 + D_{12}F$ and F replaced by $-F$, because $(C_1 + D_{12}F) - D_{12}F = C_1$ and $(A^\mathcal{V}_{B_2F})_{-B_2F} = A^\mathcal{V}$ (see Lemma 2.13).

Proof of (iv):
First of all, we note that $(i\omega - A^{\mathcal{V}})^{-1}$ is uniformly bounded for $\omega \in \mathbb{R}$, because $S(\cdot)$ is exponentially stable on $\mathcal{V}$.
The implication (4.4) $\Rightarrow$ (C.3) follows immediately from item (i), so we only need to show (C.3) $\Rightarrow$ (4.4). Suppose that $(\omega, x, u) \in \mathbb{R} \times D(A^{\mathcal{V}}) \times U$ such that $i\omega x = A^{\mathcal{V}}x + B_2u$. Then $x = (i\omega - A^{\mathcal{V}})^{-1}B_2u$ and we have that $\|x\|_{\mathcal{V}}^2 \le const\|u\|_U^2$. Since (C.3) implies that $\|C_1x + D_{12}u\|_Z^2 \ge \epsilon\|u\|_U^2$, the result follows easily.

Proof of (v):
It is clear that (C.4) implies (4.4).
Since $(A^{\mathcal{V}}, B_2)$ is exponentially stabilizable on $\mathcal{V}$, there exists an $F \in \mathcal{L}(\mathcal{V}, U)$ such that $S_{B_2F}(\cdot)$ is exponentially stable on $\mathcal{V}$. Suppose that (4.4) is satisfied. It follows from (iii) and (iv) that there exists an ϵ such that for all $u \in U$

$$\left\|\left((C_1 + D_{12}F)(i\omega I - A^{\mathcal{V}} - B_2F)^{-1}B_2 + D_{12}\right)u\right\|_Z^2 \ge \epsilon\|u\|_U^2. \qquad \text{(C.14)}$$

Defining $D_F := C_1 + D_{12}F$, we claim that there exists a $\delta > 0$ such that for all $(\omega, x, u) \in \mathbb{R} \times D(A^{\mathcal{V}}) \times U$

$$\left\| \begin{matrix} i\omega x - A^{\mathcal{V}}_{B_2F}x - B_2u \\ D_Fx + D_{12}u \end{matrix} \right\|_{\mathcal{V}\times Z}^2 \ge \epsilon(\|x\|_{\mathcal{V}}^2 + \|u\|_U^2). \qquad \text{(C.15)}$$

If (C.15) were not satisfied, there would exist a sequence $(\omega_n, x_n, u_n) \in \mathbb{R} \times D(A^{\mathcal{V}}) \times U$ with $\|x_n\|_{\mathcal{V}}^2 + \|u_n\|_U^2 = 1$ such that

$$\left\| \begin{matrix} i\omega_n x_n - A^{\mathcal{V}}_{B_2F}x_n - B_2u_n \\ D_Fx_n + D_{12}u_n \end{matrix} \right\|_{\mathcal{V}\times Z}^2 \to 0.$$

Hence, $(i\omega_n - A^{\mathcal{V}}_{B_2F})x_n - B_2u_n \overset{\mathcal{V}}{\to} 0$ and $D_Fx_n + D_{12}u_n \overset{Z}{\to} 0$ as n tends to infinity. Since $S_{B_2F}(\cdot)$ exponentially stable on $\mathcal{V}$ it follows from Lemma C.1 that there exists a $const > 0$ such that for all $\omega \in \mathbb{R}$ and for all $x \in \mathcal{V}$

$$\|D_F(i\omega I - A^{\mathcal{V}}_{B_2F})^{-1}x\|_Z \le const\|x\|_{\mathcal{V}}.$$

Therefore,

$$D_F(i\omega_n I - A^{\mathcal{V}}_{B_2F})^{-1}\left((i\omega_n - A^{\mathcal{V}}_{B_2F})x_n - B_2u_n\right) =$$

$$D_Fx_n - D_F(i\omega_n I - A^{\mathcal{V}}_{B_2F})^{-1}B_2u_n \overset{Z}{\to} 0.$$

Since also $D_Fx_n + D_{12}u_n \overset{Z}{\to} 0$ we may conclude that

$$(D_F(i\omega_n I - A^{\mathcal{V}}_{B_2F})^{-1}B_2 + D_{12})u_n \overset{Z}{\to} 0.$$

Hence, (C.14) implies that $u_n \overset{U}{\to} 0$. Finally, we shall prove that $x_n \overset{\mathcal{V}}{\to} 0$, to get a contradiction with the fact that $\|x_n\|_{\mathcal{V}}^2 + \|u_n\|_U^2 = 1$. Indeed, we have

$(i\omega_n - A^{\mathcal{V}}_{B_2F})x_n - B_2u_n \overset{\mathcal{V}}{\to} 0$ and so $x_n - (i\omega_n - A^{\mathcal{V}}_{B_2F})^{-1}B_2u_n \overset{\mathcal{V}}{\to} 0$. Now it follows from the fact that

$$(\cdot I - A^{\mathcal{V}}_{B_2F})^{-1} \in \mathcal{H}_\infty(\mathcal{L}(\mathcal{V}))$$

and $u_n \overset{U}{\to} 0$ that $x_n \overset{\mathcal{V}}{\to} 0$ and this contradicts the fact that $\|x_n\|^2_{\mathcal{V}} + \|u_n\|^2_U = 1$. Hence we have proved (C.15). Since

$$\begin{pmatrix} i\omega I - A^{\mathcal{V}} & -B_2 \\ C_1 & D_{12} \end{pmatrix} \begin{pmatrix} I & 0 \\ F & I \end{pmatrix} = \begin{pmatrix} i\omega I - (A^{\mathcal{V}} + B_2F) & -B_2 \\ (C_1 + D_{12}F) & D_{12} \end{pmatrix}$$

and $F \in \mathcal{L}(\mathcal{V}, U)$, it is clear that (C.4) follows from (C.15).

Appendix D

The relation between P, Q and $\bar{P}$

D.1

In this section we shall show how the operators P, Q and $\bar{P}$ that emerge from the $\mathcal{H}_\infty$-control problem in Chapter 5 are related. Throughout this section we shall assume that $\gamma = 1$. We start with a result that is given in [84, Lemma 4.5].

Lemma D.1
Let X and Y be separable Hilbert spaces and suppose that we have some $S \in \mathcal{L}(X, Y)$ such that $S^{-1} \in \mathcal{L}(Y, X)$. Furthermore, suppose that $T(\cdot)$ is a C_0-semigroup on X with infinitesimal generator A. Then $\tilde{T}(\cdot)$ defined by $\tilde{T}(t) = ST(t)S^{-1}$, $t \geq 0$, is a C_0-semigroup on Y and its infinitesimal generator is given by $\tilde{A} = SAS^{-1}$ with $D(\tilde{A}) = S(D(A))$. Furthermore, the growth bounds of $T(\cdot)$ and $\tilde{T}(\cdot)$ are the same.

Next we give a lemma that is related to the coupling condition of Theorem 5.4.

Lemma D.2
Let X be a real Hilbert space and suppose that $T, S \in \mathcal{L}(X)$. Then for all $\lambda \in \mathbb{C}\backslash 0$ we have

$$\lambda \in \rho(ST) \text{ if and only if } \lambda \in \rho(TS). \tag{D.1}$$

Now suppose that $P, Q \in \mathcal{L}(X)$ are nonnegative definite. Then the following are equivalent:

$$(i) \qquad r_\sigma(PQ) < 1 \tag{D.2}$$

$$(ii) \qquad 1 \in \rho(PQ) \text{ and } Q(I - PQ)^{-1} \geq 0. \tag{D.3}$$

Proof

Property (D.1) follows from [23, p.315, Problem 2]: if $1 \in \rho(ST)$ then $(I - TS)^{-1} = I + T(I - ST)^{-1}S$.

Using this result, it is easy to see that $1 \in \rho(PQ)$ if and only if $1 \in \rho(Q^{\frac{1}{2}}PQ^{\frac{1}{2}})$ and $r_\sigma(PQ) < 1$ if and only if $r_\sigma(Q^{\frac{1}{2}}PQ^{\frac{1}{2}}) < 1$. This property is crucial in the derivation of the equivalence between (D.2) and (D.3).

Proof of $(i) \Rightarrow (ii)$:

Suppose that $r_\sigma(PQ) < 1$. Of course then $1 \in \rho(PQ)$ and $Q(I - PQ)^{-1}$ is well defined. We have seen that $r_\sigma(PQ) < 1$ implies that $r_\sigma(Q^{\frac{1}{2}}PQ^{\frac{1}{2}}) < 1$. Now it is straightforward to show that $(I - Q^{\frac{1}{2}}PQ^{\frac{1}{2}})^{-1} \geq 0$ and so $Q(I - PQ)^{-1} = Q^{\frac{1}{2}}Q^{\frac{1}{2}}(I - PQ^{\frac{1}{2}}Q^{\frac{1}{2}})^{-1} = Q^{\frac{1}{2}}(I - Q^{\frac{1}{2}}PQ^{\frac{1}{2}})^{-1}Q^{\frac{1}{2}} \geq 0$.

Proof of $(ii) \Rightarrow (i)$:

Suppose that $1 \in \rho(PQ)$ and $Q(I - PQ)^{-1} \geq 0$. We have seen that also $1 \in \rho(Q^{\frac{1}{2}}PQ^{\frac{1}{2}})$. Since $Q(I - PQ)^{-1} \geq 0$, we have $< Q(I - PQ)^{-1}x, x >_X \geq 0$ for all $x \in X$ and therefore $< Qy, (I - PQ)y >_X \geq 0$ for all $y \in X$. Hence $< (I - Q^{\frac{1}{2}}PQ^{\frac{1}{2}})Q^{\frac{1}{2}}y, Q^{\frac{1}{2}}y >_X \geq 0$ for all $y \in X$. This implies that $< (I - Q^{\frac{1}{2}}PQ^{\frac{1}{2}})x, x >_X \geq 0$ for all $x \in \overline{image(Q^{\frac{1}{2}})}$ and since P and Q are bounded the same holds for all $x \in \overline{image(Q^{\frac{1}{2}})}$. It is easy to see that $< (I - Q^{\frac{1}{2}}PQ^{\frac{1}{2}})x, x >_X \geq 0$ also holds for all $x \in kernel(Q^{\frac{1}{2}}) = \overline{(image(Q^{\frac{1}{2}}))}^{\perp}$. Now it is straightforward to show that

$$< Q^{\frac{1}{2}}PQ^{\frac{1}{2}}x, x >_X \leq < x, x >_X \quad \text{for all } x \in X. \tag{D.4}$$

It follows from the above that $1 \in \rho(Q^{\frac{1}{2}}PQ^{\frac{1}{2}})$ and so (D.4) implies that $r_\sigma(Q^{\frac{1}{2}}PQ^{\frac{1}{2}}) < 1$ (see e.g. [53, Section 9.2]). Finally, this implies (D.2). ■

Lemma D.3

Let $\mathcal{W}$ and $\mathcal{V}$ be real separable Hilbert spaces such that $\mathcal{W} \hookrightarrow \mathcal{V}$ and let $S(\cdot)$ be a C_0-semigroup on these spaces with infinitesimal generators $A^{\mathcal{W}}$ and $A^{\mathcal{V}}$ respectively. Let $\{\mathcal{W}', j\}$ be a representation of $\mathcal{W}^d$ and assume that the representation of $\mathcal{V}^d$ is given by $\{\mathcal{V}', k\}$, where $\mathcal{V}' = j^{-1}\mathcal{V}^d$ with inner product given by $< x', y' >_{\mathcal{V}'} = < jx', jy' >_{\mathcal{V}^d}$ and $k = j\mid_{\mathcal{V}'}$, so that for all $(v', w) \in \mathcal{V}' \times \mathcal{W}$ we have

$$< v', w >_{<\mathcal{V}', \mathcal{V}>} = < v', w >_{<\mathcal{W}', \mathcal{W}>}$$

and $\mathcal{V}' \hookrightarrow \mathcal{W}'$ (see also (2.66)). We assume that $D(A^{\mathcal{V}}) \hookrightarrow \mathcal{W}$ so that $D((A^{\mathcal{W}})') \hookrightarrow \mathcal{V}'$ (see Theorem 2.17 item (iii)). Then for any $P \in \mathcal{L}(\mathcal{V}, \mathcal{V}')$, any $Q \in \mathcal{L}(\mathcal{W}', \mathcal{W})$ we have

$$r_\sigma^{\mathcal{W}'}(PQ) = r_\sigma^{\mathcal{V}}(Q'P'). \tag{D.5}$$

Now consider the following statements about the stabilizing solutions to the Riccati equations related to Theorem 5.4 (we have $\gamma = 1$).

(*i*) *There exists a nonnegative definite* $P \in \mathcal{L}(\mathcal{V}, \mathcal{V}')$ *satisfying the Riccati equation* (5.14) *such that the* C_0*-semigroup* $S_P(\cdot)$ *defined by* (5.15) *is exponentially stable on* $\mathcal{V}$ *and* $\mathcal{W}$.

(*ii*) *There exists a nonnegative definite* $\bar{P} \in \mathcal{L}(\mathcal{W}', \mathcal{W})$ *satisfying the Riccati equation* (5.57) *such that the* C_0*-semigroup* $S_{\bar{P}}(\cdot)$ *defined by* (5.58) *is exponentially stable on* $\mathcal{V}$ *and* $\mathcal{W}$.

(*iii*) *There exists a nonnegative definite* $Q \in \mathcal{L}(\mathcal{W}', \mathcal{W})$ *satisfying the Riccati equation* (5.16) *such that the* C_0*-semigroup* $S_Q(\cdot)$ *defined by* (5.17) *is exponentially stable on* $\mathcal{V}$ *and* $\mathcal{W}$.

Then the following hold:

a) *Suppose that* (*i*) *and* (*iii*) *hold and that* $r_\sigma^{\mathcal{W}'}(PQ) = r_\sigma^{\mathcal{V}}(QP) < 1$. *Then* $\bar{P}$ *defined by* $\bar{P} := Q(I - PQ)^{-1}$ *satisfies the conditions of* (*ii*).

b) *Suppose that* (*i*),(*ii*) *and* (*iii*) *hold. Then* $r_\sigma^{\mathcal{W}'}(PQ) = r_\sigma^{\mathcal{V}}(QP) < 1$ *and* $\bar{P} = Q(I - PQ)^{-1}$.

Proof

To prove (D.5) we note that $PQ \in \mathcal{L}(\mathcal{W}', \mathcal{V}') \subset \mathcal{L}(\mathcal{W}')$ and $Q'P' \in \mathcal{L}(\mathcal{V}, \mathcal{W}) \subset \mathcal{L}(\mathcal{V})$ and we claim that for nonzero $\lambda \in \mathbb{C}$, $\lambda \in \rho^{\mathcal{W}'}(PQ)$ if and only if $\lambda \in \rho^{\mathcal{V}}(Q'P')$.

To prove this claim we take $\lambda = 1$, without loss of generality. Suppose that $1 \in \rho^{\mathcal{W}'}(PQ)$. It follows that $(I - PQ)^{-1} \in \mathcal{L}(\mathcal{W}')$ and with $P_2 := Q(I - PQ)^{-1} \in \mathcal{L}(\mathcal{W}', \mathcal{W})$ we have

$$(I - PQ)(I + PP_2) = (I + PP_2)(I - PQ) = I.$$

Taking the adjoint of this last expression gives

$$(I + P_2'P')(I - Q'P') = (I - Q'P')(I + P_2'P') = I.$$

Hence we see that $(I - Q'P') \in \mathcal{L}(\mathcal{V})$ has an inverse $(I + P_2'P') \in \mathcal{L}(\mathcal{V})$, so that $1 \in \rho^{\mathcal{V}}(Q'P')$. The converse can be proved similarly, so we have proved our claim. Formula (D.5) follows easily from this claim.

In the rest of the proof we shall assume that

$$D_{12}'[C_1 \;\; D_{12}] = [0 \;\; I] \quad \text{and} \quad D_{21}[B_1' \;\; D_{21}'] = [0 \;\; I]. \tag{D.6}$$

These assumptions greatly simplify the formulas so that it shall be easier to understand what is going on. Of course the result holds for the general case. We note that we shall extensively use the notion of Hamiltonians (see Lemma 2.35 and Corollary 2.36). First of all, we show how the Riccati equations in items $(i)-(iii)$ and the generators of $S_P(\cdot), S'_Q(\cdot)$ and $S'_{\bar{P}}(\cdot)$ can be reformulated. Using the simplifying assumption (D.6) and Lemma 2.35, it follows that the Riccati equation for P in item (i) is equivalent to

$$PD(A^{\mathcal{V}}) \subset D((A^{\mathcal{W}})') \quad \text{and}$$

$$\left((A^{\mathcal{W}})'P + PA^{\mathcal{V}} + P(B_1B_1' - B_2B_2')P + C_1'C_1\right)x = 0 \tag{D.7}$$

for all $x \in D(A^{\mathcal{V}})$. The generator of $S_P(\cdot)$ on $\mathcal{V}$ is given by

$$D(A_P^{\mathcal{V}}) = D(A^{\mathcal{V}}) \quad \text{and} \quad A_P^{\mathcal{V}} = A^{\mathcal{V}} + (B_1B_1' - B_2B_2')P$$

(in the last formula we used the perturbation results of Lemma 2.13). Similarly, using Lemma 2.35 (or Corollary 2.36), it follows that the Riccati equation for Q in item (iii) is equivalent to

$$QD((A^{\mathcal{W}})') \subset D(A^{\mathcal{V}}) \quad \text{and}$$

$$\left(A^{\mathcal{V}}Q + Q(A^{\mathcal{W}})' + Q(C_1'C_1 - C_2'C_2)Q + B_1B_1'\right)x = 0 \tag{D.8}$$

for all $x \in D((A^{\mathcal{W}})')$. The generator of $S'_Q(\cdot)$ on $\mathcal{W}'$ is given by

$$D((A_Q^{\mathcal{W}})') = D((A^{\mathcal{W}})') \quad \text{and} \quad (A_Q^{\mathcal{W}})' = (A^{\mathcal{W}})' + (C_1'C_1 - C_2'C_2)Q.$$

Before we give the reformulation of the Riccati equation for $\bar{P}$ in item (ii), we recall from (5.46) that $S_0'(\cdot) = S'_{PB_1B_1'}(\cdot)$ and that its generator on $\mathcal{V}'$ and $\mathcal{W}'$ is given by $(A_0^{\mathcal{V}})'$ and $(A_0^{\mathcal{W}})'$, respectively. Furthermore, $D(A_0^{\mathcal{V}}) = D(A^{\mathcal{V}})$, $D((A_0^{\mathcal{W}})') = D((A^{\mathcal{W}})')$ and $(A_0^{\mathcal{W}})' = (A^{\mathcal{W}})' + PB_1B_1'$. Again from Lemma 2.35, it follows that the Riccati equation for $\bar{P}$ in item (ii) is equivalent to

$$\bar{P}D((A^{\mathcal{W}})') \subset D(A^{\mathcal{V}}) \quad \text{and}$$

$$\left(A_0^{\mathcal{V}}\bar{P} + \bar{P}(A_0^{\mathcal{W}})' + \bar{P}(PB_2B_2'P - C_2'C_2)\bar{P} + B_1B_1'\right)x = 0 \tag{D.9}$$

for all $x \in D((A^{\mathcal{W}})')$. The generator of $S'_{\bar{P}}(\cdot)$ on $\mathcal{W}'$ is given by

$$D((A_2^{\mathcal{W}})') = D((A^{\mathcal{W}})') \quad \text{and} \quad (A_2^{\mathcal{W}})' = (A_0^{\mathcal{W}})' + (PB_2B_2'P - C_2'C_2)\bar{P}.$$

To prove $a)$, we suppose that the assumptions of $a)$ hold and define $\bar{P} \in \mathcal{L}(\mathcal{W}', \mathcal{W})$ by $\bar{P} := Q(I - PQ)^{-1}$. We first show that $\bar{P}$ is well-defined and nonnegative definite. Since $PQ \in \mathcal{L}(\mathcal{W}')$ and $r_\sigma^{\mathcal{W}'}(PQ) < 1$, it follows that $QP \in \mathcal{L}(\mathcal{W})$ and $r_\sigma^{\mathcal{W}}(QP) < 1$ (for any $T \in \mathcal{L}(X)$ we have $r_\sigma^X(T) = r_\sigma^{X'}(T')$). Hence it is easy to see that $\bar{P}$ is well-defined and that $\bar{P} = \bar{P}' = (I - QP)^{-1}Q$. The representation of the dual space $\mathcal{W}^d$ is given by $\{\mathcal{W}', j\}$ and denoting $i_{\mathcal{W}}$ as the canonical isomorphism from $\mathcal{W}^d$ to $\mathcal{W}$ (see Section 2.5) we define

$$l_{\mathcal{W}} : \mathcal{W}' \to \mathcal{W} \;\; ; \;\; l_{\mathcal{W}} := i_{\mathcal{W}} j.$$

Hence $l_{\mathcal{W}}$ is an isometry from $\mathcal{W}'$ to $\mathcal{W}$ and we have

$$< w_1', w_2' >_{\mathcal{W}'} = < w_1', l_{\mathcal{W}} w_2' >_{<\mathcal{W}', \mathcal{W}>} .$$

We define

$$\left.\begin{aligned} \tilde{Q} &:= (l_{\mathcal{W}})^{-1} Q \in \mathcal{L}(\mathcal{W}') \\ \tilde{P} &:= P l_{\mathcal{W}} \in \mathcal{L}(\mathcal{W}') \\ \tilde{\bar{P}} &:= (l_{\mathcal{W}})^{-1} \bar{P} \in \mathcal{L}(\mathcal{W}'). \end{aligned}\right\} \tag{D.10}$$

It is straightforward to show that $\tilde{P}$ and $\tilde{Q}$ are both nonnegative definite on $\mathcal{W}'$ and we have $\tilde{\bar{P}} = \tilde{Q}(I - \tilde{P}\tilde{Q})^{-1}$. Since $r_\sigma^{\mathcal{W}'}(PQ) = r_\sigma^{\mathcal{W}'}(\tilde{P}\tilde{Q}) < 1$, Lemma D.2 implies that $\tilde{\bar{P}} \geq 0$ and so $\bar{P}$ is nonnegative definite.

Next, we shall prove that $\bar{P}$ is a solution of the Riccati equation (5.57) (or, equivalently, of the Riccati equation (D.9)) such that the C_0-semigroup $S'_{\bar{P}}(\cdot)$ defined by (5.58) is exponentially stable on $\mathcal{V}'$ and $\mathcal{W}'$. With H_Q representing the Hamiltonian related to the Riccati equation (D.8), i.e.

$$H_Q := \begin{pmatrix} (A^{\mathcal{W}})' & C_1'C_1 - C_2'C_2 \\ -B_1B_1' & -A^{\mathcal{V}} \end{pmatrix},$$

it follows from Corollary 2.36 that for all $x \in D((A^{\mathcal{W}})')$, $Qx \in D(A^{\mathcal{V}})$ and

$$H_Q \begin{pmatrix} I \\ Q \end{pmatrix} x = \begin{pmatrix} I \\ Q \end{pmatrix} (A_Q^{\mathcal{W}})' x. \tag{D.11}$$

We define $H_{\bar{P}}$ as the Hamiltonian corresponding to the Riccati equation (D.9), i.e.

$$H_{\bar{P}} := \begin{pmatrix} (A_0^{\mathcal{W}})' & PB_2B_2'P - C_2'C_2 \\ -B_1B_1' & -A_0^{\mathcal{V}} \end{pmatrix}.$$

Using item (i) (in fact the Riccati equation (D.7)) it follows from a straightforward calculation that for all $(x, y) \in D((A^{\mathcal{W}})') \times D(A^{\mathcal{V}})$

$$\begin{pmatrix} I & P \\ 0 & I \end{pmatrix} H_{\bar{P}} \begin{pmatrix} I & -P \\ 0 & I \end{pmatrix} \begin{pmatrix} x \\ y \end{pmatrix} = H_Q \begin{pmatrix} x \\ y \end{pmatrix}. \tag{D.12}$$

If we take

$$\begin{pmatrix} x \\ y \end{pmatrix} = \begin{pmatrix} I \\ Q \end{pmatrix} x$$

in (D.12) and premultiply (D.12) by $\begin{pmatrix} I & -P \\ 0 & I \end{pmatrix}$, it follows from (D.11) that for all $x \in D((A^{\mathcal{W}})')$

$$H_{\bar{P}} \begin{pmatrix} I - PQ \\ Q \end{pmatrix} x = \begin{pmatrix} I - PQ \\ Q \end{pmatrix} (A_Q^{\mathcal{W}})' x. \tag{D.13}$$

Now using the fact that $(I-PQ)^{-1} \in \mathcal{L}(\mathcal{W}')$ and the definition of $\bar{P}$, we have for all $x \in D((A^{\mathcal{W}})')$

$$H_{\bar{P}} \begin{pmatrix} I \\ \bar{P} \end{pmatrix} (I-PQ)x = \begin{pmatrix} I \\ \bar{P} \end{pmatrix} (I-PQ)(A_Q^{\mathcal{W}})'x. \tag{D.14}$$

Comparison of the first components of the 'vectors' on both sides of (D.14) shows that for all $x \in D((A^{\mathcal{W}})')$

$$(A_2^{\mathcal{W}})'(I-PQ)x = (I-PQ)(A_Q^{\mathcal{W}})'x. \tag{D.15}$$

Now it follows from Lemma 2.32 item (ii), that $(I-PQ)^{-1}$ maps $D((A^{\mathcal{W}})')$ into itself. Hence, we infer that $\bar{P} = Q(I-PQ)^{-1}$ satisfies $\bar{P}D((A^{\mathcal{W}})') \subset D(A^{\mathcal{V}})$. Furthermore, it follows from (D.15), the fact that $S_Q'(\cdot)$ is exponentially stable on $\mathcal{W}'$ and Lemma D.1 that $S_{\bar{P}}'(\cdot)$ is exponentially stable on $\mathcal{W}'$ and hence $S_{\bar{P}}(\cdot)$ is exponentially stable on $\mathcal{W}$. Since $S_Q(\cdot)$ is in fact exponentially stable on $\mathcal{W}$ and $\mathcal{V}$, it follows from Theorem 2.20 item (iv) that $S_{\bar{P}}(\cdot)$ is also exponentially stable on $\mathcal{V}$. Furthermore, now (D.14) and (D.15) imply that for all $x \in D((A^{\mathcal{W}})')$

$$H_{\bar{P}} \begin{pmatrix} I \\ \bar{P} \end{pmatrix} x = \begin{pmatrix} I \\ \bar{P} \end{pmatrix} (I-PQ)(A_Q^{\mathcal{W}})'(I-PQ)^{-1}x = \begin{pmatrix} I \\ \bar{P} \end{pmatrix} (A_2^{\mathcal{W}})'x$$

and the Riccati equation (D.9) follows immediately from this (consider the second components on both sides). Hence, we have shown that $\bar{P}$ satisfies all the properties of item (ii) and this completes the proof of a).

Next, we prove b), so suppose that $(i), (ii)$ and (iii) hold. We note that the corresponding Riccati equations are given by (D.7), (D.8) and (D.9). Furthermore, we define the Hamiltonians H_Q and $H_{\bar{P}}$ as in the proof of part a). It follows from the Riccati equation (D.8) and Corollary 2.36 that for all $(x,y) \in D((A^{\mathcal{W}})') \times D(A^{\mathcal{V}})$

$$\begin{pmatrix} I & 0 \\ -Q & I \end{pmatrix} H_Q \begin{pmatrix} I & 0 \\ Q & I \end{pmatrix} \begin{pmatrix} x \\ y \end{pmatrix} = \begin{pmatrix} (A_Q^{\mathcal{W}})' & C_1'C_1 - C_2'C_2 \\ 0 & -(A_Q^{\mathcal{V}}) \end{pmatrix} \begin{pmatrix} x \\ y \end{pmatrix}.$$

As in the proof of a) we have that (D.12) is satisfied and substituting this in the above equation, we see that for all $(x,y) \in D((A^{\mathcal{W}})') \times D(A^{\mathcal{V}})$

$$\begin{pmatrix} (A_Q^{\mathcal{W}})' & C_1'C_1 - C_2'C_2 \\ 0 & -A_Q^{\mathcal{V}} \end{pmatrix} \begin{pmatrix} x \\ y \end{pmatrix} =$$

$$\begin{pmatrix} I & 0 \\ -Q & I \end{pmatrix} \begin{pmatrix} I & P \\ 0 & I \end{pmatrix} H_{\bar{P}} \begin{pmatrix} I & -P \\ 0 & I \end{pmatrix} \begin{pmatrix} I & 0 \\ Q & I \end{pmatrix} \begin{pmatrix} x \\ y \end{pmatrix} =$$

$$\begin{pmatrix} I & P \\ -Q & I-QP \end{pmatrix} H_{\bar{P}} \begin{pmatrix} I-PQ & -P \\ Q & I \end{pmatrix} \begin{pmatrix} x \\ y \end{pmatrix}. \tag{D.16}$$

Furthermore, it follows from the Riccati equation (D.9) and Corollary 2.36 that for all $(x,y) \in D((A^{\mathcal{W}})') \times D(A^{\mathcal{V}})$

$$\begin{pmatrix} (A_2^{\mathcal{W}})' & PB_2B_2'P - C_2'C_2 \\ 0 & -A_2^{\mathcal{V}} \end{pmatrix} \begin{pmatrix} x \\ y \end{pmatrix} =$$

$$\begin{pmatrix} I & 0 \\ -\bar{P} & I \end{pmatrix} H_{\bar{P}} \begin{pmatrix} I & 0 \\ \bar{P} & I \end{pmatrix} \begin{pmatrix} x \\ y \end{pmatrix}. \tag{D.17}$$

In the 'finite-dimensional' proof, using the fact that $(A_Q^{\mathcal{W}})'$ and $A_2^{\mathcal{V}}$ are exponentially stable, it follows that the stable subspace of $H_{\bar{P}}$ is given by

$$image \begin{pmatrix} I - PQ \\ Q \end{pmatrix} = image \begin{pmatrix} I \\ \bar{P} \end{pmatrix}$$

and this implies the statement *b*). Since we do not have such a result about stable subspaces for infinite-dimensional systems, we adopt a different approach.

Again for simplicity of notation we introduce

$$S_1 := \begin{pmatrix} I - PQ & -P \\ Q & I \end{pmatrix} \text{ and } S_2 := \begin{pmatrix} I & 0 \\ \bar{P} & I \end{pmatrix} \tag{D.18}$$

(note that S_1, S_2, S_1^{-1} and S_2^{-1} map $D((A^{\mathcal{W}})') \times D(A^{\mathcal{V}})$ *onto* itself).
Reformulating (D.16) and (D.17) it follows that for all $(x,y) \in D((A^{\mathcal{W}})') \times D(A^{\mathcal{V}})$

$$\begin{pmatrix} (A_Q^{\mathcal{W}})' & C_1'C_1 - C_2'C_2 \\ 0 & -A_Q^{\mathcal{V}} \end{pmatrix} \begin{pmatrix} x \\ y \end{pmatrix} = S_1^{-1} H_{\bar{P}} S_1 \begin{pmatrix} x \\ y \end{pmatrix} \tag{D.19}$$

and

$$\begin{pmatrix} (A_2^{\mathcal{W}})' & PB_2B_2'P - C_2'C_2 \\ 0 & -A_2^{\mathcal{V}} \end{pmatrix} \begin{pmatrix} x \\ y \end{pmatrix} = S_2^{-1} H_{\bar{P}} S_2 \begin{pmatrix} x \\ y \end{pmatrix}. \tag{D.20}$$

We know that for all $\begin{pmatrix} x \\ y \end{pmatrix} \in D((A^{\mathcal{W}})') \times D(A^{\mathcal{V}})$, there holds $S_2^{-1} S_1 \begin{pmatrix} x \\ y \end{pmatrix} \in D((A^{\mathcal{W}})') \times D(A^{\mathcal{V}})$. So replacing $\begin{pmatrix} x \\ y \end{pmatrix}$ in (D.20) by $S_2^{-1} S_1 \begin{pmatrix} x \\ y \end{pmatrix}$, we see that for all $\begin{pmatrix} x \\ y \end{pmatrix} \in D((A^{\mathcal{W}})') \times D(A^{\mathcal{V}})$,

$$\begin{pmatrix} (A_2^{\mathcal{W}})' & PB_2B_2'P - C_2'C_2 \\ 0 & -A_2^{\mathcal{V}} \end{pmatrix} S_2^{-1} S_1 \begin{pmatrix} x \\ y \end{pmatrix} = S_2^{-1} H_{\bar{P}} S_1 \begin{pmatrix} x \\ y \end{pmatrix}. \tag{D.21}$$

Hence, combining (D.19) and (D.21), we infer that for all $(x, y) \in D((A^{\mathcal{W}})') \times D(A^{\mathcal{V}})$ we have

$$S_2^{-1}S_1 \begin{pmatrix} (A_Q^{\mathcal{W}})' & C_1'C_1 - C_2'C_2 \\ 0 & -A_Q^{\mathcal{V}} \end{pmatrix} \begin{pmatrix} x \\ y \end{pmatrix} =$$

$$\begin{pmatrix} (A_2^{\mathcal{W}})' & PB_2B_2'P - C_2'C_2 \\ 0 & -A_2^{\mathcal{V}} \end{pmatrix} S_2^{-1}S_1 \begin{pmatrix} x \\ y \end{pmatrix}, \tag{D.22}$$

where $S_2^{-1}S_1$ is given by

$$S_2^{-1}S_1 = \begin{pmatrix} I - PQ & -P \\ Q - \bar{P}(I - PQ) & I + \bar{P}P \end{pmatrix}. \tag{D.23}$$

If we take $y = 0$ in (D.22) and compare the second components of the 'vectors' on both sides of (D.22), we conclude that for all $x \in D((A^{\mathcal{W}})')$ there holds

$$(Q - \bar{P}(I - PQ))(A_Q^{\mathcal{W}})'x = -A_2^{\mathcal{V}}(Q - \bar{P}(I - PQ))x. \tag{D.24}$$

We define $T := Q - \bar{P}(I - PQ) \in \mathcal{L}(\mathcal{W}', \mathcal{W})$ and we shall prove that in fact $T = 0$. The proof is similar to that of Lemma 2.32 item (i), but not the same because $((A^{\mathcal{W}})')' = A^{\mathcal{W}} \neq A^{\mathcal{V}}$. For $x \in D((A^{\mathcal{W}})')$ and $y \in D((A^{\mathcal{V}})')$ we calculate

$$\frac{d}{dt} < TS_Q'(t)x, S_{\bar{P}}'(t)y >_{<\mathcal{W},\mathcal{W}'>} =$$

$$< T(A_Q^{\mathcal{W}})'S_Q'(t)x, S_{\bar{P}}'(t)y >_{<\mathcal{W},\mathcal{W}'>} + < TS_Q'(t)x, (A_2^{\mathcal{W}})'S_{\bar{P}}'(t)y >_{<\mathcal{W},\mathcal{W}'>} =$$

$$< T(A_Q^{\mathcal{W}})'S_Q'(t)x, S_{\bar{P}}'(t)y >_{<\mathcal{W},\mathcal{W}'>} + < TS_Q'(t)x, (A_2^{\mathcal{V}})'S_{\bar{P}}'(t)y >_{<\mathcal{W},\mathcal{W}'>} =$$

$$< T(A_Q^{\mathcal{W}})'S_Q'(t)x, S_{\bar{P}}'(t)y >_{<\mathcal{V},\mathcal{V}'>} + < TS_Q'(t)x, (A_2^{\mathcal{V}})'S_{\bar{P}}'(t)y >_{<\mathcal{V},\mathcal{V}'>} =$$

$$< T(A_Q^{\mathcal{W}})'S_Q'(t)x, S_{\bar{P}}'(t)y >_{<\mathcal{V},\mathcal{V}'>} + < A_2^{\mathcal{V}}TS_Q'(t)x, S_{\bar{P}}'(t)y >_{<\mathcal{V},\mathcal{V}'>} = 0,$$

where we used (D.24) in the last step. Hence, $< TS_Q'(t)x, S_{\bar{P}}'(t)y >_{<\mathcal{W},\mathcal{W}'>}$ is constant and since $S_Q'(\cdot)$ and $S_{\bar{P}}'(\cdot)$ are both exponentially stable on $\mathcal{W}'$ it follows that this constant equals 0. Finally, we can use the fact that $D((A^{\mathcal{W}})')$ and $D((A^{\mathcal{V}})')$ are both dense in $\mathcal{W}'$ to infer that $< Tx, y >_{<\mathcal{W},\mathcal{W}'>} = 0$ for all $x, y \in \mathcal{W}'$, which implies that

$$T = Q - \bar{P}(I - PQ) = 0. \tag{D.25}$$

Now, considering (D.23) and (D.25), we see that

$$(S_2^{-1}S_1)(S_1^{-1}S_2) =$$

$$\begin{pmatrix} I-PQ & -P \\ 0 & I+\bar{P}P \end{pmatrix} \begin{pmatrix} I+P\bar{P} & P \\ 0 & I-QP \end{pmatrix} = \begin{pmatrix} I & 0 \\ 0 & I \end{pmatrix}.$$

This shows that $(I - PQ)$ has a bounded inverse in $\mathcal{L}(\mathcal{W}')$:

$$(I - PQ)(I + P\bar{P}) = (I + P\bar{P})(I - PQ) = I.$$

Hence (D.25) implies that $\bar{P} = Q(I - PQ)^{-1}$.
To prove that $r_\sigma^{\mathcal{W}'}(PQ) < 1$ we define $\tilde{P}, \tilde{\bar{P}}$ and $\tilde{Q}$ just as in (D.10) and we note that all these operators are nonnegative definite on $\mathcal{W}'$. Furthermore, $\tilde{\bar{P}} = \tilde{Q}(I-\tilde{P}\tilde{Q})^{-1}$ and so it follows from Lemma D.2 that $r_\sigma^{\mathcal{W}'}(PQ) = r_\sigma^{\mathcal{W}'}(\tilde{P}\tilde{Q}) < 1$. ■

Bibliography

[1] B.D.O.Anderson and S.Vongpanitlerd, *Network Analysis and Synthesis*, Prentice-Hall, 1973.

[2] J.-P. Aubin, *Applied functional analysis*, Wiley, New York, 1979.

[3] A.V.Balakrishnan, *Applied functional analysis*, Springer, 1976.

[4] H.T.Banks and J.A.Burns, Hereditary control problems: numerical methods based on averaging approximations, *SIAM J. Control and Optimization*, Vol.16, pp.169-208, 1978.

[5] V.Barbu, $\mathcal{H}_\infty$-boundary control with state-feedback; the hyperbolic case, *manuscript*, Februari 1992.

[6] T.Başar and P.Bernhard, *$\mathcal{H}_\infty$-optimal control and related minimax design problems: a dynamic game approach*, Birkhäuser, Boston, 1991.

[7] A.Bensoussan and P.Bernhard, Contributions to the theory of robust control, *manuscript*, September 1991.

[8] A.Bensoussan, G.Da Prato, M.C.Delfour and S.K. Mitter, *Representation and Control of Infinite-dimensional Systems*, preprint.

[9] J.Bontsema, *Dynamic stabilization of large flexible space structures*, Ph.D. thesis, University of Groningen, The Netherlands, 1989.

[10] F.M.Callier and J.Winkin, *LQ*-optimal control of infinite-dimensional systems by spectral factorization, *Automatica*, Vol.28, No.4, pp.757-770, 1992.

[11] R.F.Curtain, A synthesis of time and frequency domain methods for the control of infinite-dimensional systems: a system theoretic approach, *Control and estimation in distributed parameter systems* (in the series Frontiers in Applied Mathematics), Ed.H.T.Banks, pp.171-224, 1992.

[12] R.F.Curtain, Robust controllers for infinite-dimensional systems, *Analysis and Optimization of Systems: State and Frequency Domain Approaches for Infinite-Dimensional Systems*, Proceedings of 10-th International Conference Sophia-Antipolis 1992, Lecture notes in control and information sciences, Vol.185, Springer Verlag, Berlin, pp.140-159, 1993.

[13] R.F.Curtain, Robust stabilizability of normalized coprime factors: the infinite-dimensional case, *International Journal of Control*, Vol.51, pp.1173-1190, 1990.

[14] R.F.Curtain and B.A.M. van Keulen, Robust control with respect to coprime factors of infinite-dimensional positive real systems, *IEEE Trans. Aut. Control*, Vol.37, No.6, pp.868-871, 1992.

[15] R.F.Curtain, H.Logemann, S.Townley and H.Zwart, Well-posedness, stabilizablity and admissibility for Pritchard-Salamon systems, to appear in the *Journal of Mathematical Systems, Estimation, and Control.*

[16] R.F.Curtain and A.J.Pritchard, *Infinite-Dimensional linear systems theory*, Lecture notes in control and information sciences, Vol.8, Springer Verlag, Berlin, 1978.

[17] R.F.Curtain and H.Zwart, *An introduction to infinite-dimensional linear systems theory, preprint.*

[18] R.Datko, Extending a theorem of A.M.Liapunov to Hilbert space, *Journal of Mathematical Analysis and Applications*, Vol. 32, pp. 610-616, 1970.

[19] R.Datko, A linear control problem in abstract Hilbert space, *Journal of differential equations*, Vol.9, pp.346-359, 1971.

[20] M.C.Delfour, The linear quadratic control problem with delays in state and control variables: a state-space approach, *SIAM J. Control and Opt.*, Vol.24, No.5, pp.835-883, 1986.

[21] M.C.Delfour, C.McCalla and S.K.Mitter, Stability and the infinite-time quadratic cost problem for linear hereditary differential systems, *SIAM J. Control and Optimization*, Vol.13, pp.48-88, 1975.

[22] C.A.Desoer and M.Vidyasagar, *Feedback systems: input-output properties,* Academic Press, New York, 1975.

[23] J.Dieudonné, *Foundations of modern analysis*, Academic press, New York, 1960.

[24] J.C.Doyle, Analysis of feedback systems with structured uncertainties, *IEE Proceedings*, Vol.129, pp.242-250, 1982.

[25] J.C.Doyle, B.A.Francis and A.R.Tannenbaum, *Feedback control theory*, Macmillan Publ. Company, 1992.

[26] J.C.Doyle, K.Glover, P.P.Khargonekar and B.A.Francis, State-space solutions to standard $\mathcal{H}_2$ and $\mathcal{H}_\infty$ control problems, *IEEE Trans. Aut. Control*, Vol.AC 34, pp.831-847, 1989.

[27] M.J.Engelhart and M.C.Smith, A four-block problem for $\mathcal{H}_\infty$-design: properties and applications, *Automatica*, Vol.27, pp. 811-818, 1991.

[28] D.S.Flamm and K.Kiplec, Numerical computation of inner factors for distributed parameter systems, *Analysis and Optimization of Systems: State and Frequency Domain Approaches for Infinite-Dimensional Systems*, Proceedings of 10-th International Conference Sophia-Antipolis 1992, Lecture notes in control and information sciences, Vol.185, Springer Verlag, Berlin, pp.598-609, 1993.

[29] D.S.Flamm and S.K.Mitter, $\mathcal{H}_\infty$-sensitivity for delay systems: Part1, *Systems and control letters*, Vol.9, pp.17-24, 1987.

[30] D.S.Flamm and H.Yang, $\mathcal{H}_\infty$-Optimal mixed sensitivity for general distributed plants, *Proc. 29th IEEE Conf. on Decision and Control*, Honolulu, 1990, pp. 134-139.

[31] C.Foias, A.Tannenbaum and G.Zames, Weighted sensitivity minimization for delay systems, *IEEE Trans Aut. Control*, AC-31, pp.763-766, 1986.

[32] C.Foias, A.Tannenbaum and G.Zames, On the $\mathcal{H}_\infty$-optimal sensitivity problem for systems with delays, *SIAM J. Control and Optimization*, Vol.25, pp.686-705, 1987.

[33] Y.Fourés and I.E.Segal, Causality and analyticity, *Trans. Amer. Math. Soc.* 78, pp.385-405, 1955.

[34] B.A.Francis, *A course in $\mathcal{H}_\infty$-control theory*, Lecture notes in control and information sciences, Vol.88, Springer Verlag, Berlin, 1987.

[35] J.Gibson, Linear quadratic control of hereditary differential systems: infinite-dimensional Riccati equations and numerical approximations, *SIAM J. Control and Opt.*, Vol.21, pp.95-139, 1983.

[36] K.Glover, D.J.N.Limebeer, J.C.Doyle, E.M.Kasenally and M.G. Safonov, A characterization of all solutions to the four block general distance problem, *SIAM J. of Control and Optimization*, Vol.29, No.2, pp. 283-324, 1991.

[37] K.Glover and J.C.Doyle, State-space formulae for all stabilizing controllers that satisfy an $\mathcal{H}_\infty$-norm bound and relations to risk sensitivity, *Systems and Control Letters*, Vol.11, pp.167-172, 1988.

[38] E.Hille and R.S.Phillips, *Functional analysis and semigroups*, AMS Publ. 31, Providence R.I., 1957.

[39] D.Hinrichsen and A.J.Pritchard, Real and complex stability radii: a survey, *Control of uncertain systems*, Proc. int. workshop Bremen, Birkhäuser Boston, pp.119-162, 1990.

[40] A.Ichikawa, $\mathcal{H}_\infty$-control and mini-max problems in Hilbert space, *manuscript*, September 1991.

[41] A.Ichikawa, Differential Games and $\mathcal{H}_\infty$-Problems, in *Recent advances in mathematical theory of systems, control, networks and signal processing I*, Eds. H.Kimura, S.Kodama, Proceedings of the international symposium MTNS-91, pp.115-120, 1992.

[42] K.Ito and H.T.Tran, Linear optimal control problem for linear systems with unbounded input and output operators: numerical approximations, *Int. Series of Numerical Mathematics*, Vol. 91, Birkhäuser, pp.171-195, 1989.

[43] C.A.Jacobson and C.N.Nett, Linear state-space systems in infinite dimensional space: the role and characterization of joint stabilizability/detectability, *IEEE Trans. on Autom. Control*, Vol.33, No.6, pp. 541-549, 1988.

[44] E.W.Kamen, P.P.Khargonekar and A.Tannenbaum, Stabilization of time-delay systems using finite-dimensional compensators, *IEEE Trans. on Autom. Control*, Vol.30, No.1, pp. 75-78, 1985.

[45] K.Kappel and D.Salamon, An approximation theorem for the algebraic Riccati equation,*SIAM J. Control and Optimization*, Vol.28, pp.1136-1147, 1990.

[46] B.A.M.van Keulen, Hankel operators for non-exponentially stabilizable infinite-dimensional systems, *Systems and Control Letters* 15, pp.221-226, 1990.

[47] B.A.M.van Keulen, The $\mathcal{H}_\infty$-problem with measurement feedback for linear infinite-dimensional systems, Report W-9103 University of Groningen The Netherlands, to appear in the *Journal of Mathematical Systems, Estimation and Control.*

[48] B.A.M.van Keulen, Redheffer's Lemma and $\mathcal{H}_\infty$-control for infinite-dimensional systems, Report W-9206 University of Groningen The Netherlands, to appear in *SIAM J. of Control and Opt.*

[49] B.A.M.van Keulen, Equivalent conditions for the solvability of the infinite-dimensional LQ-problem with unbounded input and output operators, Report W-9209 University of Groningen The Netherlands, *submitted.*

[50] B.A.M.van Keulen, $\mathcal{H}_\infty$-control with measurement-feedback for Pritchard-Salamon systems, Report W-9211 University of Groningen The Netherlands, to appear in the *International Journal on Robust and Nonlinear Control.*

[51] B.A.M.van Keulen, M.A.Peters and R.F.Curtain, $\mathcal{H}_\infty$-control with state-feedback: the infinite-dimensional case, *Journal of Mathematical Systems, Estimation, and Control*, Vol.3, No.1, pp.1-39, 1993.

[52] P.P.Khargonekar, I.R.Petersen and M.A.Rotea, $\mathcal{H}_\infty$-Optimal control with state feedback, *IEEE Trans. on Aut. Control*, Vol.33, pp.786-788, 1988.

[53] E.Kreyszig, *Introductory functional analysis with applications*, Wiley, New York, 1978.

[54] C.R.Kuiper and H.Zwart, Solutions of the ARE in terms of the Hamiltonian for Riesz-spectral systems, *Analysis and Optimization of Systems: State and Frequency Domain Approaches for Infinite-Dimensional Systems*, Proceedings of 10-th International Conference Sophia-Antipolis 1992, Lecture notes in control and information sciences, Vol.185, Springer Verlag, Berlin, pp.314-325, 1993.

[55] H.Kwakernaak, Minimax frequency domain performance and robustness optimization of linear feedback systems, *IEEE Trans. on Aut. Control*, Vol.30, pp.994-1004, 1985.

[56] H.Kwakernaak, Robust control and $\mathcal{H}_\infty$-optimization, *Automatica*, Vol.29, pp.255-274, 1993.

[57] I.LASIECKA and R.TRIGGIANI, *Differential and algebraic Riccati equations with applications to boundary/point control problems: continuous theory and approximation theory*, Lecture notes in control and information sciences, Vol.164, Springer Verlag, Berlin, 1991.

[58] K.LENZ, H.ÖZBAY, A.TANNENBAUM, J.TURI and B.MORTON, Frequency domain analysis and robust control design for an ideal flexible beam, *Automatica*, Vol.27, pp.947-962, 1991.

[59] J.L.LIONS, *Optimal control of systems governed by partial differential e-quations*, Springer, Berlin 1971.

[60] H.LOGEMANN, Stabilization and regulation of infinite-dimensional systems using coprime factorizations, *Analysis and Optimization of Systems: State and Frequency Domain Approaches for Infinite-Dimensional Systems*, Proceedings of 10-th International Conference Sophia-Antipolis 1992, Lecture notes in control and information sciences, Vol.185, Springer Verlag, Berlin, pp.102-139, 1993.

[61] J.LOUIS, *The regulator problem in Hilbert spaces and some applications to stability of nonlinear control systems*, Ph.D. thesis, Facultés Universitaires Notre-Dame de la Paix, Namur, Belgium, 1986.

[62] J.LOUIS and D.WEXLER, The Hilbert space regulator problem and operator Riccati equation under stabilizability, *Annales de la Société Scientifique de Bruxelles*, T.105, 4, pp. 137-165, 1991.

[63] D.C.MCFARLANE and K.GLOVER, *Robust controller design using normalized coprime factor plant descriptions*, Lecture notes in control and information sciences, Vol.138, 1990.

[64] C.MCMILLAN and R.TRIGGIANI, Min-max game theory for a class of boundary control problems, *Analysis and Optimization of Systems: State and Frequency Domain Approaches for Infinite-Dimensional Systems*, Proceedings of 10-th International Conference Sophia-Antipolis 1992, Lecture notes in control and information sciences, Vol.185, Springer Verlag, Berlin, pp.459-466, 1993.

[65] H.ÖZBAY, $\mathcal{H}_\infty$ Optimal controller design for a class of distributed parameter systems, to appear in *International Journal of Control.*

[66] H.ÖZBAY, Controller reduction in the 2-block $\mathcal{H}_\infty$ optimal design for distributed plants, *International Journal of Control*, Vol.54, pp.1291-1308, 1992.

[67] H.ÖZBAY and A.TANNENBAUM, A skew Toeplitz approach to the $\mathcal{H}_\infty$-optimal control of multivariable distributed systems, *SIAM J. Control and Optimization*, Vol.28, pp.653-670, 1990.

[68] H.ÖZBAY and A.TANNENBAUM, On the structure of sub-optimal $\mathcal{H}_\infty$-controllers in the sensitivity minimization problem for distributed stable systems, *Automatica*, Vol.27, pp.293-305, 1991.

[69] A.PAZY, *Semigroups of linear operators and applications to partial differential equations*, Springer, New York 1983.

[70] M.A.PETERS, A time-domain approach to $\mathcal{H}_\infty$ worst case design, Master's Thesis, University of Eindhoven, Netherlands, 1989.

[71] V.M.POPOV, *Hyperstability of Control Systems*, Springer, Berlin, 1973.

[72] A.J.PRITCHARD and D.SALAMON, The linear quadratic control problem for infinite-dimensional systems with unbounded input and output operators, *SIAM Journal of Control and Optimization*, Vol. 25, No.1, pp.121-144, 1987.

[73] A.J.PRITCHARD and D.SALAMON, The linear quadratic control problem for retarded systems with delays in control and observation, *IMA J. of Math. Control and Information*, Vol.2, pp.335-362, 1985.

[74] A.J.PRITCHARD and S.TOWNLEY, Robustness optimization for uncertain infinite-dimensional systems with unbounded inputs, *IMA J. of Math. Control and Information*, Vol.8, No.2, pp.121-134, 1991.

[75] R.RAVI, K.M.NAGPAL and P.P.KHARGONEKAR, $\mathcal{H}_\infty$-control of linear time-varying systems: a state-space approach, *SIAM J. of Control and Optimization*, Vol.29, No 6, pp.1394-1413, 1991.

[76] R.M.REDHEFFER, On a certain linear fractional transformation, *Journal of Mathematics and Physics*, Vol.39, pp. 269-286, 1960.

[77] A.A.RODRIGUEZ and M.A.DAHLEH, Weighted $\mathcal{H}_\infty$-optimization for stable infinite-dimensional systems using finite-dimensional techniques, Proc. 29-th Conference on Decision and Control, pp.1814-1820, 1990.

[78] M.ROSENBLUM and R.ROVNYAK, *Hardy classes and operator theory*, Oxford University Press, 1985.

[79] M.G.SAFONOV, D.J.N.LIMEBEER and R.Y.CHIANG, Simplifying the $\mathcal{H}_\infty$-theory via loop-shifting, matrix-pencil and descriptor concepts, *International Journal of Control*, Vol.50, No.6, pp.2467-2488, 1989.

[80] D.SALAMON, *Control and observation of neutral systems*, Vol.91, Pitman, London, 1984.

[81] D.SALAMON, Infinite-dimensional systems with unbounded control and observation: a functional analytic approach, *Trans of the Am. Math. Soc.*, Vol.300, No.2, pp.383-431, 1987.

[82] D.SALAMON, Realization theory in Hilbert space, *J. of Math. Systems Theory*, Vol.21, pp.147-164, 1989.

[83] C.SCHERER, $\mathcal{H}_\infty$-optimization without assumptions on finite or infinite zeros, *SIAM J. Control Optimization*, Vol.30, No.1, pp.143-166, 1992.

[84] H.SCHUMACHER, *Dynamic feedback in finite- and infinite-dimensional linear systems*, Math. Centre (CWI) Tracts 143 Amsterdam, 1981.

[85] M.C.SMITH, Well-posedness of $\mathcal{H}_\infty$-optimal control problems, *SIAM J.Control and Optimization*, Vol.28, pp.342-358, 1990.

[86] A.A.STOORVOGEL, *The $\mathcal{H}_\infty$-control problem: a state-space approach*, Prentice-Hall, 1992.

[87] G.TADMOR, Worst-case design in the time domain. The maximum principle and the standard $\mathcal{H}_\infty$-problem, *MCSS* 3, pp.301-324, 1990.

[88] G.TADMOR, The standard $\mathcal{H}_\infty$-problem and the maximum principle: the general linear case, to appear in *SIAM J. Control and Optimization.*

[89] S.TAKEDA and A.R. BERGEN, Instability of feedback systems by orthogonality of L_2, *IEEE Trans. Autom. Control*, AC 18, No. 6, pp. 631-636, 1973.

[90] A.R.TANNENBAUM, Frequency domain methods for the $\mathcal{H}_\infty$-optimization of distributed systems, *Analysis and Optimization of Systems: State and Frequency Domain Approaches for Infinite-Dimensional Systems*, Proceedings of 10-th International Conference Sophia-Antipolis 1992, Lecture notes in control and information sciences, Vol.185, Springer Verlag, Berlin, pp.242-278, 1993.

[91] S.B.TOWNLEY, *Robustness of infinite-dimensional systems*, PhD thesis, University of Warwick, 1987.

[92] R.TRIGGIANI, On the stabilization problem in Banach space, *J. Math. Analysis and Applications*, Vol.52, pp.383-403, 1975.

[93] E.VERMA and E.JONCKHEERE, L_∞-compensation with mixed sensitivity as a broadband matching problem, *Systems and Control Letters*, Vol.4, pp.125-130, 1984.

[94] S.WEILAND, *Theory of approximation and disturbance attenuation for linear systems*, Ph.D. thesis, University of Groningen, December 1990.

[95] G.WEISS, Admissible observation operators for linear semigroups, *Israel J.Math.*, Vol.65, pp. 17-43, 1989.

[96] G.WEISS, Admissibility of unbounded control operators, *SIAM J. Control and Optimization*, Vol.27, No.3, pp.527-545, 1989.

[97] G.WEISS, The representation of regular linear systems on Hilbert spaces, *Proc. of the Conference on Distributed Parameter Systems Vorau 1988*, International Series of Numerical Mathematics, Birkhäuser, Vol.91, pp.401-416, 1989.

[98] G.WEISS, Transfer functions of regular systems, Part 1: Characterizations of regularity, *submitted.*

[99] D.WEXLER, On frequency domain stability for evolution equations in Hilbert spaces via the algebraic Riccati equation, *SIAM J. Math Anal.*, Vol.11, No.6, pp. 969-983, 1980.

[100] J.C.WILLEMS, Least squares stationary optimal control and the algebraic Riccati equation, *IEEE Trans. Autom. Control*, AC 16, No. 6, pp.621-634, 1971.

[101] V.A.YAKUBOVICH, A frequency theorem for the case in which the state and control spaces are Hilbert spaces with an application to some problems in the synthesis of optimal controls I, *Siberian Math. Journal*, 15, pp.457-476, 1974.

[102] J.ZABCZYK, Remarks on the algebraic Riccati equation in Hilbert space, *J. of Appl. Math. and Opt.*, Vol.2, No.3, pp.251-258, 1976.

[103] G.ZAMES, Feedback and optimal sensitivity: model reference transformations, multiplicative seminorms and approximate inverses, *IEEE Trans. on Aut. Control*, Vol.23, pp.301-320, 1981.

[104] K.ZHOU and P.P.KHARGONEKAR, An algebraic Riccati equation approach to $\mathcal{H}_\infty$-optimization, Systems and Control Letters, Vol. 11, No 2, pp. 85-91, 1988.

Index

Systems & Control: Foundations & Applications

Series Editor
Christopher I. Byrnes
School of Engineering and Applied Science
Washington University
Campus P.O. 1040
One Brookings Drive
St. Louis, MO 63130-4899
U.S.A.

Systems & Control: Foundations & Applications publishes research monographs and advanced graduate texts dealing with areas of current research in all areas of systems and control theory and its applications to a wide variety of scientific disciplines.

We encourage the preparation of manuscripts in TeX, preferably in Plain or AMS TeX—LaTeX is also acceptable—for delivery as camera-ready hard copy which leads to rapid publication, or on a diskette that can interface with laser printers or typesetters.

Proposals should be sent directly to the editor or to: Birkhäuser Boston, 675 Massachusetts Avenue, Cambridge, MA 02139, U.S.A.

Estimation Techniques for Distributed Parameter Systems
H.T. Banks and K. Kunisch

Set-Valued Analysis
Jean-Pierre Aubin and Hélène Frankowska

Weak Convergence Methods and Singularly Perturbed Stochastic Control and Filtering Problems
Harold J. Kushner

Methods of Algebraic Geometry in Control Theory: Part I
Scalar Linear Systems and Affine Algebraic Geometry
Peter Falb

H^∞-Optimal Control and Related Minimax Design Problems
Tamer Başar and Pierre Bernhard

Identification and Stochastic Adaptive Control
Han-Fu Chen and Lei Guo

Viability Theory
Jean-Pierre Aubin

Representation and Control of Infinite Dimensional Systems, Vol. I
A. Bensoussan, G. Da Prato, M.C. Delfour and S.K. Mitter

Mathematical Control Theory: An Introduction
Jerzy Zabczyk

Representation and Control of Infinite Dimensional Systems, Vol. II
A. Bensoussan, G. Da Prato, M.C. Delfour and S.K. Mitter